Lecture Notes in Bioinformatics   **12558**

Subseries of Lecture Notes in Computer Science

More information about this subseries at http://www.springer.com/series/5381

João C. Setubal · Waldeyr Mendes Silva (Eds.)

Advances in Bioinformatics and Computational Biology

13th Brazilian Symposium on Bioinformatics, BSB 2020
São Paulo, Brazil, November 23–27, 2020
Proceedings

Editors
João C. Setubal (iD)
University of São Paulo
São Paulo, Brazil

Waldeyr Mendes Silva (iD)
Instituto Federal de Goiás
Formosa, Goiás, Brazil

ISSN 0302-9743   ISSN 1611-3349   (electronic)
Lecture Notes in Bioinformatics
ISBN 978-3-030-65774-1   ISBN 978-3-030-65775-8   (eBook)
https://doi.org/10.1007/978-3-030-65775-8

LNCS Sublibrary: SL8 – Bioinformatics

This Springer imprint is published by the registered company Springer Nature Switzerland AG
The registered company address is: Gewerbestrasse 11, 6330 Cham, Switzerland

Preface

The Brazilian Symposium on Bioinformatics (BSB) is an international conference that covers all aspects of bioinformatics and computational biology. This volume contains the accepted papers for BSB 2020, held virtually during November 23–27, 2020.

As in past years, the special interest group in Computational Biology (CEbiocomp) of the Brazilian Computer Society (SBC) organized the event. A Program Committee (PC) was in charge of reviewing submitted papers; this year, the PC had 46 members. Each submission was reviewed by three PC members. There were two submission tracks: full papers and short papers. One of us (Setubal) supervised the full paper track, and the other (Mendes da Silva) supervised the short paper track. In the full paper track, 20 papers were accepted; 5 papers were accepted in the short paper track. All of them are printed in this volume and were presented orally at the event. In addition to the technical presentations, BSB 2020 featured the following invited speakers, with the respective talk titles: David Roos (University of Pennsylvania, USA), "Cross-silo integration and interrogation of complex biomedical datasets: from genomics to epidemiology and back again"; Bas E. Dutilh (Utrecht University, The Netherlands), "Metagenomics: from illuminating "dark matter" to modeling the microbiome"; and Jens Stoye (Bielefeld University, Germany), "Searching for Genomic Variants in Multiple Genomes at Once".

Based on PC reviews, we chose the authors of the paper by Gabriel Oliveira et al., "COVID-19 X-ray Image Diagnostic with Deep Neural Networks", as the recipients of the BSB 2020 Best Paper Award. Although not part of the selection criteria, we were happy to see that it turned out that this paper addresses one of the many topics arising from the current COVID-19 pandemic.

BSB 2020 was made possible by the dedication and work of many people and organizations. We would like to express our thanks to all PC members, as well as to the external ad-hoc reviewers. Their names are listed in the pages that follow. We are also grateful to the local organizers and volunteers for their valuable help; to Raquel Minardi for helping out with social media; to the sponsors for making the event financially viable; to the developers of EasyChair (full papers) and JEMS (short papers), which were the systems we used to handle submissions; and Springer for agreeing to publish this volume and their staff for working with us on its production. Finally, we would like to thank all authors for their time and effort in submitting their work and the invited speakers for having accepted our invitation.

November 2020

João C. Setubal
Waldeyr Mendes Silva

Organization

Conference Chair

Daniel Cardoso Moraes de Oliveira   Fluminense Federal University, Brazil

Program Committee Chairs

João Carlos Setubal   University of São Paulo, Brazil
Waldeyr C. Mendes da Silva   Instituto Federal de Goiás, Brazil

Steering Committee

Daniel Cardoso Moraes de Oliveira   Fluminense Federal University, Brazil
João Carlos Setubal   University of São Paulo, Brazil
Luis Antonio Kowada   Fluminense Federal University, Brazil
Natália Florencio Martins   Empresa Brasileira de Pesquisa Agropecuária, Brazil
Ronnie Alves   Instituto Tecnológico Vale, Brazil
Sérgio Lifschitz   Pontifícia Universidade Católica do Rio de Janeiro, Brazil
Sérgio Vale Aguiar Campos   Federal University of Minas Gerais, Brazil
Tainá Raiol   Fundação Oswaldo Cruz, Brazil
Waldeyr Mendes   Instituto Federal de Goiás, Brazil

Program Committee

Said Sadique Adi   Federal University of Mato Grosso do Sul, Brazil
Nalvo Almeida   Federal University of Mato Grosso do Sul, Brazil
Ronnie Alves   ITV Sustainable Development, Brazil
Deyvid Amgarten   University of São Paulo, Brazil
Marilia Braga   Bielefeld University, Germany
Laurent Brehelin   Laboratoire d'Informatique, de Robotique et de Microélectronique de Montpellier, France
Marcelo Brigido   University of Brasília, Brazil
Sérgio Campos   Federal University of Minas Gerais, Brazil
Andre Carvalho   University of São Paulo, Brazil
Ricardo Cerri   Federal University of Sao Carlos, Brazil
Luís Cunha   Fluminense Federal University, Brazil
Alberto Davila   FIOCRUZ-RJ, Brazil
Ulisses Dias   University of Campinas, Brazil

Zanoni Dias	University of Campinas, Brazil
Luciano Antonio Digiampietri	University of São Paulo, Brazil
Bas E. Dutilh	Utrecht University, The Netherlands
Andre Fujita	University of São Paulo, Brazil
Ronaldo Fumio Hashimoto	University of São Paulo, Brazil
Luis Kowada	Fluminense Federal University, Brazil
Sergio Lifschitz	PUC-RJ, Brazil
Felipe Prata Lima	Federal Institute of Alagoas, Brazil
Fabrício Martins Lopes	Federal University of Technology of Paraná, Brazil
Felipe A. Louza	Federal University of Uberlândia, Brazil
Fábio Henrique Viduani Martinez	Federal University of Mato Grosso do Sul, Brazil
Natalia Martins	Embrapa Genetic Research and Biotechnology, Brazil
João Meidanis	University of Campinas, Brazil
Raquel Melo-Minardi	Federal University of Minas Gerais, Brazil
Jhonatas Monteiro	University of São Paulo, Brazil
Jefferson Morais	Federal University of Pará, Brazil
Daniel de Oliveira	Fluminense Federal University, Brazil
Sergio Pantano	Institut Pasteur de Montevideo, Uruguay
Alexandre Rossi Paschoal	Federal University of Technology of Paraná, Brazil
José Patané	Butantan Institute, Brazil
Daniel Pinheiro	University of São Paulo, Brazil
David Pires	Butantan Institute, Brazil
Rommel Ramos	Federal University of Pará, Brazil
Mariana Recamonde-Mendoza	Federal University of Rio Grande do Sul, Brazil
Marcel da Câmara Ribeiro-Dantas	Institut Curie, France
Fernando Luís Barroso Da Silva	University of São Paulo, Brazil
Thais Amaral e Sousa	Instituto Federal de Goiás, Brazil
Marcilio de Souto	University of Orleans, France
Kleber Padovani De Souza	Federal University of Pará, Brazil
Sadegh Sulaimany	University of Kurdistan, Iran
Guilherme Telles	University of Campinas, Brazil
Alessandro Varani	São Paulo State University, Brazil
Maria Emília Machado Telles Walter	University of Brasília, Brazil

Additional Reviewers

Eloi Araújo
Daniel Saad Nogueira Nunes
Lucas Oliveira

Funding

Fundação de Amparo à Pesquisa do Estado do Rio de Janeiro (FAPERJ), Brazil

Sponsor

Sociedade Brasileira de Computação (SBC), Brazil

Contents

A Classification of *de Bruijn Graph* Approaches for *De Novo* Fragment Assembly

Elvismary Molina de Armas[1] , Maristela Holanda[2] , Daniel de Oliveira[3] ,
Nalvo F. Almeida[4] , and Sérgio Lifschitz[1(✉)]

[1] Depto. Informática, PUC-Rio, Rio de Janeiro, Brazil
{earmas,sergio}@inf.puc-rio.br
[2] Depto. Ciência da Computação, UNB, Brasília, Brazil
mholanda@unb.br
[3] Instituto de Computação, UFF, Niterói, Brazil
danielcmo@ic.uff.br
[4] Faculdade de Computação, UFMS, Campo Grande, MS, Brazil
nalvo@facom.ufms.br

Abstract. Research in bioinformatics has changed rapidly since the
advent of next-generation sequencing (NGS). Despite the positive impact
on cost reduction, assembling the generated reads remains a challenge.
This paper presents in detail the main ideas related to *de novo* assembly,
the technologies involved, and theoretical concepts about the *de Bruijn*
graph structure. We also explain the existing approaches to minimize
the memory requirements for *de Bruijn* graph construction. Finally, we
propose a comparative view of several solutions, including the k-mers
codification and the data structures used to represent and persist them.

1 Introduction

The field of biological research has changed rapidly since the advent of massively parallel sequencing technologies, known as next-generation sequencing (NGS) [10,18]. Some commercial DNA sequencing platforms include the Genome Sequencer from Roche 454 Life Sciences (www.my454.com), the Genome Analyzer platform from Illumina (www.illumina.com), the SOLiD System from Applied Biosystems (www.appliedbiosystems.com), the Pacific Bioscience (PacBio) sequencers (https://www.pacb.com), and the Oxford MinION, which uses Nano Pore Sequencing technology (https://nanoporetech.com).

These platforms' vital characteristic is that they do not rely on Sanger chemistry [41] as first-generation machines did. With their arrival in the market in 2005 and the fast development since then, they have drastically lowered the cost *per* sequenced nucleotide and increased throughput by orders of magnitude [36]. Their performance dramatically increased the numbers of generated reads (many hundreds of thousands or even millions of reads) in a relatively short time [25] with good genome coverage.

© Springer Nature Switzerland AG 2020
J. C. Setubal and W. M. Silva (Eds.): BSB 2020, LNBI 12558, pp. 1–12, 2020.
https://doi.org/10.1007/978-3-030-65775-8_1

NGS has brought a relevant impact in various areas such as genomics, transcriptomics, metagenomics, proteogenomics, gene expression analysis, noncoding RNA discovery, SNP detection, and protein binding sites identification [18]. The genome assembly problem arises because it is impossible to sequence a whole genome directly in one read using current sequencing technologies.

For assemblies with no reference genome, called *de novo*, assembling a large genome (>100 Mbp) using short readings remains a challenge. Some successful approaches based on the use of *de Bruijn* graph computational data structure have been developed for *de novo* assembly. The construction and use of *de Bruijn* graph demand a large amount of main memory and execution time because of the large number of elements (nodes and edges) to process.

In the next sections, we will give more details about *de novo* assembly, the technologies involved, and theoretical concepts about the *de Bruijn* graph structure. We also explain a categorization of the existing approaches that minimize memory requirements for *de Bruijn* graph construction. Further, a comparative view of several solutions shows the codification to represent the k-mers and the data structures used to store them.

2   Genome Assembly

Genome assembly may be defined as the computational process of reconstructing a whole-genome using numerous short sequences called reads up to the chromosomal level. An assembly is a hierarchical data structure that maps the sequence data to a putative reconstruction of the target genome. However, the vast majority of sequenced genomes are made available only in draft format, having as a result only *contigs*, which are continuous stretches of DNA sequence, or *scaffolds*, which are chains of contigs with additional information about their relative positions and orientations.

There are some challenges around the use of NGS data that brings difficulties to obtain the assembly. DNA sequencing technologies share the fundamental limitation that **read lengths** are much shorter than even the smallest genomes. The process of determining the complete DNA sequence of an organism's genome overcomes this limitation by over-sampling the target genome with short reads from random positions. Also, assembly software is challenged by **repeated sequences** in the target genome. Genomics regions that share perfect repetitions may be indistinguishable, mainly if they are longer than the reads. The repetition resolution is more difficult in front of **sequencing errors**. Therefore it is necessary to further increase sequence accuracy by sequencing individual genomes in a large number of times, increasing the sequenced **reads coverage**.

In terms of computational complexity, the assembly may require High Performance Computing (HPC) platforms for large genomes and the processing of larger volumes of data. Algorithms developed for these HPC platforms are typically complex and depend on pragmatic engineering and heuristics. Heuristics help overcome complicated repetition patterns in real genomes, random and systematic errors in factual data, and real computers' physical limitations. Also, the implementations and results are tied to a suitable parameter instantiation. In case of *de novo* assembly, using a k-mer based algorithms (see Sect. 3), the selection of k value is vital.

3 The *de Bruijn* Graph *De Novo* Assembly Approach

There are two main categories of NGS assemblers can: Overlap-Layout-Consensus approach (OLC) [35], based on an overlap graph, and *de Bruijn* Graph (DBG) approach, which relies on some form of k-mers as vertices of the graph. The *de Bruijn* graph was defined outside the realm of DNA sequencing to represent strings from a finite alphabet. The nodes represent all possible fixed-length strings. The edges represent suffix-to-prefix perfect overlaps [34]. In the context of genome assembly, the *de Bruijn* graph could be defined as follows:

Definition 1. *A read r with length m represents a genome substring, over the alphabet $\Sigma = \{A, T, C, G\}$. Each character of the alphabet represents one of the four nitrogenous bases present in DNA: adenine (A), guanine (G), cytosine (C) and thymine (T).*

Definition 2. *A k-mer is a substring over a read with specific k length.*

The k-mer is a string whose length is $k, 1 < k < m$. k defines the minimum length of a substring that two reads must share to define an overlap, linking two reads in the graph traverse. Using a larger k value involves more accuracy to discover repeated regions in the genome, but also increases the chances of loose overlaps in reads, causing the loss of links in the graph. Consequently, it is not easy to estimate the right k value for the best assembly. The total number of k-mers present in one read is equal to $m - k + 1$, while the total number of k-mers present in n reads is $(m - k + 1)n$. The unique k-mers space for k value is 4^k.

Definition 3. *A de Bruijn graph $G_k(V, E)$ represents overlaps between k-mers, in which:*

- *The set of vertices is defined by $V = S = \{s_1, s_2, ..., s_p\}$, where S is a set of unique k-mers over a given set of reads.*
- *The set of edges is defined by $E = \{e_1, e_2, ..., e_q\}$, where $e = (s_i, s_j)$ if and only if the $k - 1$-th suffix of s_i matches exactly the $k - 1$-th prefix of s_j. s_i and s_j must be adjacent k-mers in at least one read.*

The life cycle of DBG for genome assembly can be summarized in two steps. First, construction involves the generation of all k-mers to generate a node per distinct k-mer and an edge between two nodes if these k-mers have a $k - 1$ overlapped in at least one read. In the second step, the processing is carried on by simplifying the graph and traverse it to generate contiguous genome regions called *contigs*.

4   *de Bruijn Graph* Based Assemblers

There are some assemblers that use DBG approach, for example Velvet [48,49], ALLPATHS [5], ABySS [45], SOAPdenovo [29], SPAdes [1] and Contrail (https://sourceforge.net/projects/contrail-bio/). However, the *de Bruijn* approach has as a drawback that *de Bruijn* graph may require an enormous amount of memory (several gigabytes of RAM). Besides, the construction and analysis of a *de Bruijn* graph are not easily parallelized [42]. As a result, *de Bruijn* assemblers such as Velvet and ALLPATHS, which have been used successfully on bacterial genomes, do not scale to larger genomes. For example, these programs would require several terabytes of RAM for human-sized genomes to store their *de Bruijn* graphs [42] and memory requirements may be higher for more complex genome organisms, as is the case of many plants.

This high memory consumption problem is expected to worsen in the future because the NGS data generation rate has exceeded expectations based on Moore's law [23], meaning that the amount of raw data is expected to grow much faster than the capacity of available memory [25].

4.1   Main Classification of Approaches

There are several techniques to reduce the memory footprint for the assembly process. These approaches can be divided into two general groups. The first group considers solutions that reduce the amount of data as a cost of **not exact representations of DBG**. The goal is to reduce the data to process as much as possible by removing or sampling until it is possible to execute the assembly with available resources. However, it may affect the final quality of the assembly. In that group, we also include those approaches that use probabilistic data structures as Bloom Filters (BF) [2].

In that way, they can use a fixed amount of memory, independently on the number of items to be processed. However, there is no accurate measure to guide the reduction of the data. The memory consumption during the assembly process is highly sensitive to data and the value of k. Thus the success of the reduction is only validated when the assembly is achieved with the available memory. The second group considers solutions that increase the memory resources for the same amount of data for **an exact representations of DBG**, through some techniques like partitioning or distributing, in main or external memory. We discuss these approaches in detail in the following sections.

4.2   General Strategies to Reduce Memory Footprint for DBG Construction

More specifically, alternating between the aforementioned groups we have some techniques to reduce the memory requirements for the assembly process, which can be examined through the following categories:

Pre-processing Techniques: First, there are pre-processing techniques such as Diginorm in [46], Quake [24], ALLPATHS-LG error corrector [22], which try to reduce the input size removing redundant information and errors before the assembly process itself start.

Optimized Data Structures for Graph Representation: To minimize the memory requirements during the k-mer unique identification process, indexes for identifying duplicate k-mers may be used. Hash tables in memory have also been successfully used for many assemblers, such as in [29,45,48], to identify duplicate k-mers. However, for a large amount of NGS data, hash tables do not work well because they may not fit in memory. Suffix-array is a data structure also used to compute overlaps. The FM-index [44] has also been used to allow the compressed representation of input reads and fast computation of overlaps in string graph (equivalent to overlap graph), but it is not tested yet in the construction of *de Bruijn* graph.

Succinct data structures have also been explored to represent *de Bruijn* graph [3,4,11]. In [11] a succinct bitmap is also used to compress the representation of *de Bruijn* graph, but overall its need for space will continue to increase as the graph becomes "bigger". Other approaches are based on the idea of sparseness in genome assembly [47], where only a subset of k-mers present in the dataset is stored. Bloom Filters (BF) have been arduously explored as solution to deal with DBG computationally demanding for a not exact representations of DBG [9,33,50]. They are used to store vertices (k-mers), while the edges are implicitly deduced by querying the Bloom filter for the membership of all possible extensions of a k-mer. However, this approach does not correspond exactly to the edges contained in the reads. Some works have been focused on mechanisms that avoid false positives using Bloom Filters [9,40].

An extra-compacted *de Bruijn* Graph structure is introduced in [12]. It represents intermediate states of the DBG during its generation through a series of iterations in which the number of k-mers to be processed is iteratively reduced. The extra-compacted *de Bruijn* Graph nodes represents the unique dk-mers with length equal or less than d, while edges corresponds to unique edges formed by adjacent dk-mers, whose sharing $k-1$-length overlap in at least one read.

Extending the Computational Resources: Some solutions proposed the use of cloud-based resources to overcome the memory requirements limitations. In [25] were designed a set of assembler experiments using the GAGE datasets and a group of program assemblers in virtual machines in the Amazon AWS environment. The financial analysis reveals that the cost for assembly increases as the complex of genomes to be assembled, because such an operation requires more expensive virtual machines, and the assembly may be executed for several hours. Other solutions are based or combined with parallelization techniques as BCALM2 [8], and the use of GPU and other memory systems like Gerbil [19] and k-mer counter FPGA-based solution [32].

External Memory Approaches: Partition assembly algorithms were also proposed for external processing. For example, the Minimum Substring Partitioning (MSP) [28] technique allows us to split the input reads into subsequences and distributes then into this disk partitions, then processing one disk partition at a time. Moreover, the k-mers partitions can be processed in a distributed manner, as well as the Contrail proposed to avoid memory bottleneck. Moreover, BCALM1 [7] and BCALM2 [8] propose a construction of a compacted DBG using partition and disk distribution approaches, such as DSK k-mer counter. Among the techniques used for partitioning were found hash functions over the k-mer *minimum substring* of p length and *minimizers*.

The construction of the graph embedded into a relational database management system (RDBMS) is another type of approach explored in [13–15,43]. Some indexes configuration based on B+-tree, hash over k-mer [15]) and k-mer p-minimum substring [13] were tested for textitk-mer mapping process as part of the DBG construction, in junction with an *ad-hoc* cost model to measure the performance gained [14]. Although some optimizations are needed to improve the execution time given by the index evaluation, the case study implemented with PostgreSQL based on the Velvet assembly algorithm shows the feasibility of using DBMS to manage I/O operations in the k-mer process mapping, allowing incremental processing without reprocessing and recovery from failures [43].

The external memory approaches presented so far process from the beginning the total number of k-mers, to following obtain the vertices of the graph (unique k-mers) and corresponding edges. These imply carrying on the process with the high level of redundancy present in those k-mers, which significantly impacts the amount of memory needed, and the number of I/O operations.

An approach presented in [12] combines an external memory and optimized data structures approach. Unlike the above techniques, it proposes the construction of an exact representation of *de Bruijn* Graph without the necessity of process all k-mers. Through an iterative sequence of reductions, it is possible to process the graph as much as possible in the main memory, and only when the available main memory becomes insufficiently, will be using an external memory solution. Then, large duplicate regions had already been identified, avoiding processing a significant amount of duplicated k-mers in external memory, reducing the number of I/O operations.

4.3   Specific Strategies to Reduce the Large Memory Consumption

There are a series of specific elements on which the studied solutions have focused to reduce the large memory consumption in terms of implementation. As can be seen, some of them correspond directly to one of the general approaches.

One of those elements is using data structures for fast lookup with lower overhead to store the graph elements, especially for the set of unique k-mers that correspond to the nodes. The most frequent data structure used is hash tables, followed by search-trees data structure variations. Another essential element to taking into account is the codification of k-mer for less memory per element.

We classified the approaches available in literature into two groups, those based on a lossless compression, which is an exact representation of DBG, the other based on lossy data compression. In the first group, we observed the classical 2-bits k-mer codification. For example, in DSK, Velvet, and ABYSS, they remove parts of $p1$ and $p2$ length prefixes such as in KMC1, and some forms of hash optimization as in Jellyfish with the use of a bijective function and Meraculous, which use a lightweight hash (a combination of the hash family). In the second group, we have seen best represented by Bloom Filters (BF) with false positives (FP) or specialized probabilistic data structures based on BF.

Table 1. *de Bruijn* graph solutions comparison. Abbreviation used for the approach classification (Cl.): KC (K-mer Counter), A (Assembler), S (Space efficient solution) and E (external approach by disk distribution)

Approach	Cl.	Description of the approach	Data structure to store k-mers
Jellyfish [31]	KC	Main memory. Hash table merging in disk if not memory available	Hash table using a quadratic reprobing function
BFCounter [33]	KC	BF in RAM. Two pass algorithm to correct false positives	BF. 4-bits per k-mer, Google sparsehash library
Meracolous [6]	A	Reliance on the linear U-U component of the graph	Novel lightweight hash
DSK [39]	KC	Disk distribution based on hashing	Hash table
KMC1 [16]	KC	Disk distribution based on prefixes. Sorting using the least-significant-digit (LSD) radix sort	Bins (disk files) according on prefixes $p1$ and $p2$ of k-mers
Minia [9]	S	BF in RAM + set of FP on disk with a fixed amount of memory	BF for nodes. cPF structure for a set of critical FP
khmer [50]	KC	Count-Min Sketch in RAM. Not error correction is used	Count-Min Sketch to storing the frequency distributions of distinct k-mers
Cascading BF [40]	A	BF + set of FP	Cascading BFs to store FP k-mers
BCALM [7]	E	DSK + BCALM algorithm	DBGFM to store no-branch path, codified as FM index
KMC2 [17]	KC	Disk distribution based on k-mer *signatures*. Sorting using LSD radix sort	Bins (disk files) according to the related canonical minimizer. HT for counting
KCMBT [30]	KC	Trie-based in-memory algorithm. Three phases algorithm	Multiple Burst Trees (KCMBT)
BCALM2 [8]	E	Based on BCALM + parallelization. Three stages algorithm	Union-find data structure for partitions with a minimal perfect hash function based indexing
KMC3 [26]	KC	Disk distribution. Better balance of bin sizes and fast radix sort	Bins to store k-mers
HaVec [38]	S	BF + the quotient of the hash function division to verify FP	Vector to store the k-mers along with their neighbor information
Squeakr [37]	KC	Based upon a counting filter data structure CQFs	CQF to store an approximation of a multiset S of k-mers, maintaining a false positive rate
FastEtch algorithm [20, 21]	S	Based upon Count-Min sketch probabilistic data structure	Count-Min sketch stores an approximated DBG with the subset of nodes that are most likely to contribute to the *contig* generation step

Table 1 summarizes several approaches present in the literature, showing a brief description of each one and classification based on the type of application: k-mer counter, assembler, space-efficient solutions for DBG and disk distribution as an external approach variant. It also presents the data structure used to represent the DBG storing the k-mers with a specific codification.

Some variants for k-mer codification are used to reduce the amount of memory needed to represent and store each k-mer. DSK (Minia assembler) [39] and ABYSS [45], for example, use a classical 2 bits representation for each nucleotide base in k-mers. With the use of probabilistic data structures like Bloom Filters, Minia [9], for example, allows approximated $1.44 \log_2(16k/2.08) + 2.08$ bits/k-mer.

Other solutions use strategies based on hash tables, for example Jellyfish [31] codifying a part of k-mer as the index of the table using a bijective hash function. Meracolous [6] uses a recursive collision strategy with multiple hash functions to avoid explicitly storing the k-mer themselves. khmer [50] uses a Count-Min Sketch storing only counts, while k-mers must be retrieved from the original data set, and HaVec [38] uses 5 bytes for each index in the hash table plus $2k$ bits for k-mer. As another strategy, KMC1 [16], for example, stores k-mer without the $p1$ and $p2$ prefixes.

4.4 k-mers Counters

The DBG construction implies a subroutine to identify distinct k-mers and get their multiplicity. Identify distinct k-mers problem also has been touched by counting k-mer tools [16, 17, 27, 31, 33, 39]. Although k-mer counter tools aim at generating histograms over k-mers distributions, their processes has some similarities to the ones that get the vertices set of the DBG.

Identifying distinct k-mers have been approached by sorting [16, 17], hashing [27, 31, 39] or using Bloom Filters [33], combined sometimes with parallel approaches to speed up the process [16, 17]. Some of them [16, 17, 27, 39] have been focused on distribution the k-mers in disk partitions to counter them before, loading in main memory each partition at time.

It is valid to note that k-mer counters have not the notion of vertices and edges. Besides, to reduce the amount of data, they make some assumptions such as do not count the k-mers with frequencies smaller or more significant than a given value, which is not appropriate for the DBG construction.

5   Conclusions

Next-generation sequencing (NGS) data has significantly impacted several fields of bioinformatics, greatly reducing costs. However, the genome assembly continues to be a challenge for genomic research since no technology is capable of sequencing the whole-genome. Also, some aspects of NGS data makes the assembly difficult, such as the error profiles for each NGS platform, the non-uniform coverage of the target, hampering the resolution of genome repetitions.

The *de Bruijn* graph (DBG) approach for the *de novo* assembly is used when there is no reference genome. The construction of DBG needs a high main memory space and is responsible for the high computational cost. The most critical parameter in the DBG is the k value, which impacts the assembly's accuracy, the number of vertices and edges, and the memory requirements.

Several techniques have been proposed to construct a DBG. This paper classifies those approaches into two main categories, one for an exact and complete graph, another for non-exact DBG representation. Considering the solutions based on probabilistic data structures used for an approximate DBG representation, the most explored data structure are the bloom filters. This data structure allows vertices to be stored independently of their number, but at the cost of false positives.

By revisiting the literature, we could find some solutions using external memory to construct a graph with an exact representation. They process all the k-mers in external memory. A new algorithm building an accurate representation without the necessity of process all k-mers have also been found.

This paper presented in detail the main theoretical and practical aspects related to *de novo* assembly, particularly the *de Bruijn* graph structure. We have enumerated the existing approaches to reduce the memory requirements for DBG construction. Finally, we have proposed a comparative view of the existing solutions in the literature.

Acknowledgments. The authors are partially supported by grants from FUNDECT, FAPERJ, CNPq, CAPES and INCA, Brazilian Public Funding Agencies and Research Institutes.

References

1. Bankevich, A., et al.: Spades: a new genome assembly algorithm and its applications to single-cell sequencing. J. Comput. Biol.: J. Comput. Mol. Cell Biol. **19**(5), 455–477 (2012). https://doi.org/10.1089/cmb.2012.0021. https://pubmed.ncbi.nlm.nih.gov/22506599

2. Bloom, B.H.: Space/time trade-offs in hash coding with allowable errors. Commun. ACM **13**(7), 422–426 (1970). https://doi.org/10.1145/362686.362692

3. Boucher, C., Bowe, A., Gagie, T., Puglisi, S., Sadakane, K.: Variable-order de Bruijn graphs. In: Data Compression Conference Proceedings 2015 (2014). https://doi.org/10.1109/DCC.2015.70

4. Bowe, A., Onodera, T., Sadakane, K., Shibuya, T.: Succinct de Bruijn graphs. In: Raphael, B., Tang, J. (eds.) WABI 2012. LNCS, vol. 7534, pp. 225–235. Springer, Heidelberg (2012). https://doi.org/10.1007/978-3-642-33122-0_18

5. Butler, J.: ALLPATHS: De novo assembly of whole-genome shotgun microreads. Genome Res. **18**(5), 810–820 (2008). https://doi.org/10.1101/gr.7337908

6. Chapman, J.A., Ho, I., Sunkara, S., Luo, S., Schroth, G.P., Rokhsar, D.S.: Meraculous: de novo genome assembly with short paired-end reads. PLoS ONE **6**(8), e23501 (2011). https://doi.org/10.1371/journal.pone.0023501

7. Chikhi, R., Limasset, A., Jackman, S., Simpson, J.T., Medvedev, P.: On the representation of de Bruijn graphs. In: Sharan, R. (ed.) RECOMB 2014. LNCS, vol. 8394, pp. 35–55. Springer, Cham (2014). https://doi.org/10.1007/978-3-319-05269-4_4

8. Chikhi, R., Limasset, A., Medvedev, P.: Compacting de Bruijn graphs from sequencing data quickly and in low memory. Bioinformatics **32**(12), i201 (2016). https://doi.org/10.1093/bioinformatics/btw279

9. Chikhi, R., Rizk, G.: Space-efficient and exact de Bruijn graph representation based on a Bloom filter. Algorithms Mol. Biol. **8**(1), 22 (2013). https://doi.org/10.1186/1748-7188-8-22

10. Claros, M.G., Bautista, R., Guerrero-Fernández, D., Benzerki, H., Seoane, P., Fernández-Pozo, N.: Why assembling plant genome sequences is so challenging. Biology **1**(2), 439 (2012). https://doi.org/10.3390/biology1020439

11. Conway, T.C., Bromage, A.J.: Succinct data structures for assembling large genomes. Bioinformatics **27**(4), 479–486 (2011). https://doi.org/10.1093/bioinformatics/btq697

12. de Armas, E.M., Castro, L.C., Holanda, M., Lifschitz, S.: A new approach for de Bruijn graph construction in de novo genome assembling. In: 2019 IEEE International Conference on Bioinformatics and Biomedicine, pp. 1842–1849 (2019)

13. de Armas, E.M., Ferreira, P.C.G., Haeusler, E.H., de Holanda, M.T., Lifschitz, S.: K-mer mapping and RDBMS indexes. In: Kowada, L., de Oliveira, D. (eds.) BSB 2019. LNCS, vol. 11347, pp. 70–82. Springer, Cham (2020). https://doi.org/10.1007/978-3-030-46417-2_7

14. de Armas, E.M., Haeusler, E.H., Lifschitz, S., de Holanda, M.T., da Silva, W.M.C., Ferreira, P.C.G.: K-mer Mapping and de Bruijn graphs: the case for velvet fragment assembly. In: 2016 IEEE International Conference on Bioinformatics and Biomedicine (BIBM), pp. 882–889 (2016). https://doi.org/10.1109/BIBM.2016.7822642

15. de Armas, E.M., Silva, M.V.M., Lifschitz, S.: A study of index structures for K-mer mapping. In: Proceedings Satellite Events of the 32nd Brazilian Symposium on Databases. Databases Meet Bioinformatics Workshop, pp. 326–333 (2017)

16. Deorowicz, S., Debudaj-Grabysz, A., Grabowski, S.: Disk-based k-mer counting on a PC. BMC Bioinform. **14**(1) (2013). https://doi.org/10.1186/1471-2105-14-160

17. Deorowicz, S., Kokot, M., Grabowski, S., Debudaj-Grabysz, A.: KMC 2: fast and resource-frugal k-mer counting. Bioinformatics **31**(10), 1569 (2015). https://doi.org/10.1093/bioinformatics/btv022

18. El-Metwally, S., Hamza, T., Zakaria, M., Helmy, M.: Next-generation sequence assembly: four stages of data processing and computational challenges. PLoS Comput. Biol. **9**(12), 1–19 (2013). https://doi.org/10.1371/journal.pcbi.1003345

19. Erbert, M., Rechner, S., Müller-Hannemann, M.: Gerbil: a fast and memory-efficient k-mer counter with GPU-support. Algorithms Mol. Biol. **12**(1), 9:1–9:12 (2017). https://doi.org/10.1186/s13015-017-0097-9

20. Ghosh, P., Kalyanaraman, A.: A fast sketch-based assembler for genomes. In: Proceedings of the 7th ACM International Conference on Bioinformatics, Computational Biology, and Health Informatics. BCB '16, pp. 241–250. ACM, New York, NY, USA (2016). https://doi.org/10.1145/2975167.2975192

21. Ghosh, P., Kalyanaraman, A.: FastEtch: a fast sketch-based assembler for genomes. IEEE/ACM Trans. Comput. Biol. Bioinform. **16**(4), 1091–1106 (2019). https://doi.org/10.1109/TCBB.2017.2737999

22. Gnerre, S., et al.: High-quality draft assemblies of mammalian genomes from massively parallel sequence data. Proc. Natl. Acad. Sci. U.S.A. **108**(4), 1513–1518 (2011). https://doi.org/10.1073/pnas.1017351108. 21187386[pmid]

23. Jackman, S.D., Birol, I.: Assembling genomes using short-read sequencing technology. Genome Biol. **11**(1), 202 (2010). https://doi.org/10.1186/gb-2010-11-1-202. https://www.ncbi.nlm.nih.gov/pubmed/20128932, 20128932[pmid]

24. Kelley, D.R., Schatz, M.C., Salzberg, S.L.: Quake: quality-aware detection and correction of sequencing errors. Genome Biol. **11**(11), R116 (2010). https://doi.org/10.1186/gb-2010-11-11-r116

25. Kleftogiannis, D., Kalnis, P., Bajic, V.B.: Comparing memory-efficient genome assemblers on stand-alone and cloud infrastructures. PLoS ONE **8**(9) (2013). https://doi.org/10.1371/journal.pone.0075505

26. Kokot, M., Dlugosz, M., Deorowicz, S.: KMC 3: counting and manipulating k-mer statistics. Bioinformatics **33**(17), 2759–2761 (2017). https://doi.org/10.1093/bioinformatics/btx304

27. Li, Y., Yan, X.: MSPKmerCounter: a fast and memory efficient approach for K-mer Counting. arXiv e-prints (2015)

28. Li, Y., Kamousi, P., Han, F., Yang, S., Yan, X., Suri, S.: Memory efficient minimum substring partitioning. PVLDB **6**(3), 169–180 (2013). https://doi.org/10.14778/2535569.2448951

29. Luo, R., et al.: SOAPdenovo2: an empirically improved memory-efficient short-read de novo assembler. GigaScience **1**(1), 1–6 (2012). https://doi.org/10.1186/2047-217X-1-18

30. Mamun, A.A., Pal, S., Rajasekaran, S.: KCMBT: a k-mer counter based on multiple burst trees. Bioinformatics **32**(18), 2783 (2016). https://doi.org/10.1093/bioinformatics/btw345

31. Marcais, G., Kingsford, C.: A fast, lock-free approach for efficient parallel counting of occurrences of k-mers. Bioinformatics **27**(6), 764–770 (2011). https://doi.org/10.1093/bioinformatics/btr011

32. McVicar, N., Lin, C., Hauck, S.: K-Mer counting using bloom filters with an FPGA-attached HMC. In: 25th IEEE Annual International Symposium on Field-Programmable Custom Computing Machines, FCCM 2017, Napa, CA, USA, 30 April–2 May 2017, pp. 203–210 (2017). https://doi.org/10.1109/FCCM.2017.23

33. Melsted, P., Pritchard, J.K.: Efficient counting of k-mers in DNA sequences using a bloom filter. BMC Bioinform. **12**(1), 333 (2011). https://doi.org/10.1186/1471-2105-12-333

34. Miller, J.R., Koren, S., Sutton, G.: Assembly algorithms for next-generation sequencing data. Genomics **95**(6), 315–327 (2010). https://doi.org/10.1016/j.ygeno.2010.03.001. 20211242[pmid]

35. Myers, E.W.: Toward simplifying and accurately formulating fragment assembly. J. Comput. Biol. **2**(2), 275–290 (1995). https://doi.org/10.1089/cmb.1995.2.275. pMID: 7497129

36. Niedringhaus, T.P., Milanova, D., Kerby, M.B., Snyder, M.P., Barron, A.E.: Landscape of next-generation sequencing technologies. Anal. Chem. **83**(12), 4327–4341 (2011). https://doi.org/10.1021/ac2010857

37. Pandey, P., Bender, M.A., Johnson, R., Patro, R.: deBGR: an efficient and near-exact representation of the weighted de Bruijn graph. Bioinformatics **33**(14), i133–i141 (2017). https://doi.org/10.1093/bioinformatics/btx261

38. Rahman, M.M., Sharker, R., Biswas, S., Rahman, M.: HaVec: an efficient de Bruijn graph construction algorithm for genome assembly. Int. J. Genom. **2017**, 1–12 (2017). https://doi.org/10.1155/2017/6120980

39. Rizk, G., Lavenier, D., Chikhi, R.: DSK: k-mer counting with very low memory usage. Bioinformatics **29**(5), 652–653 (2013). https://doi.org/10.1093/bioinformatics/btt020

40. Salikhov, K., Sacomoto, G., Kucherov, G.: Using cascading bloom filters to improve the memory usage for de Bruijn graphs. Algorithms Mol. Biol.: AMB **9**, 2 (2014). https://doi.org/10.1186/1748-7188-9-2

41. Sanger, F., Coulson, A., Barrell, B., Smith, A., Roe, B.: Cloning in single-stranded bacteriophage as an aid to rapid DNA sequencing. J. Mol. Biol. **143**(2), 161–178 (1980). https://doi.org/10.1016/0022-2836(80)90196-5

42. Schatz, M.C., Delcher, A.L., Salzberg, S.L.: Assembly of large genomes using second-generation sequencing. Genome Res. **20**(9), 1165–1173 (2010). https://doi.org/10.1101/gr.101360.109

43. Silva, M.V.M., de Holanda, M.T., Haeusler, E.H., de Armas, E.M., Lifschitz, S.: VelvetH-DB: Persistência de Dados no Processo de Montagem de Fragmentos de Sequências Biológicas. In: Proceedings Satellite Events of the 32nd Brazilian Symposium on Databases. Databases Meet Bioinformatics Workshop, pp. 334–341 (2017)

44. Simpson, J.T., Durbin, R.: Efficient construction of an assembly string graph using the FM-index. Bioinformatics (Oxford, England) **26**(12), i367–i373 (2010). https://doi.org/10.1093/bioinformatics/btq217

45. Simpson, J.T., Wong, K., Jackman, S.D., Schein, J.E., Jones, S.J., Birol, I.: ABySS: a parallel assembler for short read sequence data. Genome Res. **19**(6), 1117–1123 (2009). https://doi.org/10.1101/gr.089532.108

46. Titus Brown, C., Howe, A., Zhang, Q., Pyrkosz, A.B., Brom, T.H.: A reference-free algorithm for computational normalization of shotgun sequencing data. arXiv e-prints arXiv:1203.4802 (2012)

47. Ye, C., Sam Ma, Z., Cannon, C., Pop, M., Yu, D.: Exploiting sparseness in de novo genome assembly. BMC Bioinform. **13**(Suppl. 6), S1 (2012). https://doi.org/10.1186/1471-2105-13-S6-S1

48. Zerbino, D.: Velvet software. EMBL-EBI. https://www.ebi.ac.uk/zerbino/velvet/ (2016). Accessed 15 June 2019

49. Zerbino, D.R., Birney, E.: Velvet: algorithms for de novo short read assembly using de Bruijn graphs. Genome Res. **18**(5), 821–829 (2008). https://doi.org/10.1101/gr.074492.107

50. Zhang, Q., Pell, J., Canino-Koning, R., Howe, A.C., Brown, C.T.: These are not the K-mers you are looking for: efficient online K-mer Counting Using a Probabilistic Data Structure. PLoS ONE **9**(7), 1–13 (2014). https://doi.org/10.1371/journal.pone.0101271

Redundancy Treatment of NGS Contigs in Microbial Genome Finishing with Hashing-Based Approach

Marcus Braga[1]([✉]) [ID], Kenny Pinheiro[2] [ID], Fabrício Araújo[2] [ID],
Fábio Miranda[2] [ID], Artur Silva[2] [ID], and Rommel Ramos[2] [ID]

[1] Campus Paragominas, Universidade Federal Rural da Amazônia,
Paragominas, Pará, Brazil
marcusbbraga@yahoo.com.br

[2] Instituto de Ciências Biológicas, Universidade Federal do Pará, Belém, Pará,
Brazil

Abstract. Repetitive DNA sequences longer than reads' length produce assembly gaps. In addition, repetition can cause complex and misassembled rearrangements that creates branches in assembler graphs. Algorithms must decide which way is the best. Incorrect decisions create false associations, called chimeric contigs. Reads coming from different copies of a repetitive region on genome may be wrongly assembled as a unique contig, a repetitive contig. Furthermore, the growth of hybrid assembling approaches using different sequencing platforms data, different fragment sizes or even data from distinct assemblers are responsible for significantly increasing in the amount of generated contigs and therefore subsequent redundancy on data. Thus, this work presents a hybrid computational method to detect and eliminate redundant contigs from microbial genome assemblies. It consists of two Hashing-Based techniques: a Bloom Filter to detect duplicated contigs and a Locality-Sensitive Hashing (LSH) to remove similar contigs. The redundancy reduction facilitates downstream analysis and diminishes the required time to finishing and curate genomic assemblies. The hybrid assembly of GAGE-B dataset was performed with SPAdes (De Bruijn Graph) assembler and Fermi (OLC) assembler. The proposed pipeline was applied to the resulting contigs and the performance compared to other similar tools such as HSBLASTN, Simplifier and CD-HIT. Results are presented.

Keywords: NGS contigs · Redundancy detection · Genome finishing · Bloom filter · LSH

1 Introduction

DNA sequencing is routinely used in various fields of biology. When Whole Genome Sequencing is performed, the DNA is fragmented and the nucleotides are sequenced. High-Throughput Sequencing Technology, also known as Next Generation Sequencing (NGS), allowed the parallelization of the sequencing process,

© Springer Nature Switzerland AG 2020
J. C. Setubal and W. M. Silva (Eds.): BSB 2020, LNBI 12558, pp. 13–24, 2020.
https://doi.org/10.1007/978-3-030-65775-8_2

generating much more data than previous methods [1]. From the data generated by NGS technologies, several new applications have emerged. Many of these analyzes begin with the computational process of sequence assembly [2]. NGS sequence assembly consists of grouping a set of sequences generated in sequencing, producing longer contiguous sequences, called contigs. These contigs are joined together to form even larger known sequences, the scaffolds [3].

There are two general approaches to assembling NGS fragments: reference-based and *de novo* approaches. In the first approach, a reference genome of a related species is used as a guide to align the reads. *De novo* assembly is based only on the overlapping reads to generate contigs [4]. These, in turn, may contain gaps (regions not represented in the assembly).

New hybrid strategies have been developed to take advantage of each type of assembly [5,6]. For example, hybrid strategies can combine reads and assemblies from different sequencing technologies and different assembly algorithms, or use assemblies generated by different assemblers, combining the results (contigs and/or scaffolds) produced by those tools to produce a new sequence [7].

1.1 DNA Repetitions and Contigs Redundancy

Repetitive sequences of DNA are present in all genomes. Repetitions have always been technical challenges for sequence mapping and assembly tools. NGS sequencing with short reads and high throughput made these challenges more difficult. From a computational perspective, repetitions create ambiguities in alignment and assembly which can lead to distortions and errors in the interpretation of results [8].

Repetitions can include from only two copies to millions of copies, can range in size from one to two bases (mono and dinucleotide) to millions of bases. The best documented example of interspersed repetitions in the human genome is the Alu repetitive element class, which covers approximately 11% of the genomes [9].

Repetitions that are sufficiently divergent do not present many computational problems, therefore, for this work, a repetition is defined as a sequence of at least 100 bp in length that occurs two or more times in the genome and exhibits over 97% of identity for at least one other copy of itself. This definition excludes many repetitive sequences, but includes those that present the main computational challenges [10].

Repetitions may create gaps in *de novo* assembly. Besides creating gaps, repetitions can erroneously collapse over one another and cause complex rearrangements and assembly errors [11]. The degree of difficulty, in terms of correctness and contiguity, that repetitions cause during genome assembly largely depends on the read's length.

Several new assemblers have emerged that address this problem, in particular overlay-based and de Bruijn-based graphs. Both create graphs of different types from the sequencing data. Repetitions cause ramifications in these graphs and assemblers must guess which branch to follow. Incorrect decisions create false associations, generating chimeric contigs and wrong copy numbers. If the assembler is more conservative, it will break the assembly at these junction points, leading to precise but fragmented assembly with very small contigs.

The central problem with the repetitions is that the assembler is unable to distinguish from each other, which means that the regions flanking these repetitions can be mistakenly assembled. It is common for an assembler to create a chimera by joining two regions that are not close in the genome and, in the end, the reads eventually align with the wrongly assembled genome [8].

There is a combination of strategies for solving repetitive DNA problems, including the use of varying fragment libraries [12], post-processing software to detect wrong assemblies [11], analysis of coverage statistics to detect and resolve entanglements in DBGs.

NGS Technologies remain unable to generate one single sequence per chromosome, instead, they produce a large and redundant set of reads, each read being a fragment of the complete genome [13].

Assembly algorithms explore graphs through heuristics, selecting and traversing paths and generating sequences as output. The set of contigs is hardly satisfactory and usually needs to be postprocessed, most of the time to discard very short contigs or contigs contained within other contigs. This procedure is performed to reduce redundancy [14].

Specific errors inherent in sequencing platforms also affect the quality of the generated assembly. In the case of the 454 and Ion Torrent platforms, for example, errors in identifying homopolymer sequences may affect the construction of contigs in the de Bruijn graph since k-mers derived from these regions may not show agreement, resulting in a greater fragmentation of the assembly. Thus, it is important to consider using different assembly tools and to give preference to those that have greater tolerance to errors observed in the platform of interest.

The point is that when using a hybrid assembly approach, the problem of redundant contigs persists and even increases. When assembling with different assemblers using different methods, the contigs resulting from these assemblies are merged, generating a much larger amount. These contigs, in general, represent different regions of the genome and lead to different gaps. The large number of contigs generated by these hybrid assembly approaches require considerable computational and human resources for analysis, especially for the identification of assembly errors and the elimination of bases with higher probability of contiguous edge error, which prevents the extent of overlap due to mismatch errors [15].

1.2 Computational Methods for Redundancy Detection in Sequences

There are a large number of computational methods that can identify redundancies in text sequences, such as biological sequences. String matching, or pattern matching, or sequence matching, is a classic computational problem. Algorithms of this nature are used to find matches between a standard input string and a specified string. These methods can locate all occurrences of a character pattern within another sequence [16].

Several methods have been proposed to find similarities between sequences. Some of these look for exact matches between sequences while others allow character insertions, deletions or substitutions trying to find the best possible alignment [17].

These algorithms can be divided into two main categories: exact string matching and approximate string matching methods, which are a generalization of the previous approach, which looks for similar or approximate patterns (sub-chains) in a string [18].

Similarity Search. One form of searching for similarities is the Nearest Neighbor Search and consists of locating data items whose distances to a query item are the smallest in a large data set [19]. The search for similarities has become a primary computational task in many areas, including pattern recognition, data mining, biological databases, multimedia information retrieval, machine learning, data compression, computer vision and statistical data analysis. The concept of exact string matching rarely has meaning in these environments, while concepts such as proximity, distance (similarity/dissimilarity) are much more beneficial for this type of search [20].

Several methods have been developed to solve this kind of problem and many efforts have been devoted to the approximate search. Hashing techniques have been widely studied, especially Locality Sensitive Hashing, as a useful concept for these categories of applications.

In general terms, hashing is an approach where it turns the data item into a small numeric representation or, equivalently, a short code consisting of a sequence of bits [21]. Hashing the nearest neighbor can be done in two ways: by indexing data items through hash tables that store items with the same code in the same hash bucket or by approximating the distance using what was calculated with short codes [22].

The hashing approach to the approximate search aims to map the query items to the destination items so that the approximate search by the nearest neighbor can be performed efficiently and accurately using the destination items and possibly a small subset of the raw query items. Target items are called hash codes (also known as hash values) [23].

Locality Sensitive Hashing (LSH). The term Locality Sensitive Hashing was introduced in 1988 [24] to designate a randomized structure able to efficiently search for the nearest neighbor in large spaces. It is based on the definition of LSH functions, a family of hash functions that maps similar input items with the same hash code to a much higher probability than for different items. The first specific LSH function, minHash, was invented by Broder [25] for the detection and grouping of nearly duplicate web pages and is still one of the most studied and applied LSH methods [20].

In conventional hashing, close items, which are similar, can be mapped/scattered to different positions after hashing, but in LSH similar items remain close even after hashing.

In LSH, we call candidate pairs to item pairs that have been mapped to the same compartment. When the banding technique is applied, comparisons are performed only on candidate pairs rather than on all pairs, as in a linear search. If the goal is to find an exact match, techniques used to process data such as MapReduce [26], Twitter Storm [27] can be used. These techniques are based on parallelism, resulting in reduced time, however, these methods require additional hardware. In most cases, only the most similar pairs are desired. The level of similarity is defined by some threshold and the desired result is what is known by searching for the nearest neighbor, and in these cases LSH model is the best option. To apply LSH in different applications, it needs to be developed according to the application domain [28].

Another important consideration is that false negatives and false positives should be avoided as far as possible. It is said that there is a false negative when the most similar pairs are not mapped to the same compartment. A false positive happens when different pairs are mapped to the same compartment [29].

Bloom Filter. Bloom Filter is a probabilistic data structure that uses multiple hash functions to store data in a large array of bits. It was introduced in 1970 by Burton H. Bloom and is used in applications that perform membership queries on a large set of elements [30].

A Bloom Filter (BF) is, therefore, a simple data structure that uses bit arrays to represent a set and determine whether or not an element is present in it. False positives are allowed, that is to say, with high probability, if an element is in the set. On the other hand, false negatives are not possible, that is, it is possible to know exactly if an element does not belong to the set. The achieved space savings can often overcome this disadvantage of false positives if the probability of error is controlled [31].

BF uses hash functions to map elements in an array, the filter. The membership is tested by comparing the mapping results with the potential members of the vector. An element is considered part of the set if and only if a hash function maps that location to a key [32].

1.3   Contribution of This Work

Repetitions in the genome associated with small reads length are known to be one of the most common reasons for fragmentation of a consensus sequence, as reads that come from different copies of a repetitive region in the genome end up not being properly identified and assembled in the same contig, known as repetitive contigs. This problem is already addressed with a set of strategies such as using miscellaneous fragment libraries, postprocessing software to detect assembly errors, and coverage statistics analysis to detect and resolve entanglements in DBGs, however, none of these mechanisms completely solves the problem. On the other hand, sequencing errors also affect the assembly, either by generating further fragmentation of the assembly, or by resulting in abrupt ends in graph pathways and ultimately contigs. As a result, hybrid assembly approaches are often used to try to achieve greater error tolerance. However, in these approaches, a new problem arises, the production of an even larger amount of redundant contigs.

In this paper, we present a computational method for detecting and eliminating redundant contigs from microbial assemblies based on Bloom Filter and LSH combination, allowing to minimize the computational effort in the genome finishing step.

2   Biological Dataset and Assembly

The GAGE-B dataset [33] is used to evaluate large-scale genomic assembly algorithms. GAGE-B data is originated from the genome sequencing of eight bacteria, with size between 2.9 and 5.4 Mb and GC content between 33 and 69%. This data is publicly available for Illumina sequencing. Some genomes were included which HiSeq and MiSeq data were available also, resulting in 12 datasets (Table 1). All GAGE-B datasets had underwent preprocessing steps such as adapter removal and q10 quality trimming using the Ea-utils package [34].

The *de novo* assembly of the GAGE-B datasets were performed with SPAdes [35] and Fermi [36] assemblers on an AMD Opteron (TM) Processor 6376 computer with 64 CPUs, 1TB RAM, operating system CentOS release 5.10 Linux OS version 2.6.18371.6.1.el5.

Table 1. Microbial genomes and sequenced reads from GAGE-B.

Species	Genome size (MB)	% GC content	NGS platform	Read size (pb)	Frag. size (pb)	Coverage
A.hydrophila SSU	4.7	65	HiSeq	101	180	250x
B.cereus VD118	5.4	35	HiSeq	101	180	100–300
B.cereus ATCC 10987	5.4	35	MiSeq	250	600	100x
B.fragilis HMW 615	5.3	43	HiSeq	101	180	250x
M.abscessus 6G-0125-R	5.1	64	HiSeq	100	335	115x
M.abscessus 6G-0125-R	5.1	64	MiSeq	250	335	100x
R.sphaeroides 2.4.1	4.6	69	HiSeq	101	220	210x
R.sphaeroides 2.4.1	4.6	69	MiSeq	251	540	100x
S.aureus M0927	2.9	33	HiSeq	101	180	250x
V.cholerae CO1032(5)	4.0	48	HiSeq	100	335	110x
V.cholerae CO1032(5)	4.0	48	MiSeq	250	335	100x
X.axonopodis pv. Manihotis UA323	2.9	33	HiSeq	101	400	250x

*Note: All datasets used paired-end reads from both ends of each fragment.

Prior to assembling with SPAdes, KmerGenie software [37] was used to estimates the best k-mer length for each assembly. To predict the best K value, the raw reads of each sample were used. First, we merged the forward and reverse reads into a single file. Then the file containing the forward and reverse reads was submitted to KmerGenie. For two samples, the assembly was performed with K values arbitrated by SPAdes itself.

After the assemblies, the contigs generated by both assemblers were submitted to the QUAST software [38] to measure the assembly performance. In addition to the number of contigs and the N50 value, the largest contig, completeness (genome fraction -%), misassemblies and number of genes (complete and partial) metrics were also computed.

3   The Proposed Hybrid Model

In this application, a standard Bloom Filter was implemented on a bit array to represent a set of fixed length contigs. For illustrative purposes, we assume that the size of the BF is 20. All positions in the filter are started with the value 0. At first, the contigs are presented to the BF (insertion operation). For each contig, BF uses hash functions (in this case 3) to determine positions along the array. Each hash function generates hash codes, which will be the corresponding positions in the filter. Since the functions are conventional, a uniform distribution of the positions generated along the structure occurs. Finally, each position indicated by a hash code has its value changed from 0 to 1. The contig does not have to be stored in the structure, only the indication of whether it is present or not, through the value of bit 0 (missing) or 1 (present) in the set. This makes BF fast and low on memory space usage.

To determine if a contig belongs to the set (test operation), the contig is submitted to the same hash functions that generate the positions in the vector. If the values in all generated positions are equal to 1, then that contig already belongs to the set. In this case we have a duplication, i.e., an equal contig has been detected and must be eliminated from the set. BFs are probabilistic structures, that is, if all positions correspond to 1, it can be said with high probability the contig is part of the set. BFs are susceptible to false positives. Otherwise, if the contig is submitted to BF and at least one of the generated hash codes maps to a position with a 0 value, it can be stated with absolute certainty that the contig does not belong to the set, since the BFs are not susceptible to false negatives. These membership queries are possible because of uniform hash functions properties.

The LSH approach used in this hybrid model to detect similar contigs is also used to find similar items in documents and is divided into three steps: Shingling, where contigs are divided into smaller k-size sequences (k-shingles or k-grams); MinHashing, where smaller sequences are converted to small numeric values (signatures) without losing their original similarity information; Locality-Sensitive Hashing, where the similarity of signature (candidate) pairs is verified.

The proposed computational application, BFLSH, was developed in Java and consists of two main modules. The first module implements a standard

Bloom Filter with MD5, CRC32, and SHA-1 hash functions to do an exact string match. Duplicate contigs are deleted and a new file is generated with the unique sequences only. Then, an LSH approach performs approximate string matching to identify similar contigs. The LSH function chosen was MinHash, which behaves well with the Jaccard distance. The LSH method was implemented as described in [39]. The contigs generated in the GAGE-B assemblies were separated into different files in FASTA format and submitted to the pipeline for detection and removal of redundant copies.

To compare and measure the performance of the proposed model, three other methods for reducing redundancy were implemented or used. The first, not yet published, is based on HS-BLASTN [40], which uses the Burrows-Wheeler transform as a basis for sequence alignment. The second was the Simplifier [13] and finally the CD-HIT [41] was applied.

All programs were performed on an AMD Opteron (TM) Processor 6376 computer, 64 CPUs, 1TB RAM, CentOS release 5.10 operating system OS Linux version 2.6.18371.6.1.el5. The similarity percentage used as a reference to consider contigs as similar was 75%.

4   Results and Discussion

BFLSH was applied to the 12 GAGE-B datasets to reduce redundant contigs and the results produced are shown in Table 2.

Table 2. BFLSH applied to the contigs of GAGE-B.

Species	# Contigs SPAdes	# Contigs BFLSH (%reduction)	# Contigs Fermi	# Contigs BFLSH (%reduction)
A.hydrophila (HiSeq)	350	277 (20,85%)	1.600	**371 (76,81%)**
B.cereus (HiSeq)	4.248	3.484 (17,98%)	1.993	**1.009 (49,37%)**
B.cereus (MiSeq)	347	153 (55,90%)	386	**129 (66,58%)**
B.fragilis (HiSeq)	191	155 (18,84%)	1.028	**538 (47,66%)**
M.abscessus (HiSeq)	349	343 (1,71%)	1.204	**246 (79,56%)**
M.abscessus (MiSeq)	2.769	**1.430 (48,35%)**	11.172	10.897 (2,46%)
R.sphaeroides (HiSeq)	514	321 (37,54%)	4.482	**1.863 (58,43%)**
R.sphaeroides (MiSeq)	4.031	4.031 (0%)	1.572	**727 (53,75%)**
S.aureus (HiSeq)	702	655 (6,69%)	343	**119 (65,30%)**
V.cholerae (HiSeq)	247	130 (47,36%)	1.866	**553 (70,36%)**
V.cholerae (MiSeq)	4.731	**2.584 (45,38%)**	7.461	7.276 (2,47%)
X.axonopodis pv. Manihotis (HiSeq)	305	212 (30,49%)	9.900	**4.508 (54,46%)**

* Bold and italic values represent for which assembly the method reduced the most contigs per organism, proportionally.

In percentage terms, BFLSH achieved better results in eliminating redundant contigs in SPAdes assemblies for the *M.abscessus* (MiSeq) and *V.cholerae* (MiSeq) data. On the other hand, the method performed better on *A.hydrophila* (HiSeq), *B.cereus* (HiSeq), *B.cereus* (MiSeq), *B.fragilis* (Hi-Seq), *M.abscessus*

(HiSeq), *R.sphaeroides* (HiSeq), *R.sphaeroides* (MiSeq), *S.aureus* (HiSeq), *V.cholerae* (HiSeq) and *X.axonopodis pv. Manihotis* (HiSeq). The results of the comparison between BFLSH and HSBLASTN, Simplifier and CD-HIT) in reducing SPAdes-generated contigs are shown in Table 3.

Table 3. Comparison between redundancy remove methods. SPAdes data.

Species	Raw contigs	BFLSH	HS-BLASTN	Simplifier	CD-HIT
A.hydrophila (HiSeq)	350	277	299	346	275
B.cereus (HiSeq)	4.248	3.484	3.677	4.102	3.353
B.cereus (MiSeq)	347	153	215	321	120
B.fragilis (HiSeq)	191	155	178	190	156
M.abscessus (HiSeq)	349	343	344	349	346
M.abscessus (MiSeq)	2.769	1.430	1.632	1.783	780
R.sphaeroides (HiSeq)	514	321	366	490	303
R.sphaeroides (MiSeq)	4.031	4.031	3.130	2.310	1.088
S.aureus (HiSeq)	702	655	680	698	649
V.cholerae (HiSeq)	247	130	167	222	120
V.cholerae (MiSeq)	4.731	2.584	2.671	3.522	1.780
X.axonopodis pv. Manihotis (HiSeq)	305	212	251	295	186

For the data generated by SPAdes, the most efficient method for decreasing contiguity redundancy was CD-HIT, having obtained the largest reduction for 10 out of 12 datasets (Table 3, however BFLSH achieved the second lowest number of contigs for nine organisms Table 3. For *R.sphaeroides* (MiSeq), BFLSH could not detect any contiguous redundancy in SPAdes data, while the other methods found and considerably reduced the number of contiguous sequences. When BFLSH was applied to the contigs generated with Fermi, there was a significant elimination of redundancy (53.75%), compatible with the result obtained by the other compared methods (HS-BLATN, Simplifier and CD-HIT).

It can be inferred that, for this organism and in the specific case of assembling with SPAdes, the proposed Hashing-based technique failed to properly map the actual similarities between the contigs into the Bloom Filter, that is, it was not able to efficiently translate the proximity between the elements of the set of contigs using the Jaccard similarity used in the LSH method.

Another hypothesis would be the usage of inappropriate parameters during the application of the method for this isolated case. Bloom Filter's parameter variance, such as the number of hash functions, the size of the BF, or even the amount of difference allowed to consider two items as similar can affect the engine output. LSH methods may be more or less stringent depending on the parameter used for similarity. In the case of this work, a value of 75% was used as default for all experiments. One last possibility can still be raised: the difference in the assembly approach used by SPAdes and Fermi. Since SPAdes is based on Bruijn

Table 4. Comparison between redundancy removal methods. Fermi Data.

Species	Raw contigs	BFLSH	HS-BLASTN	Simplifier	CD-HIT
A.hydrophila (HiSeq)	1.600	371	1.344	1.506	266
B.cereus (HiSeq)	1.993	1.009	1.298	1.689	777
B.cereus (MiSeq)	386	129	351	346	119
B.fragilis (HiSeq)	1.028	538	873	954	401
M.abscessus (HiSeq)	1.204	246	1.021	1.168	229
M.abscessus (MiSeq)	11.172	10.897	11.004	10.437	9.191
R.sphaeroides (HiSeq)	4.482	1.863	1.989	4.299	1.631
R.sphaeroides (MiSeq)	1.572	727	1.402	1.550	621
S.aureus (HiSeq)	343	119	290	253	99
V.cholerae (HiSeq)	1.866	553	1.514	1.788	425
V.cholerae (MiSeq)	7.461	7.276	7.384	7.014	6.291
X.axonopodis pv. Manihotis (HiSeq)	9.900	4.508	4.999	9.453	987

graphs, its internal way of generating contigs uses kmers, while Fermi, being from the OLC family of assemblers, does not. This factor may have in some way affected the nature of the contigs generated by SPAdes for this organism, which, combined with the other factors, may have led to unexpected yield for this case.

The results of the comparison between BFLSH and other methods for reducing Fermi-generated contigs are shown in Table 4.

In the comparison performed for the data generated by Fermi, the most efficient method to reduce contigs redundancy was CD-HIT, having obtained the largest reduction for the 12 datasets. However, BFLSH achieved the second lowest number of contigs for twelve data sets in absolute terms.

5   Conclusion

A computational application has been developed capable of efficiently detecting and eliminating redundant NGS contigs generated by *de novo* assemblers in microbial genomes. The problem of redundant contigs generated in NGS data assembly has been minimized with BFLSH.

For efficiency evaluation, the BFLSH was compared with three other distinct methods and proved to be efficient. Bloom Filter and LSH techniques can effectively be used to find similar items in biological sequences. The probabilistic nature of hash functions makes the possibility of false negatives possible, but these can be minimized with proper techniques.

One way to consider in the future is to adopt a parallelization approach, which can significantly increase the time and memory efficiency, especially of the LSH step.

Another alternative implementation for future thinking would be the use of other LSH functions, such as Super-Bit, which uses the Cosine similarity, among others [5].

References

1. Ambardar, S., Gupta, R., Trakroo, D., Lal, R., Vakhlu, J.: High throughput sequencing: an overview of sequencing chemistry. Indian J. Microbiol. **56**(4), 394–404 (2016)
2. Nagarajan, N., Pop, M.: Sequence assembly demystified. Nat. Rev. Genet. **14**(3), 157–167 (2013)
3. El-Metwally, S., Hamza, T., Zakaria, M., Helmy, M.: Next-generation sequence assembly: four stages of data processing and computational challenges. PLoS Comput. Biol. **9**(12), e1003345 (2013)
4. Martin, J.A., Wang, Z.: Next-generation transcriptome assembly. Nat. Rev. Genet. **12**(10), 671–682 (2011)
5. Goswami, M., et al.: Distance sensitive bloom filters without false negatives. In: Proceedings of the Twenty-Eighth Annual ACM-SIAM Symposium on Discrete Algorithms, SODA '17, USA, 2017, pp. 257–269. Society for Industrial and Applied Mathematics (2017)
6. Tang, L., Li, M., Fang-Xiang, W., Pan, Y., Wang, J.: MAC: Merging assemblies by using adjacency algebraic model and classification. Front. Genet. **10**, 1396 (2020)
7. de Sousa Paz, H.E.: reSHAPE : montagem hibrida de genomas com foco em organismos bacterianos combinando ferramentas de novo. Dissertacao (2018)
8. Treangen, T.J., Salzberg, S.L.: Repetitive DNA and next-generation sequencing: computational challenges and solutions. Nat. Rev. Genet. **13**(1), 36–46 (2011)
9. Batzer, M.A., Deininger, P.L.: Alu repeats and human genomic diversity. Nat. Rev. Genet. **3**(5), 370–379 (2002)
10. Zavodna, M., Bagshaw, A., Brauning, R., Gemmell, N.J.: The accuracy, feasibility and challenges of sequencing short tandem repeats using next-generation sequencing platforms. PLoS ONE **9**(12), e113862 (2014)
11. Phillippy, A.M., Schatz, M.C., Pop, M.: Genome assembly forensics: finding the elusive mis-assembly. Genome Biol. **9**(3), R55 (2008)
12. Wetzel, J., Kingsford, C., Pop, M.: Assessing the benefits of using mate-pairs to resolve repeats in de novo short-read prokaryotic assemblies. BMC Bioinform. **12**(1), 95 (2011)
13. Bradnam, K.R., et al.: Assemblathon 2: evaluating de novo methods of genome assembly in three vertebrate species. GigaScience **2**(1), 2047–217X (2013)
14. Nagarajan, N., Pop, M.: Parametric complexity of sequence assembly: theory and applications to next generation sequencing. J. Comput. Biol. **16**(7), 897–908 (2009)
15. Ramos, R.T.J., Carneiro, A.R., Azevedo, V., Schneider, M.P., Barh, D., Silva, A.: Simplifier: a web tool to eliminate redundant NGS contigs. Bioinformation **8**(20), 996–999 (2012)
16. Galil, Z., Giancarlo, R.: Data structures and algorithms for approximate string matching. J. Complex. **4**(1), 33–72 (1988)
17. Pandiselvam, P., Marimuthu, T., Lawrance, R.: A comparative study on string matching algorithm of biological sequences (2014)
18. Al-Khamaiseh, K., ALShagarin, S.: A survey of string matching algorithms. Int. J. Eng. Res. Appl. **4**, 144–156 (2014)

19. Wang, J., Shen, H.T., Song, J., Ji, J.: Hashing for similarity search: a survey (2014)
20. Chauhan, S.S., Batra, S.: Finding similar items using lsh and bloom filter. In: 2014 IEEE International Conference on Advanced Communications, Control and Computing Technologies, pp. 1662–1666 (2014)
21. Bender, M.A., et al.: Don't thrash: how to cache your hash on flash. Proc. VLDB Endow. **5**, 1627–1637 (2012)
22. Slaney, M., Casey, M.: Locality-sensitive hashing for finding nearest neighbors [lecture notes]. Signal Process. Mag. IEEE **25**, 128–131 (2008)
23. Baluja, S., Covell, M.: Learning to hash: forgiving hash functions and applications. Data Min. Knowl. Disc. **17**(3), 402–430 (2008)
24. Indyk, P., Motwani, R.: Approximate nearest neighbors: towards removing the curse of dimensionality. In: Conference Proceedings of the Annual ACM Symposium on Theory of Computing, pp. 604–613 (2000)
25. Broder, A.Z.: On the resemblance and containment of documents. In: Proceedings of the Compression and Complexity of SEQUENCES 1997 (Cat. No.97TB100171), pp. 21–29 (1997)
26. Jain, R., Rawat, M., Jain, S.: Data optimization techniques using bloom filter in big data. Int. J. Comput. Appl. **142**, 23–27 (2016)
27. Stephens, Z.D., et al.: Big data: astronomical or genomical? PLoS Biol. **13**(7), e1002195 (2015)
28. Andoni, A., Indyk, P.: Near-optimal hashing algorithms for approximate nearest neighbor in high dimensions. Commun. ACM **51**(1), 117–122 (2008)
29. Ding, K., Huo, C., Fan, B., Xiang, S., Pan, C.: In defense of locality-sensitive hashing. IEEE Trans. Neural Netw. Learn. Syst. **29**(1), 87–103 (2018)
30. Bloom, B.H.: Space/time trade-offs in hash coding with allowable errors. Commun. ACM **13**(7), 422–426 (1970)
31. Broder, A., Mitzenmacher, M.: Survey: Network applications of bloom filters: a survey. Internet Math. **1**, 11 (2003)
32. Naor, M., Yogev, E.: Tight bounds for sliding bloom filters. Algorithmica **73**(4), 652–672 (2015)
33. Magoc, T., et al.: GAGE-B: an evaluation of genome assemblers for bacterial organisms. Bioinformatics **29**(14), 1718–1725 (2013)
34. Aronesty, E.: Comparison of sequencing utility programs. Open Bioinform. J. **7**(1), 1–8 (2013)
35. Bankevich, A., et al.: SPAdes: a new genome assembly algorithm and its applications to single-cell sequencing. J. Comput. Biol. **19**(5), 455–477 (2012)
36. Li, H.: Exploring single-sample SNP and INDEL calling with whole-genome de novo assembly. Bioinformatics (Oxford, England) **28**(14), 1838–1844 (2012)
37. Chikhi, R., Medvedev, P.: Informed and automated k-mer size selection for genome assembly. Bioinformatics **30**(1), 31–37 (2013)
38. Gurevich, A., Saveliev, V., Vyahhi, N., Tesler, G.: QUAST: quality assessment tool for genome assemblies. Bioinformatics **29**(8), 1072–1075 (2013)
39. Leskovec, J., Rajaraman, A., Ullman, J.D.: Mining of Massive Datasets. Cambridge University Press, Cambridge (2014)
40. Chen, Y., Ye, W., Zhang, Y., Yuesheng, X.: High speed BLASTN: an accelerated MegaBLAST search tool. Nucleic Acids Res. **43**(16), 7762–7768 (2015)
41. Li, W., Godzik, A.: CD-HIT: a fast program for clustering and comparing large sets of protein or nucleotide sequences. Bioinformatics **22**(13), 1658–1659 (2006)

Efficient Out-of-Core Contig Generation

Julio Omar Prieto Entenza[iD], Edward Hermann Haeusler[iD], and Sérgio Lifschitz$^{(\boxtimes)}$ [iD]

Departamento de Informática - (PUC-Rio), Rio de Janeiro, Brazil
sergio@inf.puc-rio.br

Abstract. Genome sequencing involves splitting a genome into a set *reads* that are assembled into *contigs* that are eventually ordered and organized as *scaffolds*. There are many programs that consider the use of the *de Bruijn* Graph (dBG) but they must deal with a high computational cost, mainly due to internal RAM consumption. We propose to use an external memory approach to deal with the *de Bruijn* graph construction focusing on contig generation. Our proposed algorithms are based on well-known I/O efficient methods that identify unitigs and remove errors such as tips and bubbles. Our analytical evaluation shows that it becomes feasible to generate *de Bruijn* graphs to obtain the needed contigs, independently of the available memory.

1 Introduction

Genome sequencing is the process that determines the order of nucleotides within a DNA molecule. Modern instruments splits a genome into a set of many short sequences (*reads*) that are assembled into longer contiguous sequences, *contigs*, followed by the process of correctly ordering *contigs* into *scaffolds* [18].

We may associate *genome sequencing* with the problem of finding a Hamiltonian Cycle through an Overlap Layout Consensus (OLC) assembly method. Alternatively, it can be modeled as the problem of finding a Eulerian Cycle considering the *de Bruijn Graph* (dBG) based methods [14]. The latter can be seen as a breakthrough for the research on genome assembly. This is due to the fact that to find a Hamiltonian Cycle is an NP-complete problem [14].

When we handle actual dBGs, we must consider the existence of errors that appear due to high-frequencies distortions on Next-generation Sequencing (NGS) platforms. These errors induce the dBG size to be more prominent than the *overlap* graph used in the OLC genome sequencing method using the same reads.

To remove the errors, we need to have some data structure representation of the dBG. Current real-world datasets induce challenging problems as they have already reached high volumes and will continue to grow as sequencing technologies improve [19]. As a consequence, the dBG increases the complexity leading to tangles and topological structures that are difficult to resolve [16]. Also, the graph has a high memory footprint for large organisms (*e.g.,* sugarcane plants) and it becomes worse due to the increase of the genome datasets. Therefore, there are research works that focus on dealing with the ever-growing graph sizes

© Springer Nature Switzerland AG 2020
J. C. Setubal and W. M. Silva (Eds.): BSB 2020, LNBI 12558, pp. 25–37, 2020.
https://doi.org/10.1007/978-3-030-65775-8_3

for dBG-based genome assembly [3,15]. Any proposed solution must be aware of the fact that the dBG is not entirely built before error pruning. After the splitting of the genome in the first phase and, for a natural number k chosen based on an empirical criterion, for each *read* of size m we will have $m - k + 1$ possible k-mers, which correspond to the nodes of the dBG. The number of k-mers depends on the adjacency determined by the *read* itself. The splitting error introduced in a given *read* may large increase the size of the subgraph it induces.

Roughly speaking, the dBG is a set of k-mers (subsequences of length k) [23] linked to each other, according to information provided by their *reads*. Some k-mers can come from different *reads* and the information about adjacency supplied by any other *read* is processed only at dBG constructing phase. Thus, error pruning should happen at this particular stage. The memory needed is so significant that we must use external memory to accomplish the dBG construction.

The different approaches that deal with the dBG size aim to design lightweight data structures to reduce the memory requirements and to fit the assembly graph into the main memory. Although it might be efficient, the amount of memory increases according to the size of the dataset and the DNA of the organism. While bacteria genomes currently take only a few gigabytes of RAM, species like mammalian and plants require over tens to hundreds of gigabytes. For instance, approximately 0.5 TB of memory is required to build a hash table of 75-mers for the human genome [19].

We propose algorithms to simplify and remove errors in the *de Bruijn* graph using external memory. As a result, it will be able to generate contigs using a fixed amount of RAM, independently of the read dataset size. There are other works addressing *de Bruijn* graph processing using external memory [8,11], but they focus only on the constructing of large *de Bruijn* graphs efficiently with no error prune considerations. To the best of our knowledge, this is the first proposal of using an external memory approach focusing on the dBG simplification and errors removal for contig generation during dBG construction.

We show an algorithm that provides out-of-core *contraction of unambiguous paths* with an I/O cost of $O(|E|/B)$, where E is the set of edges of the dBG and B is the size of the partition loaded to the RAM each time. With the overhead for creating the new partitions, the overall I/O complexity is $O((sort(|E|) + |E|/B) \log Path)$, where $Path$ is the length of the longest unambiguous path in the dBG. For a machine with memory M, and a dBG satisfying $|E| < M^2/4B$ $sort(E)$ is performed with I/O complexity $O(|E|/B)$ [9]. Summing up the I/O complexity of this out-of-core contraction of paths is $O((|E|/B) \log Path)$. The out-of-core *graph cleaning* phase, by removing *tips* and *bubbles*, is performed with a similar I/O complexity $O((|E|/B + sort(|E|)) log Path)$. The *creation of contigs* is performed by a full scan of the graph with I/O complexity $O(|V|/B)$.

2   De Novo Assembly Using *de Bruijn* Graph

Given a *de Bruijn* Graph $G = (V, E)$ for genomic sequence assembly each node holds unique k-mers, and the edges reflect consecutive nodes if they overlap by

$k-1$ characters. The assembly aims to construct a set of contigs from the dBG G. Given a dBG as input, to generate contigs is equivalent to output all contiguous sequences which represent unambiguous paths in the graph.

The use of the dBG to generate contigs consists of a pipeline: nodes enumeration, compaction, and graph cleaning. In the first step, a set of distinct k-length substrings (k-mers) is extracted from the reads. Each k-mer becomes a graph node. Next, all paths with all but the first vertex having in-degree 1 and all but the last vertex having out-degree 1 (unitigs) are compacted into a single vertex. Finally, the last step removes topological issues from the graph due to sequencing errors and polymorphism [5].

The number of nodes in the graph can be huge. For instance, the size of the genome of white spruce is 20 Gbp and generates 10.7×10^9 k-mers (with k = 31) and needs 4.3 TB of memory [5]. Also, the whole genome assembly of 22 Gbp (bp - base pairs) loblolly pine generates 13×10^9 k-mers and requires 800GB of memory [5]. Theoretically speaking, a 1,000 Genomes dataset with 200 Terabytes of data can generate about 2^{47} or nodes, 64–128 times larger than the problem size of the top result in the Graph 500 list [15].

Next-generation sequencing platforms do not provide comprehensive read data from the genome sequences. Hence, the produced data is distorted by high frequencies of sequencing errors and genomic repeats [18]. Sequencing errors compound this problem because each such error corrupts the correct genomic sequence into up to k erroneous k-mers. These erroneous k-mers introduce new vertices and edges to the graph, significantly expanding its size and creating topological artifacts as tips and bubbles [23].

Different solutions have been proposed to address the memory issues in genome assembly problem. One approach samples the entire k-mer set and performs the assembly process over the selected k-mers [21]. Another approach address to encode the dBG into efficient data structures such as light-weight hash tables [3], succinct data structures [2] or bloom filters[6,17]. There are research works based on distributed memory systems for processing power and memory demanding resources [3,7,15].

Although their apparent differences, all of these approaches are based exclusively on in-memory systems. Consequently, if the size of the graph exceeds the amount of memory available, it will be necessary to increase the RAM. As in the next future, the size of datasets will increase dramatically, and this situation will stress the different systems [20]. There is a need for new approaches to process all of this massive amount of information in a scalable way. We propose in this work to use an external memory approach to process the dBG. To increase the amount of RAM does not guarantee that the graph will always fit.

3   Overview of Our Proposed Approach

Our basic pipeline of *de novo* genome assembly could be divided into five basic operations [23]: 1) *dBG construction*, which constructs a dBG from the DNA reads; 2) *Contraction of unambiguous paths*, which merges unambiguous vertices into unitigs; 3) *Graph cleaning*, which filters and removes errors such as tips and bubbles; 4) *Contigs creation*, which create a first draft of the contigs and 5) *Scaffolds*, which joins the previous contigs together to form longer contigs. In this work, we face steps 2, 3, and 4 using an external memory approach. We assume a *de Bruijn* graph exists, and it is persisted as an edge-list format.

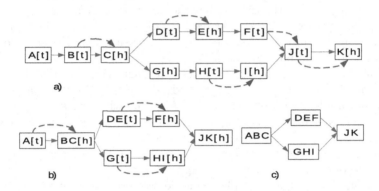

Fig. 1. Graph Contraction. Dashed arcs represent the messages and the label between brackets indicates if a vertex is a tail (t) or a head (h). a) A flipped coin choose which node will be t or h. If a tail vertex has out-degree = 1 then it sends a message to its neighbour. b) If a h vertex receives a message and its in-degree = 1 then both vertices are merged. c) Shows the result after some repeated steps.

The **Contraction of unambiguous paths** simplifies the graph by merging some nodes in the graph. Whenever a node with only one outgoing directed edge points out to another node with only one incoming directed edge, these two nodes are merged. These single nodes are called *unitigs* and are maximal if they extend in either direction. Thus, the problem of compacting a *de Bruijn* graph is to report the set of all maximal *unitigs*. Figure 1 shows how we simplify the dBG into a compacted graph. All the remained nodes are maximal *unitigs*.

The algorithmic solution to this problem is straightforward in the in-memory context. Let's assume that each path representing a unitig is a linked list where the head and the tail can be branching nodes (see Fig. 1a). Then, to obtain a maximal unitig, we only need to visit each u_c node and merge them with its successor node into the new one.

Although the in-memory algorithm is straightforward, the use of this approach is not efficient in the external memory because the number of I/O accesses is linear concerning the number of nodes. Finding all the maximal unitigs is analogous to apply the well-known graph edge contraction technique in external memory on all the unambiguous paths. Given a graph G, the contraction of an edge (u, v) is the replacement of u and v with a single vertex such that edges incident to the new vertex are the edges other than (u, v) that were incident with u or v [22]. As in our case, this algorithm can only apply on unambiguous paths. Therefore, we divided our proposal into two steps: 1) select branching nodes u_a

as head/tail of each unitig paths; and 2) apply the graph contraction technique over these paths until all nodes are maximal unitigs.

In *Graph cleaning* we aim to remove short dead-end divergences, called **tips**, from the main path. One strategy consists of testing each branching node for all possible path extensions up to a specified minimum length. If the length of the path is less than a certain threshold (set by the user) then the nodes belonging to this path are removed from the graph [7,23].

The tips removing process is analogous to traversal the paths from a branching node, u_a, to a dead-end node u_e. The graph does not fit into RAM, even after the unitig process. We need to traverse the dBG in an I/O efficient way to find and remove all tips. Our algorithm is based on an external memory list ranging from the u_e to u_a nodes. However, we have to make two significant modifications: (i) the ranking is represented by each edge/node's coverage to decide which path will be removed, and (ii) as two of more dead-ends could reach the same u_a node, we need to keep in RAM a data structure to make the traversal backward. This way, we eliminate the selected path from the branching node.

Bubbles are paths that diverge from a node then converge into another. The process of fixing bubbles begins by detecting the divergence points in the graph [23]. For each point, all paths from it are detected by tracing the graph forward until a convergence point is reached. Some assemblers restricts the size of the bubble to n nodes where $k \leq n \leq 2k$ [7], others use a modified version of Dijkstra's algorithm [23]. To simulate the different external memory approaches, we need to identify all branching nodes u_a. We execute an I/O-efficient breathfirst search (BFS) from u_a until we find a visited node. It means that there is a bubble at some point in the search (v_b). Then, we select the branch that will be kept and start another BFS in a backward direction (the start node is v_b). Finally, we remove the other paths until we find back u_a. After the execution of steps 2 and 3, the *Contigs creation* step involves the output of all the contigs represented by nodes.

Processing Out-of-Core Graphs. Many graph engines implement a vertex-centric or "think like a vertex" (TLAV) programming interface. This paradigm iteratively executes a user-defined program over vertices of a graph. The vertex program is designed from the vertex's perspective, receiving as input the data from adjacent vertices and incident edges. Execution halts after a specified number of iterations, called supersteps, are completed. It is important to note that each vertex knows the global supersteps. There is no other knowledge about the overall graph structure but its immediate neighbors and the messages that are exchanged along their incident edges [13].

The computation graph engines proceed in supersteps. For every superstep, it loads one or more partitions p based on available RAM. Then, it processes the vertices and edges that belong to p and saves the partitions back to disk. A different subset of partitions is then loaded and processed until all of them have been treated for the given superstep. The process is then repeated for the next superstep until there are no remaining vertices to visit (see Algorithm 1

Algorithm 1. Out-of-core graph processing taken from GraphChi [9]

Input:
    $G = (V, E)$: a general graph
    UpdateFunction: a user-defined program to apply over the vertices
1: **procedure** OUTOFCORE(G, *UpdateFunction*)
2:    **for each** superstep $\in$ Algorithm **do**   ▷ superstep is a global variable
3:        **for each** p $\in$ Partition **do**
4:            p.load()   ▷ Load partition to RAM
5:            **parallel for each** v $\in$ p.vertices() **do** ▷ Apply UpdateFunction
6:                UpdateFunction(G,v)
7:            p.save()   ▷ Save partition to disk

---

from GraphChi [9]). If the machine has sufficient RAM to store the graph and metadata, all partitions can be kept in RAM, and there is no disk access [9].

Because *all the operations related to partitions and parallel vertices processing are fixed*, from now we will only highlight the *UpdateFunction* and the number of supersteps. For simplicity, *UpdateFunction(G,u)* means that we apply the function over the vertex u in the graph G.

4   Contig Generation

The graph contraction algorithm (Algorithm 2) is based on I/O-efficient list ranking algorithm based on graph contraction [4] but using a TLAV [12]. The output is the distance of each node from the beginning of the list. In this case, it corresponds to k-mer concatenation. Thus, in our output, the beginning node will have the unitig concatenation. Initially, we assign the k-mer as the rank of each vertex. We then continue recursively: first, in one superstep, we find a maximal independent set among all vertices that belong to a unitig (line 11). For each of these, we flip a fair coin and vertices that flipped "tails" pick a neighbor that flipped "heads" (if any) to contract with [1]. Later the vertices identified as "tails" with precisely one outgoing edge send the required information to its neighbors (lines 12 and 13). In the next superstep, the vertices marked as head, with only one in-going edge, are merged with tail vertices, and the edge information is updated (lines 17 to 22). The function *PreprocessNewGraph* creates new partitions from the removed and added vertices and edges (line 9). This step implies merging nodes, and remove duplicate edges and update the graph partitioning. Next, we recursively continue coarsening the graph until all unambiguous nodes are collapsed.

I/O Analysis. Lines 11–22 can be done with a full scan over all the graph partitions. We load a partition into RAM, we update the vertices and edge values, and then write them to the disk. Then, we load the next partition and its vertices and edges according to the saved partition. So, the I/O cost is $O(|E|/B)$.

On the other hand, *PreprocessNewGraph* (line 9) creates new partitions from superstep $i - 1$ to be processed in the superstep i. To create the new

Algorithm 2. Graph Contraction

Input: $G = (V, E)$: a contracted dBG
Output: G': a graph with all no ambiguous paths contracted.
 1: **procedure** CONTRACTGRAPH(G)
 2:     $superstep \leftarrow 0$ and $merge \leftarrow true$
 3:     **while** $merge = true$ **do**
 4:         $merge \leftarrow false$
 5:         $MergeNodes(G, u)$ ▷ External Memory Context
 6:         $superstep \leftarrow superstep + 1$
 7:         **if** $superstep$ is odd **then**
 8:             $G \leftarrow PreprocessNewGraph(G)$

 9: **procedure** MERGENODES(u)
10:     $flag \leftarrow T \vee H$ in randomly way
11:     **if** superstep is even **then**
12:         **if** $flag = true \wedge d^+(u) = 1$ **then**
13:             $send(\{flag, seq, neighbors, id\}, j)$ ▷ message to outgoing edge
14:     **else**
15:         **if** $d^-(u) = 1$ **then**   ▷ merge nodes from incoming neighbor
16:             $m \leftarrow receive(\{flag, seq, i\})$
17:             **if** $m.flag \neq flag$ **then**
18:                 $add\_edge(u, m.neighbors)$
19:                 $seq \leftarrow glue(seq, m.seq)$
20:                 $delete(m.id)$
21:                 $merge \leftarrow true$

partitions, first, it is necessary to divide the nodes by their ID and later sort all the edges based on their destination vertex ID [9]. Thus, the I/O cost of the process for the created graph is $O(sort(|E|))$.

Finally, at each superstep, Algorithm 2 contracts a constant fraction of the vertices per iteration by graph contraction [22]. It expected $O(\log Path)$ iterations with high probability, where $Path$ is the longest unambiguous path in the graph [1]. Hence, the overall expected I/O cost for simplifying the graph is $O((sort(|E|) + |E|/B) \log Path)$. If we assume $|E| < M^2/(4B)$ then $sort(|E|) = O(|E|/B)$. This condition can be satisfied with a typical value of M, say 8 GB, B in the order of kilobytes and a graph size smaller than a petabyte [10]. On these conditions the I/O cost is $O((|E|/B) \log Path)$ (Fig. 2).

To **remove tips**, we design a straightforward procedure in few supersteps (Algorithm 3). At this point, all nodes are contracted. Thus, all terminal nodes are potential tips, and they may be removed. In the first superstep, the vertices having an in-degree or out-degree of zero and sequence's length less than $2k$ are identified as potential tips (line 10), and a message is sent to their neighbors (line 11). In the next superstep (lines 14–17), the vertices that received the messages, search for the maximal multiplicity among all neighbors and remove those potential tips with multiplicity less than the maximal value. Removing the tips generates new linear paths in the graph that will be contracted. Note

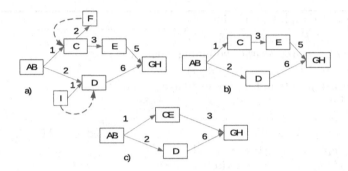

Fig. 2. Removing tips. Dotted lines show the sent messages. a) F and I are marked as potential tips. Later, they send a message to their neighbors. b) C and D receive the messages and check their multiplicity to eliminate the real tips. So, H and I are removed. c) The graph is compressed to obtain the final result.

that once the initial set of tips are removed, it could produce other tips. Most assemblers execute the removal tip algorithm a fixed number of times.

Algorithm 3. Tips removal

Input: $G = (V, E)$: a dBG
Output: G': another graph with tips removed.
1: **procedure** TIPS(G)
2:     $superstep \leftarrow 0$
3:     **while** $tips = true$ **do**
4:         $tips \leftarrow false$
5:         $RemoveTips(G, u)$
6:         $superstep \leftarrow superstep + 1$
7:     $ContractGraph(G)$)

8: **procedure** REMOVETIPS(u)   ▷ External Memory Context
9:     **if** $superstep = 0$ **then**
10:         **if** u is $u_e \wedge |seq| \leq 2k$ **then**   ▷ Identify all potential tips
11:             $send(id)$ ▷ Send a message to its neighbor
12:     **else**
13:         $maximal \leftarrow \max(\forall u.neighbors.multip)$ ▷ Get max multiplicity
14:         **for all** $m \in receive(id)$ **do** ▷ Identify the real tips and remove them
15:             **if** $(u, m.id).mult < maximal$ **then**
16:                 $delete\_node(m.id)$
17:                 $tips \leftarrow true$

I/O Analysis. The function *RemoveTips* needs only two supersteps to remove any tip given a graph: one to identify all tips and other to removes all of them (lines 8–17). This can be carried out with at most two full scans over all the graph partitions. Thus, the I/O cost is $O(|E|/B)$. Although *RemoveTips* only executes two supersteps, the graph contraction dominates the I/O cost (line 7). The I/O cost is $O((|E|/B + sort(|E|)) \log Path)$, where $Path$ represents the longest path created after the tips are removed.

The primary approach to **identify and remove bubbles** is based on BFS (breadth-first search) algorithm. As bubbles consist of paths with very different multiplicities, those paths with low multiplicity are deleted and use the path with the highest multiplicity to represent the bubble. In our algorithm, each vertex manages its history, which makes it easy to control the different paths and to pick up the right one.

The proposed algorithm has two stages: forward and backward. In the forward stage (Algorithm 4) identifies all paths that form a bubble and select one of them. On the other hand, the backward stage, (Algorithm 5), removes the redundant paths and compacts the graph. Due to space limitations, we will illustrate and explain both algorithms by examples. See Fig. 3 for the forward stage and Fig. 4 for the backward stage.

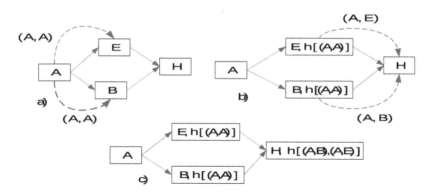

Fig. 3. Forward Bubble Detection. The figures only show the id par in the sent messages. a) A is a potential bubble beginning, so it sends messages to the outgoing neighbors. b) B and E keep it and forward new messages updating the vertices IDs. c) When H, in-degree ≥ 2, receives a message it selects that E belongs to a bubble. Later, it marks the node for the backward stage.

I/O Analysis. In the forward stage, *SelectPath* (line 4) iterates through the bubbles in a constant number of times depend on a *limit* value. Also, only a small number of messages are present in the graph, each one originating from any ambiguous vertex whose out-degree is greater than 2. Moreover, each message will be passed along an edge exactly once, as notifications are only sent

---

Algorithm 4. Forward stage bubble detection

Input:
    $G = (V, E)$: a compressed dBG and *limit*: max. length of the bubble path
Output: G': a graph with all bubbles selected.
1: **procedure** FINDBUBLE(G)
2:    $superstep \leftarrow 0$
3:    **while** $superstep \leq limit$ **do**
4:        $SelectPath(G, u))$
5:        $superstep \leftarrow superstep + 1$
6: **procedure** SELECTPATH(G,u)   ▷ External Memory Context
7:    **if** $superstep = 0 \land d^+(vertex) \geq 2$ **then** ▷ Identify possible bubble start
8:        $outgoing.add(id, id, seq)$
9:    **else**
10:        $outgoing \leftarrow \{\}$ and $incomming \leftarrow receive(id1, id2, seq)$
11:        **for each** $m \in incomming$ **do**
12:            **if** $m.id \notin history$ **then**
13:                $m.seq \leftarrow glue(seq, m.seq)$ and $m.id2 \leftarrow id$
14:                $history.add(m)$ and $outgoing.add(m)$
15:            **else**
16:                **for each** $h \in history$ **do**   ▷ All possible bubble ends
17:                    **if** $h.id1 = m.id1$ **then**
18:                        $apply\_heuristic(m, h)$
19:                      **if** m is bubble **then**
20:                          mark $m.id2$ and $history.add(m)$
21:                      **else**
22:                          mark $h.id2$
23:    $send(outgoing)$ ▷ send message to all out-edges

---

along outgoing edges. This means that only in-edges are read, and the out-edges are written. As this algorithm implies an external BFS traversal, the I/O cost is $O(BPath(|V| + |E|)/B) \approx O(BPath|E|/B)$ where $BPath$ is the longest length among all bubbles and $|E| = O(|V|)$ because a dBG is a sparse graph. Then the total I/O cost is $O(limit * BPath * |E|/B) = O(BPath * |E|/B)$. On the other hand, the backward stage uses another BFS but in the opposite direction to the forward phase, so the I/O cost is the same. Additionally, it iterates through the set of vertices (lines 6–8) and executes a contraction on the resulting graph. Therefore, the I/O cost of the backward stage is $O((|E|/B + sort(|E|)) \log BPath)$.

At this point, the graph should be formed by contracted vertices. As each vertex represents a contig, we can output them using a full scan over the graph. Thus, the I/O complexity is $O(|V|/B)$, where $|V|$ is the number of contigs.

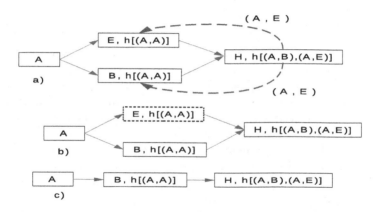

Fig. 4. Backward Bubble Elimination. a) H is a bubble ending, so it sends messages to the incoming neighbors communicating that E is marked. b) When E receives the message, it is marked and does not send more messages because it detects that the next node is the bubble. B does not send any message because its history does not have the node E. c) The graph is simplified after all bubbles are removed.

Algorithm 5. Backward step bubble detection

Input:
$G = (V, E)$:compressed dBG and *limit*: max. length of the bubble path
Output: G': a graph with all bubbles nodes marked.

```
 1: procedure BACKTRACKPATH(G)
 2:     superstep ← 0
 3:     while superstep ≤ limit do
 4:         RemovePath(G, u)
 5:         superstep ← iter + 1
 6:         for each u ∈ V do                    ▷ External Memory Context
 7:             if u.mark = true then
 8:                 delete(v)
 9:         CompressGraph(G))
10: procedure REMOVEPATH(G,u)               ▷ External Memory Context
11:     if superstep = 0 ∧ d⁻(u) ≥ 2 then   ▷ Identify all possible bubble ends
12:         for each m ∈ history do
13:             if m is marked then
14:                 outgoing.add(m.id, m.id2)
15:     else
16:         outgoing ← {}; incomming ← receive(id1, id2)
17:         for each m ∈ incomming do    ▷ Find the closest common ancestor
18:             if m.id2 = id then
19:                 dst ← history.pop(m.id)
20:                 if dst ≠ id then
21:                     outgoing.add(m) and mark = true
22:     send(outgoing)
```

5   Conclusions

In this paper, we have proposed out-of-core algorithms for dealing with contig generation, one of the most critical steps of fragment assembly methods based on *de Bruijn* graphs. Besides presenting these algorithms, we have made an I/O analytical evaluation that shows that it becomes feasible to generate *de Bruijn* graphs to obtain the needed contigs, independently of the available memory.

The I/O cost studies show that graph simplification is one of the most expensive steps. Actually, we could expect it because this phase involves a more significant number of vertices and edges. To deal with that, one could choose to do the assembly without simplifying the graph. In this condition, a tip is a branch with low coverage and not just a vertex. Hence, it will be necessary to apply a list ranking from the dead-end to branching nodes.

Among other issues, we may cite that it is hard to estimate the number of executions related to each phase. Primarily it depends on the number of nodes in the graph, which itself depends on the properties of the read's datasets. As these values vary from one dataset and sequencing technology to another, the assembly algorithms execute each step, a fixed and empirical number of times. As future work, we may cite the evaluation of other graph simplification approaches targeting erroneous and non-recognizable structures, such as X-cuts.

Acknowledgments. This work is partially supported by grants from CNPq, CAPES and FAPERJ, Brazilian Public Funding Agencies.

References

1. Anderson, R.J., Miller, G.L.: A simple randomized parallel algorithm for list-ranking. Inf. Process. Lett. **33**(5), 269–273 (1990)
2. Bowe, A., Onodera, T., Sadakane, K., Shibuya, T.: Succinct de Bruijn Graphs. In: Raphael, B., Tang, J. (eds.) WABI 2012. LNCS, vol. 7534, pp. 225–235. Springer, Heidelberg (2012). https://doi.org/10.1007/978-3-642-33122-0_18
3. Chapman, J.A., et al.: Meraculous: de novo genome assembly with short paired-end reads. PLoS One **6**(8), e23501 (2011)
4. Chiang, Y.J., et al.: External-memory graph algorithms. Procs. ACM/SIAM Symp. Discr. Algorithm. (SODA) **95**, 139–149 (1995)
5. Chikhi, R., Limasset, A., Medvedev, P.: Compacting de Bruijn graphs from sequencing data quickly and in low memory. Bioinformatics **32**(12), i201–i208 (2016)
6. Chikhi, R., Rizk, G.: Space-efficient and exact de Bruijn graph representation based on a bloom filter. Algorithm. Mol. Biol. **8**(1), 22 (2013)
7. Jackman, S.D., et al.: Abyss 2.0: resource-efficient assembly of large genomes using a bloom filter. Genome Res. **27**(5), 768–777 (2017)
8. Kundeti, V.K., et al.: Efficient parallel and out of core algorithms for constructing large bi-directed de Bruijn graphs. BMC Bioinf. **11**(1), 560 (2010)
9. Kyrola, A., Blelloch, G., Guestrin, C.: Graphchi: large-scale graph computation on just a PC. In: USENIX Symposium on Operating Systems Design and Implementation (OSDI), pp. 31–46 (2012)

10. Kyrola, A., Shun, J., Blelloch, G.: Beyond synchronous: new techniques for external-memory graph connectivity and minimum spanning forest. In: Gudmundsson, J., Katajainen, J. (eds.) Experimental Algorithms — SEA 2014. Lecture Notes in Computer Science, vol. 8504, pp. 123–137. Springer, Cham (2014). https://doi.org/10.1007/978-3-319-07959-2_11

11. Li, Y., Kamousi, P., Han, F., Yang, S., Yan, X., Suri, S.: Memory efficient minimum substring partitioning. Proc. VLDB Endow. 6(3), 169–180 (2013)

12. Malewicz, G., et al.: Pregel: a system for large-scale graph processing. In: Process ACM SIGMOD Intl. Conf. on Manage. Data, pp. 135–146 (2010)

13. McCune, R.R., Weninger, T., Madey, G.: Thinking like a vertex: a survey of vertex-centric frameworks for large-scale distributed graph processing. ACM Comput. Surv. (CSUR) 48(2), 25:1–25:39 (2015)

14. Medvedev, P., Georgiou, K., Myers, G., Brudno, M.: Computability of models for sequence assembly. In: Giancarlo, R., Hannenhalli, S. (eds.) Algorithms in Bioinformatics — WABI 2007. Lecture Notes in Computer Science, pp. 289–301. Springer, Cham (2007). https://doi.org/10.1007/978-3-540-74126-8_27

15. Meng, J., Seo, S., Balaji, P., Wei, Y., Wang, B., Feng, S.: Swap-assembler 2: optimization of de novo genome assembler at extreme scale. In: Proceedings of the 45th ICPP, pp. 195–204. IEEE (2016)

16. Miller, J.R., Koren, S., Sutton, G.: Assembly algorithms for next-generation sequencing data. Genomics 95(6), 315–327 (2010)

17. Salikhov, K., Sacomoto, G., Kucherov, G.: Using cascading bloom filters to improve the memory usage for de Brujin graphs. Algorithm. Mol. Biol. 9(1), 2 (2014)

18. Simpson, J.T., Pop, M.: The theory and practice of genome sequence assembly. Ann. Rev. Genomics Hum. Genet. 16, 153–172 (2015)

19. Sohn, J., Nam, J.W.: The present and future of de novo whole-genome assembly. Briefings Bioinf. 19(1), 23–40 (2016)

20. Stephens, Z.D., et al.: Big data: astronomical or genomical? PLoS Bio. 13(7), e1002195 (2015)

21. Ye, C., Ma, Z.S., Cannon, C.H., Pop, M., Douglas, W.Y.: Exploiting sparseness in de novo genome assembly. BMC (BioMed Central) Bioinf. 13, S1 (2012)

22. Zeh, N.: I/o-efficient graph algorithms. In: EEF Summer School on Massive Data Sets (2002)

23. Zerbino, D.R., Birney, E.: Velvet: algorithms for de novo short read assembly using de Bruijn graphs. Gen. Res. 821–829 (2008)

In silico Pathogenomic Analysis
of *Corynebacterium Pseudotuberculosis*
Biovar *Ovis*

Iago Rodrigues Blanco(ID), Carlos Leonardo Araújo(ID),
and Adriana Carneiro Folador(✉) (ID)

Laboratory of Genomics and Bioinformatics, Center of Genomics and Systems Biology,
Institute of Biological Sciences, Federal University of Pará, Belém, Pará, Brazil
adrianarc@ufpa.br

Abstract. *Corynebacterium pseudotuberculosis* is a pathogenic bacterium that may transmit caseous lymphadenitis, veterinary infection that severely attacks animals such as goats and sheep. It is known that the toxin Phosholipase D is the major virulence factor associated with this disease. However, genomic computational studies can reveal further information concerning pathogenicity mechanisms of bacteria. Through sequence analysis tools, it is possible to assess the genomic bases of these mechanisms and to analyze similarities among the different strains of this species. Nitrate reductase-negative bacteria are classified in the biovar *ovis*, able to transmit the infection. Thus, we developed an *in silico* comparative pathogenomic analysis with genomes of 33 strains of *C. pseudotuberculosis* biovar *ovis* strains, which cause caseous lymphadenitis. Looking for the identification of pathogenicity-related genes, virulence factors and composition of pathogenicity islands, it was possible to computationally predict pathogenicity potentials of target proteins and their respective biological processes during infection, besides identification of prophage genome elements and prediction of protein protein interactions.

Keywords: Pathogenomics · *Corynebacterium pseudotuberculosis* ·
Pathogenicity islands · Virulence factors

1 Introduction

1.1 Corynebacterium Pseudotuberculosis

The bacterial genus *Corynebacterium* is a member of the CMNR group (*Corynebacterium, Mycobacterium, Nocardia* and *Rhodococcus*), composed of microorganisms that share structural and genetic characteristics, such as cell wall organization and G + C content varying from 47% to 74% [1]. In this genus, species can be classified as pathogenic and non-pathogenic. Within the first group, some species stand out, such as *C. pseudotuberculosis, C. diphtheriae* and *C. ulcerans*, responsible for the transmission of caseous

© Springer Nature Switzerland AG 2020
J. C. Setubal and W. M. Silva (Eds.): BSB 2020, LNBI 12558, pp. 38–49, 2020.
https://doi.org/10.1007/978-3-030-65775-8_4

lymphadenitis (CLA), ulcerative lymphangitis and diphtheria. These bacteria are capable of producing virulence factors (VF) and pathogenicity-related proteins, responsible for causing diseases [2].

C. *pseudotuberculosis* can be divided into two biovars: *equi* and *ovis*, causers of serious infections. This discrimination depends on the pathogen's ability to reduce nitrate: positive nitrate reductase bacteria are members of biovar *equi*, responsible for causing ulcerative lymphangitis in buffaloes and horses, and negative nitrate reductase bacteria are classified in biovar *ovis,* etiological causers of CLA, forming wounds and abscesses in internal organs of small ruminants, as goats and sheep [3–5].

1.2 Comparative Pathogenomics

After the advent of next generation sequencing (NGS) technologies, there has been an exponential growth in studies related to the investigation of new organisms and their biological processes. Consequently, the time and cost applied to the techniques were reduced and a revolution was made in genomics [6]. The assessment of bacterial genomes through worldwide available databases allows the investigation of genetic factors through *in silico* researches – those that focus on computational predictions. Thus, it is possible to reveal pathogenicity mechanisms and other biological processes underlying infection by studying the genome of the pathogen [7, 8].

Several biological processes relate with pathogenicity in *Corynebacterium,* as adherence, iron uptake, secretion of toxins, invasion and colonization in the host [9]. It is possible to predict genes related with these functions by obtaining amino acid sequences and aligning them in *online* databases. Pathogenomic analyses seek to identify important genetic determinants of pathogenicity, as VF and their roles in pathogenesis through analysis of the pathogen's genetic repertoire [10]. Comparative genomic analyses are increasingly fast and promising, highlighting its recent advances.

The analysis of determinants of pathogenicity in bacteria is useful to determine which genes are responsible for pathogenic functions and how they work. We can observe the conservation of these genes among different strains of a same species, or even among different species of a same genus [11]. Comparative pathogenomics can point out important genetic factors among determined genomes. The identification of VF is so important once significant antimicrobial resistance has been observed in pathogenic bacteria, what is associated to bacterial adaptation factors and its ability to modify biological mechanisms adapting to the host's defense system [12]. Therefore, the identification of VF can reveal notable information about this pathogen [13, 14].

1.3 Determinants of Pathogenicity

Genes acquired through horizontal transfer are frequently found in the genomes of pathogenic bacteria, which may present different genetic characteristics from the rest of the genome, such as low G + C content and pathogenic function of the genic repertoire [15]. These genes are frequently clustered in genomic regions called "pathogenic islands" (PAI), which may suggest characteristics related to the pathogen's lifestyle and its infection mechanisms. The composition of PAI can reveal new genes related to pathogenicity and protein-protein interactions among closely related proteins [9].

Through application of *in silico* approaches, it is possible to predict the composition of these regions, as well as features of proteins constituting them. It is also feasible to predict pathogenic potential of molecules, by means of computational calculation based on amino acid sequences, evincing their roles in pathogenicity.

Other VF in *C. pseudotuberculosis* include toxic lipids associated to the cell wall, which can mediate bacterial resistance to phagocyte attacks, besides neuroaminidases and endonucleases, which play pivotal role in pathogenic processes [16]. Also, the insertion of prophage DNA in bacterial genomes is sometimes observed in pathogenic bacteria, directly related to phage encoded virulence genes, what can be identified through *in silico* tools.

2   Methods

2.1   Pan-Genomic Analysis of *C. Pseudotuberculosis* Biovar *Ovis*

Complete genomes of 33 strains of biovar *ovis* were downloaded and standardized for nucleotides, amino acids and function files (.nuc,.pep and.function). The strains were: 1002, 1002B, 12C, 226, 267, 29156, 3995, 4202, C231, E55, E56, FRC41, I19, MEX1, MEX25, MEX29, MIC6, N1, P54B96, PA01, PA02, PA04, PA05, PA06, PA07, PA08, PAT10, PO22241, AN902, T1, VD57 and ft_2193-67. The software PGAP [11] generated orthologous gene clusters among strains. The Gene Family method was used with minimum score of blastall 40, 80% identity, 90% coverage and e-value of 0.00001.

2.2   Prediction of Virulence Factors in *Corynebacterium*

The Virulence Factor Database (VFDB) provides an online platform to identify VF and their conservation in various bacterial genomes. In order to identify VF, their classes and sequences, we have used the VFAnalyzer tool, which resorts genomic data from the NCBI database [17]. The genomes analyzed were compared to the reference genome of the non-pathogenic strain *C. glutamicum* ATCC 13032.

2.3   Composition of Pathogenicity Islands (PAI)

GIPSy software was used to predict the existence and composition of PAI. It aligns with complementary databases, such as VFDB and NCBI. The genome of *C. pseudotuberculosis* 12C was considered as query, using as reference (non-pathogenic) the genome of *C. glutamicum* ATCC 13032. GIPSy provides composition, position of genes, G + C content and G + C deviation of PAI [18].

2.4   Synteny in *C. Pseudotuberculosis* Genomes

The online tool SimpleSynteny [19] was used to compare genomes of *C. pseudotuberculosis* 12C, FRC41 and 1002, generating images of genomic *loci*. The tool requires FASTA sequences of the genomes and target genes, showing their conservation among genomes through BLAST, and their positions. We performed gapped-type BLAST with e-value threshold of 0.001 and minimum query cutoff of 50%.

2.5 In Silico Prediction of Pathogenicity Potentials

We applied computational prediction to the obtained genes in order to predict whether they have pathogenic potential or not. This analysis was performed through the MP3 tool [20], able to predict the pathogenic ability of genomes. MP3 uses the highly reliable methods Support Vector Machine (SVM) and Hidden Markov Model (HMM).

Through each mentioned method, a score is calculated for proteins, classifying them as pathogenic (P) or non-pathogenic (NP). To have reliable prediction, we considered as reference for our analysis the proteins predicted as pathogenic by both methods.

2.6 Protein-Protein Interactions

VF sequences were submitted to the STRING database [21], which predicts direct and indirect interactions among proteins through computational approach, generating a reliability score. We analyzed those proteins with a predicted score above 0.7 and maximum of 10 interactions. The networks were generated based on the conserved gene neighborhood, co-ocurrence of genes and protein homology. *C. pseudotuberculosis* strain E19 was used as reference for the prediction by protein homology.

2.7 Identification of Prophages

The presence of phage DNA in the genome of *C. pseudotuberculosis* 12C was performed with the web tool PHAST (Phage Search Tool), which contains an extensive set of genes identified in prophages. The input genome must be inserted as a *fasta* file, generating homologous annotated hits through BLAST with different databases and a graphic visualization of the prophage genetic content [22].

3 Results and Discussion

3.1 Identification of Adherence Factors

The pathogenomic analysis of the genus revealed 4 major classes of VF, comprising 12 types of them. The complete analysis showed 36 genes associated with VF in *Corynebacterium*. Regarding their conservation, we noticed different results for the different strains analyzed.

The first class of VF groups the adherence proteins, which are divided into Collagen-Binding Proteins; SpaA-type pili; SpaD-type pili; SpaH-type pili and Surface-anchored pilus proteins. For our considered dataset, however, collagen-binding proteins were not conserved in any of the strains [23].

Pili are filamentary cellular structures that provide adhesive and invasive functions to the pathogen after infection [24]. In *C. diphtheriae, pili* are formed by the sortase machinery, being also present in other pathogenic genera such as *Streptococcus, Actinomyces* and *Clostridium*. The three Sortase-mediated Pilus Assembly structures SpaA, SpaD and SpaH have similar composition, and are essential for the attachment to host's tissues after infection. SpaA pilus is responsible for the adherence of bacteria to human pharyngeal epithelial cells, while SpaB and SpaC are responsible for binding to pharyngeal cells [25].

Of the surface-anchored pilus proteins SapA, SapD and SapE, initially characterized as VF in *C. jeikeium* K411 genome [26], only SapA was conserved in the genomes of biovar *ovis* and *C. ulcerans* FRC11. Some adherence-related VF in *Corynebacterium* are present only in *C. diphtheriae*, as *spaA*, *spaB*, *spaE*, *spaF*, *spaG*, *spaH*, *srtD*, *srtE*, *sapD* and *sapE*. It is suggestible that some of these adherence genes may have been horizontally transferred to *C. pseudotuberculosis* [15]. Genes *srtA*, *spaD*, *srtB*, *srtC*, *sapA* are conserved in most of strains, while *spaC* is shared among nearly half of them (Table 1).

Table 1. Conservation of VF in genomes of *C. pseudotuberculosis* biovar *ovis*

VF Class	VF	Gene	Cp 12C (CP011474)	Cp FRC 41 (NC_014329)	Cp 1002 (CP001809)	N. of strains
Adherence	SpaA-type pili	*spaC*	Cp12C_2000; Cp12C_2001; Cp12C_2002	cpfrc_01902; cpfrc_01903; cpfrc_01901	Cp1002_1899; Cp1002_1900; Cp1002_1901	16 13 12
		srtA	Cp12C_2004	cpfrc_01905	Cp1002_1901	33
	SpaD-type pili	*spaD*	Cp12C_1973	cpfrc_01874	Cp1002_1872	33
		srtB	Cp12C_1974	cpfrc_01875	Cp1002_1874	31
		srtC	Cp12C_1971	cpfrc_01873	Cp1002_1870	33
	Surface-anchored pilus proteins	*sapA*	Cp12C_1968	cpfrc_01870	Cp1002_1867	30
Iron uptake	ABC transporter	*fagA*	Cp12C_0036	cpfrc_00032	Cp1002_0030	30
		fagB	Cp12C_0035	cpfrc_00031	Cp1002_0029	31
		fagC	Cp12C_0034	cpfrc_00030	Cp1002_0028	33
		fagD	Cp12C_0037	cpfrc_00033	Cp1002_0031	33
	ABC-type heme transporter	*hmuT*	Cp12C_0482	cpfrc_00455	Cp1002_0451	33
		hmuU	Cp12C_0483	cpfrc_00456	Cp1002_0452	33
		hmuV	Cp12C_0484	cpfrc_00457	Cp1002_0453	31
	Iron uptake and siderophore biosynthesis	*ciuA*	Cp12C_1041	cpfrc_00987	Cp1002_0981	33
		ciuB	Cp12C_1042	cpfrc_00988	Cp1002_0982	33
		ciuC	Cp12C_1043	cpfrc_00989	Cp1002_0983	33
		ciuD	Cp12C_1044	cpfrc_00990	Cp1002_0984	30
		ciuE	Cp12C_1045	cpfrc_00991	Cp1002_0985	33
Regulation	Diphtheria toxin repressor	*dtxR*	Cp12C_1290	cpfrc_01219	Cp1002_1213	33
Toxin	Phospholipase D	*pld*	Cp12C_0033	cpfrc_00029	Cp1002_0027	33

3.2 Identification of Iron Uptake Factors

The second class of VF groups the iron uptake proteins, which can be divided into ABC transporters; ABC-type heme transporters; iron uptake and siderophore biosynthesis proteins and siderophore-dependent iron uptake systems. ATP-binding cassette (ABC) transport systems are known to be related to iron and manganese uptake in *Corynebacterium* [27], while ABC hemin transport proteins are known to be related to regulation of VF in *Yersinia pestis, Escherichia coli, Vibrio colerae* and *Shigella dysenteriae* [28].

An operon associated with virulence in this genus is *fagABCD*, present in PAI in *C. pseudotuberculosis* together with PLD, determinant factor for CLA [9]. All of these orthologs were conserved in the prediction for these strains. In addition, the ABC-type heme transporters, which encode ABC heme proteins in *Corynebacterium* are also exhibited as VF in our analysis as the genes *hmuT, hmuU* and *hmuV*.

Genes related to iron uptake are often found in pathogenic bacteria, once they require this element for living. Iron can be located in human cells and binds to heme and heme proteins. After infection, pathogenic bacteria acquire host iron sources, especially heme, heme proteins, transferrin and lactoferrin [29].

The iron uptake gene cluster *ciuABCDE* is related to ABC-type iron transport, biosynthesis and siderophore production. It is reported that this gene cluster exhibits iron-dependent manner regulation [30]. In our analysis, the entire cluster was found to be present in all analyzed strains, indicating essential function of these genes on the biological mechanisms of bacteria. Further, an additional group of genes in *C. diphtheriae* also relates to siderophore-dependent iron uptake: the *irp6* operon, which acts as promoter and may be repressed by DtxR metal ions, the diphtheria toxin repressor [31].

3.3 Identification of Regulation Factors

For *Corynebacterium*, the diphtheria toxin repressor DtxR stands out as a major regulation factor. It is a metal ion-activated molecule that acts as a transcriptional regulator and is conserved in *C. efficiens, C. glutamicum, C. jeikeium, C. diphtheriae* and all of our analyzed strains. It is responsible for regulation of virulence genes in various pathogenic bacteria [32]. It is suggested that *dtxR* genes in distinct species of *Corynebacterium* have been acquired through horizontal gene transfer and having different functional roles in distinct species, an indicative of selection for differing functions [33].

3.4 Identification of Toxin Factors

Only one toxin factor was identified in *C. pseudotuberculosis* genomes: PLD, the major VF in this species, conserved in all biovar *ovis* strains. It is the main responsible for CLA and is placed in PAI. It causes host macrophage cell death and may be regulated by several environmental factors [34]. Further, *C. ulcerans* and *C. diphtheriae* have a second VF in their genomes, DT, responsible for the different pathogeny diphtheria.

3.5 Prediction of Pathogenicity Islands

We noted pathogenicity-related genes in these genomes. Thus, we also identified the complete genomic content of PAI. We were able to identify 11 PAI, with genes annotated as VF, iron uptake proteins, toxins, etc. Some genes are still classified as hypothetical proteins, emphasizing the need for further studies on the pathogen's genome. In addition, the conservation of major VF in three genomes is showed in Fig. 1.

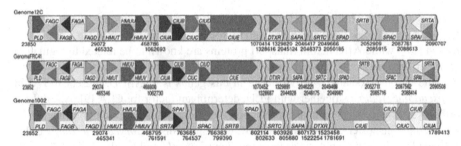

Fig. 1. Synteny analysis of main *C. pseudotuberculosis* VF in the genomes of strains 12C, FRC41 and 1002. Strains are compared regarding to conservation and position of important VF clusters in their genomes.

In PAI 1, some genes stand out, as those inserted in the *fagABCD* operon, followed by the toxin *pld* gene. In PAI 3, some genes show relation with manganese uptake and ABC transport, as mentioned. In PAI 4, many genes present association with iron ABC transport, as the *ciuABCDE* cluster, experimentally proven to have virulent function [9]. In PAI 5, a genomic region is conserved, highlighting the virulent function of the PTS sugar transporter pfoS, responsible for regulation of toxins in *Clostridium perfringens*. Aditionally, the proline iminopeptidase protein was found, essential for virulence of *Xanthomonas campestris* and conserved in *C. pseudotuberculosis* [9]. In PAI 7, an adhesin-related protein is found, which may relate to the role of adherence in the pathogenic process, once it is essential for the invasion of the bacteria in host cells.

Further, in PAI 8 we noted two adherence-related molecules: the collagen-binding surface protein and the sortase *srtA1*. In PAI9, the iron dicitrate transport proteins phuC and fecD are placed, as well as the sortase srtA2, conserving virulent function [23].

3.6 Prediction of Pathogenicity Potentials

The sequences inserted in PAI were submitted to the MP3 tool to predict their probable pathogenic roles. Pathogenicity potentials for all coding sequences of the complete genome *C. pseudotuberculosis* 12C, composed by 2,220 coding sequences, were calculated. 683 were classified as pathogenic by both prediction methods (HMM and SVM), adding an average value of 30.3% of proteins predicted as pathogenic *in silico*.

The number of pathogenic proteins in PAI, however, was different. The composition of these *loci* is formed by low G + C content sequences, what has implied in a higher pathogenic content prediction for most of these islands. We found a mean value of 48.6% of pathogenic proteins, 18.3% more than the complete genome. The pathogenic content varied significantly, showing standard deviation of 16.3% among PAI regarding to pathogenic protein content (Table 2).

3.7 Identification of Prophages

In order to complement the analysis, we looked for prophage sequences in the genome of the pathogen. Three incomplete prophage regions have been identified. One of

Table 2. Pathogenic composition of PAI in *C. pseudotuberculosis* 12C genome

Genomic region	N. of genes	Pathogenic genes	G + C deviation
Complete genome	2,220	683 (30.7%)	10%
PAI 1	13	7 (53.8%)	23%
PAI 2	12	4 (33.3%)	33%
PAI 3	21	9 (42.8%)	9%
PAI 4	14	11 (78.5%)	35%
PAI 5	15	7 (46.6%)	23%
PAI 6	7	3 (42.8%)	14%
PAI 7	7	2 (28.5%)	14%
PAI 8	17	7 (41.1%)	35%
PAI 9	18	13 (72.2%)	5%
PAI 10	6	2 (33.3%)	16%
PAI 11	8	5 (62.5%)	12%

the prophage regions was found to be common in *Micobacterya*, *Burkholderia* and *Synechococcus*. It is composed of six coding sequences, with annotations similar to enzymatic, structural and ABC transport functions.

Another prophage region, composed by 10 coding sequences shows homology to two ribonucleotide reductase proteins, a 50S ribosomal protein and three hypothetical proteins. The third prophage region shows homology to a plasmid partitioning protein, an inner membrane translocase and some bacillus-related hypothetical proteins.

Especially, one of these proteins showed homology to phage tail genes, which are reported to have multiple roles, and work as adhesion molecules for bacterial attachment in the host tissue. These tail molecules interact with the cell surface and cell wall of the host cell and are subjected to strong adaptative selection pressure [35].

3.8 Prediction of Protein-Protein Interactions

As observed, most of VF exhibit some relation regarding to biological mechanisms. We tried to obtain a visual representation of the interaction among the main pathogenicity determinants in *C. pseudotuberculosis*. Thus, we generated a PPI network for main VF, including proteins of the ciuABCDE, hmuTUV and fagABCD clusters. Further, a putative ABC transporter "AIG10663.1" is placed among them.

Despite being a computational prediction, we note through this method experimentally-proved interactions through homology-based biochemical pathways database data, conservation of structural protein domains and existing reports in the literature [9, 36–39], what may be confirmed through experimental support (Fig. 2).

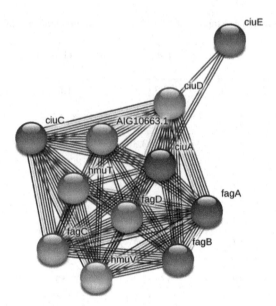

Fig. 2. PPI network of major VF in *C. pseudotuberculosis*

An experimental report had suggested the interaction between *ciuABCDE* and *fagABCD* cluster genes for iron uptake [40]. Our study is in concordance with this finding, suggesting an additional role of *hmuTUV* in the regulation of this mechanism and pathogenicity. The identification of associations among pathogenic proteins is directly associated to their role. Our results showed a high-conserved pathogenic network, indicating they may work together in the pathogenicity mechanism, what might be supported by complementary molecules [3].

4   Conclusion

This study proposed the investigation of pathogenic determinants of *C. pseudotuberculosis* biovar *ovis* to elucidate mechanisms of VF, genetic elements, PAI and orthology among the strains. Thus, we have made use of *in silico* analyses to reveal these data,.

Comparative pathogenomics revealed four main classes of *Corynebacteria* VF, which sum up to 36 genes. However, only 20 of these genes were observed in the analyzed strains. These VF are essential for invasion, propagation and survival of the pathogen after infection in the host cell. Not all of the *Corynebacterium* VF are present in the analyzed strains, indicating different acquisition of them among strains.

Additionally, we showed the synteny of these VF, genomic *loci* and the presence and composition of PAI in the reference genome of *C. pseudotuberculosis* 12C. The majority of VF described are found in these regions and organized in gene clusters, however in different genomic positions among distinct strains. Proteins composing PAI were analyzed *in silico* for potentials of pathogenicity, finding a 18.6% higher potential in these low G + C content regions in comparison to the complete genome.

Once the presence of mobile genetic elements is often found to be related with pathogenicity in bacterial genome, we looked for presence of phage DNA in the *C. pseudotuberculosis* genome, identifying three prophage regions, highlighting one phage tail gene, which is known to be related with pathogen attachment to host cell after infection. Finally, the main VF identified in this research were analyzed looking for PPI networks, exhibiting high-confidence associations among ciuABCDE, hmuTUV and clusters, suggesting combined roles in pathogenic function.

References

1. Dorella, F., Pacheco, L., Oliveira, S., et al.: *Corynebacterium pseudotuberculosis*: microbiology, biochemical properties, pathogenesis and molecular studies of virulence. Vet. Res. **37**(2), 201–218 (2016)
2. Araújo, C., Alves, J., Lima, A., et al.: The Genus *Corynebacterium* in the Genomic Era. Basic Biology and Applications of Actinobacteria, Shymaa Enany, IntechOpen (2018)
3. Araújo, C., Blanco, I., Souza, L., et al.: *In silico* functional prediction of hypothetical proteins from the core genome of *Corynebacterium pseudotuberculosis*. PeerJ **8**, e9643 (2020)
4. Biberstein, E., Knight, H., Jang, S.: Two biotypes of *Corynebacterium pseudotuberculosis*. The Vet. Rec. **89**, 691–692 (1971)
5. Williamson, L.: Caseous lymphadenitis in small ruminants. The Vet. Clin. North Am. Food Anim. Pract. **17**(2), 359–371 (2001)
6. Van Dijk, E., Jaszczyszyn, Y., Naquin, D., et al.: The third revolution in sequencing technology. Trends Genet. **34**(9), 666–681 (2018)
7. Guedes, M., Souza, B., Sousa, T., et al.: Infecção por *Corynebacterium pseudotuberculosis* em equinos: aspectos microbiológicos, clínicos e preventivos. Pesquisa Veterinária Brasileira **35**(8), 701–708 (2015)
8. Weerasekera, D.: Characterization of virulence factors of *Corynebacterium diphtheriae* and *Corynebacterium ulcerans*. thesis (2019)
9. Ruiz, J., D'afonseca, V., Silva, A., et al.: Evidence for reductive genome evolution and lateral acquisition of virulence functions in two Corynebacterium pseudotuberculosis strains. PLoS ONE **6**(4), e18551 (2011)
10. Pallen, M., Wren, B.: Bacterial pathogenomics. Nature **449**(7164), 835–842 (2007)
11. Zhao, Y., Wu, J., Yang, J., et al.: PGAP: Pan-genomes analysis pipeline. Bioinformatics **28**(3), 416–418 (2012)
12. Quiroz-Castañeda, R.: Pathogenomics and molecular advances in pathogen identification. In: Farm Animals Diseases, Recent Omic Trends and New Strategies of Treatment. IntechOpen (2018)
13. Cassiday, P., Pawloski, L., Tiwari, T., et al.: Analysis of toxigenic *Corynebacterium ulcerans* strains revealing potential for false-negative real-time PCR results. J. Clin. Microbiol. **46**(1), 331–333 (2007)
14. Lo, B.: Diptheria. Medscape. https://emedicine.medscape.com/article/782051-print. Accessed on 24 08 2020
15. Guimarães, L., Soares, S., Trost, E., et al.: Genome informatics and vaccine targets in *Corynebacterium urealyticum* using two whole genomes, comparative genomics, and reverse vaccinology. BMC Genom. **16**, 5 (2015)
16. Collin, M., Fischetti, V.: A novel secreted endoglycosidase from *Enterococcus faecalis* with activity on human immunoglobulin G and ribonuclease B. J. Biol. Chem. **279**(21), 22558–22570 (2004)

17. Liu, B., Zheng, D., Jin, Q., et al.: VFDB 2019: a comparative pathogenomic platform with an interactive web interface. Nucleic Acids Res. **47**(1), 687–692 (2019)

18. Soares, S., Geyik, H., Ramos, R., et al.: GIPSy: genomic island prediction software. J. Biotechnol. **232**, 2–11 (2016)

19. Veltri, D., Wight, M., Crouch, J.: SimpleSynteny: a web-based tool for visualization of microsynteny across multiple species. Nucleic Acids Res. **44**, 41–45 (2016)

20. Gupta, A., Kapil, R., Dhakan, D.B., et al.: MP3: a software tool for the prediction of pathogenic proteins in genomic and metagenomic data. PLoS ONE **9**, 4 (2014)

21. Szklarczyk, D., Gable, A., Lyon, D., et al.: STRING v11: protein–protein association networks with increased coverage, supporting functional discovery in genome-wide experimental datasets. Nucleic Acids Res. **47**(D1), 607–613 (2019)

22. Zhou, Y., Liang, Y., Lynch, K., et al.: PHAST: a fast phage search tool. Nucleic Acids Res. **39**, 347–352 (2011)

23. Ton-That, H., Marraffini, L., Schneewind, O.: Sortases and pilin elements involved in pilus assembly of *Corynebacterium diphtheriae*. Mol. Microbiol. **53**, 251–261 (2004)

24. Ton-That, H., Schneewind, O.: Assembly of pili on the surface of *Corynebacterium diphtheriae*. Mol. Microbiol. **50**(4), 1429–1438 (2003)

25. Mandlik, A., Swierczynski, A., Das, A., et al.: *Corynebacterium diphtheriae* employs specific minor pilins to target human pharyngeal epithelial cells. Mol. Microbiol. **64**, 111–124 (2007)

26. Hansmeier, N., Chao, T., Daschkey, S., et al.: A comprehensive proteome map of the lipid-requiring nosocomial pathogen Corynebacterium jeikeium K411. Proteomics **7**(7), 1076–1096 (2007)

27. Tauch, A., Kaiser, O., Hain, T., et al.: Complete genome sequence and analysis of the multiresistant nosocomial pathogen *Corynebacterium jeikeium* K411, a lipid-requiring bacterium of the human skin flora. J. Bacteriol. **187**(13), 4671–4682 (2005)

28. Schmitt, M., Drazek, E.: Construction and consequences of directed mutations affecting the hemin receptor in pathogenic *Corynebacterium* species. J. Bacteriol. **183**, 1476–1481 (2001)

29. Stojiljkovic, I., Perkins-Balding, D.: Processing of heme and heme-containing proteins by bacteria. DNA Cell Biol. **21**(4), 281–295 (2002)

30. Kunkle, C., Schmitt, M.: Analysis of a DtxR-regulated iron transport and siderophore biosynthesis gene cluster in *Corynebacterium diphtheriae*. J. Bacteriol. **187**(2), 422–433 (2005)

31. Qian, Y., Lee, J., Holmes, R.: Identification of a DtxR-regulated operon that is essential for siderophore-dependent iron uptake in *Corynebacterium diphtheriae*. J. Bacteriol. **184**(17), 4846–4856 (2002)

32. D'Aquino, J., Tetenbaum-Novatt, J., White, A., et al.: Mechanism of metal ion activation of the diphtheria toxin repressor DtxR. Proc. Nat. Acad. Sci. U.S. Am. **102**(51), 18408–18413 (2005)

33. Oram, D., Avdalovic, A., Holmes, R.: Analysis of genes that encode DtxR-like transcriptional regulators in pathogenic and saprophytic corynebacterial species. Infect. Immun. **72**(4), 1885–1895 (2004)

34. McKean, S., Davies, J., Moore, R.: Expression of phospholipase D, the major virulence factor of *Corynebacterium pseudotuberculosis*, is regulated by multiple environmental factors and plays a role in macrophage death. Microbiology **153**, 2203–2211 (2007)

35. Brüssow, H.: Impact of phages on evolution of bacterial pathogenicity. In: Bacterial Pathogenomics, ASM Press (2007)

36. Guerrero, J., de Oca Jiménez, R., Dibarrat, J., et al.: Isolation and molecular characterization of *Corynebacterium pseudotuberculosis* from sheep and goats in Mexico. Microbial Pathog. **117**, 304–309 (2018)

37. de Sá, M.A., Gouveia, G., Krewer, C.: Distribution of PLD and FagA, B, C and D genes in *Corynebacterium pseudotuberculosis* isolates from sheep and goats with caseous lymphadenitis. Genet. Mol. Biol. **36**(2), 265–268 (2013)
38. Galvão, C., Fragoso, S., de Oliveira, C., et al.: Identification of new *Corynebacterium pseudotuberculosis* antigens by immunoscreening of gene expression library. BMC Microbiol. **17**(1), 202 (2017)
39. Silva, W., Folador, E., Soares, S., et al.: Label-free quantitative proteomics of *Corynebacterium pseudotuberculosis* isolates reveals differences between Biovars *ovis* and *equi* strains. BMC Genom. **18**(1), 451 (2017)
40. Raynal, J., Bastos, B., Vilas-Boas, P., et al.: Identification of membrane-associated proteins with pathogenic potential expressed by *Corynebacterium pseudotuberculosis* grown in animal serum. BMC Res. Notes **1**, 11 (2018)

Assessing the Sex-Related Genomic Composition Difference Using a k-mer-Based Approach: A Case of Study in *Arapaima gigas* (Pirarucu)

Renata Lilian Dantas Cavalcante[1]([⊠])(iD), Jose Miguel Ortega[2](iD),
Jorge Estefano Santana de Souza[1], and Tetsu Sakamoto[1](iD)

[1] Bioinformatics Multidisciplinary Environment – BioME, IMD, Federal University of Rio Grande do Norte – UFRN, Natal, Brazil
`renatalilian@ufrn.edu.br, tetsu@imd.ufrn.br`
[2] Lab Biodados, Department of Biochemistry and Immunology, Federal University of Minas Gerais – UFMG, Belo Horizonte, Brazil

Abstract. *Arapaima gigas* is the largest freshwater bony fish in the world, in which adults could weigh 200 kg and measure 3 m in length. Due to its large size and its low-fat meat, *Arapaima gigas* has quickly become a species of special interest in fish-farming. One challenge faced during their production is the lack of an efficient sexing methodology, since their sexual maturation occurs late (around the third to the fifth year) and the genetic mechanisms linked to their sex determination system are not known yet. For a more sustainable management, it is of paramount importance to seek an effective and non-invasive method to differentiate sexually juvenile individuals of *Arapaima gigas*. For this, the establishment of genetic markers associated with sexual differentiation would be an advantageous tool. In this study, we proposed a k-mer based approach to identify genome features with sex-determining properties. For this purpose, we used genomic data from six adult representatives of *Arapaima gigas*, three males and three females, and counted the k-mers comprising them. As result, we found k-mers from repetitive regions with high difference and disproportion in the count among individuals of the opposite sex. These differences in the k-mer-based genomic composition could indicate the existence of genetic factors involved in the sexing of individuals in *Arapaima gigas*.

Keywords: Osteoglossiformes · Sexual differentiation · Amazon fish · Repetitive sequence · Molecular marker · Fish-farming

1 Introduction

Arapaima gigas, commonly known as "Pirarucu" or "Paiche", is the largest bony freshwater fish in the world. It belongs to the bonytongues (Order Osteoglossiformes) Arapaimidae family, and has a natural habitat in the Amazon Basin

© Springer Nature Switzerland AG 2020
J. C. Setubal and W. M. Silva (Eds.): BSB 2020, LNBI 12558, pp. 50–56, 2020.
https://doi.org/10.1007/978-3-030-65775-8_5

[1]. Adult specimens may weigh around 200 kg and measure about 3 m [2,3]. Due to its large size, its flesh containing low-fat and low fishbone, along with its physiological characteristic of emerging to the surface at intervals of 15 min to assimilate oxygen, *Arapaima gigas* became a vulnerable species to overfishing in the Amazon region [4] leading to greater surveillance of the marketing of Pirarucu by the Brazilian government in the early 2000s [5,6].

Some studies show that the use of Pirarucu in intensive fish farming is facilitated, in part, by the physiological characteristics of the animal that guarantee the rusticity of the species [7]. For example, the obligate air-breathing causes this species to tolerate environments with low concentrations of dissolved oxygen in the water [8]. In addition to the facility for captive management, attributes such as the low content of fat, combined with the rapid growth of the species, which weighs an average of 10 kg in its first year of life, add value for an intensification in the commercial exploitation of the *Arapaima gigas* [5,9,10].

One of the problems related to its fishing exploitation and fish-farming is that the genetic mechanisms linked to sex-differentiation in *Arapaima gigas* [11] is not know yet. Since its sexual maturation occurs late, around the third to the fifth year of life, and sexual dimorphism is not a strong feature of the species [10], its sexing is yet performed using laborious procedures (e.g. laparoscopy and transrectal ultrasound). In recent decades, the creation of *Arapaima gigas* in captivity has been increasingly stimulated, either to develop research to better know the particularities of the species or to exploit its economic potential [12,13]. For more sustainable management, both for captive breeding and for the study of the species, it is of paramount importance to seek an effective and non-invasive method to differentiate sexually juvenile individuals.

For this, the establishment of a molecular genetic marker related to sexual differentiation would be an advantageous tool. Previous analysis of the *Arapaima gigas* genome found no genes associated with the identification of the sex determination system of these individuals [14,15]. And chromosomal characterization studies could not distinguishes cytologically a sex chromosome in *Arapaima gigas* [16,17]. In this study, we proposed to asses the genomic composition of *Arapaima gigas* using a k-mer-based approach to identify regions in excess or missing in one of the sexes.

2   Materials and Methods

2.1   Sequencing and Data Processing

In this study, we used genomic data from six adult representatives of *Arapaima gigas*, three males (M15, M20 and M25) and three females (F15, F20 and F25). Four samples (M15, M20, F15 and F20) were collected from Bioproject PRJEB22808 available in National Center for Biotechnology Information (NCBI) database [14]. And the other samples (M25 and F25) were acquired from another Bioproject nominated as PRJNA540910, also available in NCBI database [15].The quality of the reads was verified with the help of FastQC (v. 0.11.4) [18], and low-quality reads were trimmed with the help of the Sickle paired-end (v. 1.33) [19]. Data sets were processed in the sagarana HPC cluster, CEPAD-ICB-UFMG.

2.2   k-mer Analysis

This analysis was performed in four steps: (1) k-mer count, (2) k-mer count normalization, (3) k-mer filtering for repetitive regions and (4) k-mer count comparison. The k-mer count was performed with the help of the tool K-mer Counter (KMC, v.3.1.1) [20], a free software written in C++, whose premise is to count k-mers (sequences of consecutive k nucleotide) in a given genome sequencing file. The trimmed fastq files of the 6 representatives *Arapaima gigas* were submitted as input data to KMC algorithm using the parameter -k, on the k-mer size, as 23.

After KMC step, in order to normalize the data Quantile Normalization (QN), a global arrangement method, which consists of a non-parametric methods that makes two or more distributions identical on statistical properties [21], was performed using an in-house Perl script (v.5.16.3). Lastly, to compare the average of the normalized counts between male and female samples, we performed a T-Test in R to identify which k-mers presented small p-value ($p \leq 0.05$). For further procedures, we used a in-house Perl script to select k-mers which were extracted from a repetitive region with a repeat unit up to 8 nucleotides. K-mer with the same repeat unit were merged and had their normalized count summed up.

3   Results and Discussion

After trimming low-quality sequences of genome sequencing data of 6 samples of *Arapaima gigas*, we had per each file 9.1 million, 39 million, 38.1 million, 38.5 million, 14 million, 38 million reads remained for F15, F20, F25, M15, M20 and M25, respectively. The KMC result of k-mer count showed a number of total k-mers counted of 6 billion, 26 billion, 32 billion, 25 billion, 9 billion and 32 billion k-mers for F15, F20, F25, M15, M20 and M25 samples, respectively (See Table 1).

Table 1. Stats of average of reads and total number of k-mer in six samples of *Arapaima gigas*.

Samples	Reads average	Total no. of k-mers
F15	9.170.700	6.036.151.500
F20	39.041.211	26.057.501.877
F25	38.168.487	32.523.525.409
M15	38.590.767	25.269.745.617
M20	14.141.274	9.380.791.863
M25	38.030.714	32.336.285.757

Because of the difference on the number of reads among samples, we used the Quantile Normalization method (QN) to make them comparable. For further analysis, we considered only k-mers comprised of repeating units of size 1 to 8 bp. The average comparison of normalized k-mers count between individuals of the opposite sex and the same sex showed some repeats which were more abundant in one of the sex (Table 2).

Table 2. Repeat units with significant difference (p-value < 0.05) on the average of the normalized k-mer count between female and male samples of *Arapaima gigas*.

Repeat unit	F15	F20	F25	M15	M20	M25	p-value
AAAAC	10626.84	9914.83	10002.67	7261.5	7390.34	8410.68	0.007
AACAGCTG	162	139.16	161.17	84.85	108	110.67	0.009
AGAGCGG	71.49	78.33	66.51	87.51	92.01	96.99	0.011
AAGGC	31540.49	28319.83	17831.17	7457.32	9015.83	58.5	0.019
ACGTC	3440.67	4072.83	3104.49	1584.17	2348.67	2398.16	0.021
ACCACCAG	50.16	48.66	40.16	28.34	32.34	34.5	0.023
AAGC	11656.51	9888.33	9237.18	6352.5	6537.66	5921.67	0.025
ACACATCC	0	0	0	0.85	0.83	1.34	0.026
AAGGCC	2552.17	2221.5	1767.66	638.18	975.49	1480.83	0.026
ACCAGT	241.17	295.5	249.84	306.66	360.32	354.17	0.030
AAAAG	5775.17	5536.67	4985.32	3803.83	3697.84	4719.84	0.031
AGGG	11899.16	12294.5	10126.67	8220	8186.33	9562.84	0.031
ACATATC	88.34	75	86.83	71.17	54	53.67	0.033
AAACATT	95.32	100.32	99.18	113.68	110	104.34	0.035
AACC	10210.34	9663.99	10725.17	6843.5	6845.99	8788.84	0.035
AAAGTCAC	32.67	28.5	36.83	24.33	21.5	21.83	0.039
ACTGGC	0	0	81.33	135.01	166	96.99	0.039
ACCCAGGT	32.67	30.33	36	26.17	29	24.5	0.040
AACAACCC	78.99	72	61.5	90.68	92.01	95.67	0.040
AAAGTAAT	18.68	21.32	9.85	0	0	0	0.040
AACAC	12706	12649.66	11450.5	7673.34	9504	10389.66	0.042
AAGACATT	23.83	23	24.67	26.17	29	26.5	0.042
AATGATG	61	60	60.66	48.66	54	54.34	0.043

Repeat units considered as promising candidates for sex differentiation are highlighted.

Arapaima gigas plays an important role in the economy of the north region of Brazil [12,13]. Understand the mechanisms of sexual determination in fish are essential for a sustainable management of ichthyofauna, either for commercial or conservation purposes [22]. But, elucidating these mechanisms in fish is challenging, especially in *Arapaima gigas* case, because this species do not have a typical sex chromosome in their genome [16,17] and the difference of the genome sequences between samples of opposite sex seems to be minimal. Both *Arapaima gigas* genome assemblies [14,15] did not find significant differences between the genomic content of male and female samples.

In this context, we explored other genomic features in *Arapaima gigas* to find some clues about the genetic factors involved in the sex determining system. For this, we analysed the genomic composition of Pirarucu using k-mer based approach. In this study, we have noticed the existence of k-mers from repeat regions over or underrepresented in one of the sexes, indicating potential differences in the genetic composition between males and females of *Arapaima gigas*.

The difference is not so expressive, which corroborates the reports that estimate 0.01% [14] to 0.1% [15] of the genome of this species as linked to the sexual determination. The sequences reported in this work are all part of repetitive sequences. Despite of their low complexity, repetitive regions have been reported to have important role on sex determination [23]. In medaka, which has the XY system, there is a large stretch of repetitive regions on the Y-specific regions [24]. The chromosome Y of Pacific salmon also bears a specific repetitive regions (OtY1) that is used as genetic marker to differentiate sex [25].

In this context, the repetitive sequences found in this study could be a component that could be used to determine sex in individuals of *Arapaima gigas*. We recognize, however, the necessity to perform analyses with a greater number of samples to obtain a better statistical support for our results, as well as to suggest bench trials for the validation of the in-silico analyses. Despite of that, the k-mer-based method applied on this work has demonstrated to be an interesting strategy to help us discover the sex-determination system in Pirarucu specimens and can be extended to other species.

4   Conclusions

This short paper reports a few repetitive genome sequences that can be differentiated in quantity in male and female of *Arapaima gigas*. Indicating that k-mer-based methods is an interesting approach to assist us in unraveling the sex-determination system in *Arapaima gigas*.

Acknowledgments. This work was supported by CAPES (Brazilian Federal Agency for the Support and Evaluation of Graduate Education). To Bioinformatics Multidisciplinary Environment - BioME and CEPAD-ICB-UFMG for the computational structure for the accomplishment of this work. To Dr. Manfred Schartl from University of Würzburg - Germany that met our request to publish the reads of the *Arapaima gigas* genome from Bioproject PRJNA540910. And to Dr. Sidney Emanuel Batista dos Santos from Federal University of Pará - UFPA for contributing in the discussions of this project.

References

1. Reis, R.E., Kullander, S.O., Ferraris, C.J.: Check List of the Freshwater Fishes of South and Central America. Edipucrs, Porto Alegre (2003)

2. Bezerra, R.F., Soares, M.C.F., Maciel Carvalho, E.V.M., Coelho, L.C.B.B.: Pirarucu, Arapaima Gigas, the Amazonian Giant Fish is Briefly Reviewed. Fish, Fishing and Fisheries Series, p. 41. Nova Science, New York (2013)
3. Nelson, J.S., Grande, T.C., Wilson, M.V.H.: Fishes of the World. John Wiley & Sons, Hoboken (2016)
4. Goulding, M., Smith, N.J.H., Mahar, D.J.: Floods of Fortune: Ecology and Economy Along the Amazon. Columbia University Press, New York (1996)
5. Bayley, P.B., Petrere, M., Jr.: Amazon Fisheries: Assessment Methods, Current Status and Management Options. Canadian Special Publication of Fisheries and Aquatic Sciences/Publication Speciale Canadienne des Sciences Halieutiques et Aquatiques (1989)
6. Wilson, D.E., Burnie, D.: Animal: The Definitive Visual Guide to the World's Wildlife. Dorling Kindersley, London (2001)
7. de Oliveira, E.G., et al.: Effects of stocking density on the performance of juvenile pirarucu (arapaima gigas) in cages. Aquaculture **370**, 96–101 (2012)
8. Bard, J., Imbiriba, E.P.: Piscicultura do pirarucu, arapaima gigas. Embrapa Amazônia Oriental-Circular Técnica (INFOTECA-E) (1986)
9. de M. Carvalho, L.O.D., do Nascimento, C.N.B.: Engorda de pirarucus (arapaima gigas) em associação com búfalos e suínos. Embrapa Amazônia Oriental-Circular Técnica (INFOTECA-E) (1992)
10. Almeida, I.G., Ianella, P., Faria, M.T., Paiva, S.R., Caetano, A.R.: Bulked segregant analysis of the pirarucu (arapaima gigas) genome for identification of sex-specific molecular markers. Embrapa Amazônia Oriental-Artigo em periódico indexado (ALICE) (2013)
11. Hrbek, T., Farias, I.P.: The complete mitochondrial genome of the pirarucu (arapaima gigas, arapaimidae, osteoglossiformes). Genet. Mol. Biol. **31**(1), 293–302 (2008)
12. Fontenele, O.: Contribuicao para o conhecimento da biologia do pirarucu arapaima gigas (cuvier), em cativeiro (actinopterygii, osteoglossidae) [Ceara, Brasil] (1955)
13. Imbiriba, E.P.: Potencial de criação de pirarucu, arapaima gigas, em cativeiro. Acta Amazonica **31**(2), 299 (2001)
14. Vialle, R.A., et al.: Whole genome sequencing of the pirarucu (arapaima gigas) supports independent emergence of major teleost clades. Genome Biol. Evol. **10**(9), 2366–2379 (2018)
15. Kang, D., et al.: The genome of the arapaima (arapaima gigas) provides insights into gigantism, fast growth and chromosomal sex determination system. Sci. Rep. **9**(1), 5293 (2019)
16. Marques, D.K., Venere, P.C., Galetti Jr., P.M.: Chromosomal characterization of the bonytongue arapaima gigas (osteoglossiformes: Arapaimidae). Neotropical Ichthyol. **4**(2), 215–218 (2006)
17. de Oliveira, E.A., et al.: Cytogenetics, genomics and biodiversity of the South American and African arapaimidae fish family (teleostei, osteoglossiformes). PLoS ONE **14**(3), e0214225 (2019)
18. Wingett, S.W., Andrews, S.: Fastq screen: a tool for multi-genome mapping and quality control. F1000Research, p. 7 (2018)
19. Joshi, N.A., Fass, J.N.: Sickle: a sliding-window, adaptive, quality-based trimming tool for fastq files (version 1.33) [software] (2011)
20. Kokot, M., Długosz, M., Deorowicz, S.: KMC 3: counting and manipulating k-mer statistics. Bioinformatics **33**(17), 2759–2761 (2017)
21. Hicks, S.C., Irizarry, R.A.: Quantro: a data-driven approach to guide the choice of an appropriate normalization method. Genome Biol. **16**(1), 117 (2015)

22. Martínez, P., Viñas, A.M., Sánchez, L., Díaz, N., Ribas, L., Piferrer, F.: Genetic architecture of sex determination in fish: applications to sex ratio control in aquaculture. Front. Genet. **5**, 340 (2014)
23. Ezaz, T., Deakin, J.E.: Repetitive sequence and sex chromosome evolution in vertebrates. Adv. Evol. Biol. **2014**, 9 (2014)
24. Kondo, M., et al.: Genomic organization of the sex-determining and adjacent regions of the sex chromosomes of medaka. Genome Res. **16**(7), 815–826 (2006)
25. Devlin, R.H., Biagi, C.A., Smailus, D.E.: Genetic mapping of Y-chromosomal DNA markers in pacific salmon. Genetica **111**(1–3), 43–58 (2001)

COVID-19 X-ray Image Diagnostic with Deep Neural Networks

Gabriel Oliveira, Rafael Padilha$^{(\boxtimes)}$, André Dorte, Luis Cereda, Luiz Miyazaki, Maurício Lopes, and Zanoni Dias

Institute of Computing, University of Campinas, Campinas, SP, Brazil
`rafael.padilha@ic.unicamp.br`

Abstract. The COVID-19 pandemic impacted all spheres of our society. The outbreak increased the pressure on public health systems, urging the scientific community to develop and evaluate methods to reliably diagnose patients. Driven by their effectiveness in medical imaging analysis, deep neural networks have been seen as a possible alternative to automatically diagnose COVID-19 patients from chest X-rays. Despite promising initial results, most analyses so far have been performed in small and under-represented datasets. Considering this, in this work, we evaluate state-of-the-art convolutional neural network architectures proposed in recent years by the deep learning field on images from COVIDx [24], a dataset consisting of 13,975 chest X-ray from COVID-19, pneumonia, and healthy patients. In our experiments, we investigate the effect of data pre-processing steps and class unbalancing for this task. Our best model, an ensemble of several networks, achieved an accuracy above 93% in the testing set, showing promising results in a challenging dataset.

Keywords: COVID-19 diagnostic · Chest X-ray image analysis · Deep learning

1 Introduction

With its impact on society, public health, and economy, the outbreak of COVID-19 (SARS-CoV-2) has shaped how the year 2020 will be remembered. The scientific community has devoted its efforts to trace the origin of the disease, studying its effects on the human body, as well as evaluating treatment methods [7,10,28]. Amidst the crisis, society directed its resources onto identifying how artificial intelligence could aid the fight against the pandemic [13], such as predicting mortality and growth rates [23,25], analyzing the virus genome [18], and discovering possible drugs [2].

As the increase in COVID-19 cases overwhelms healthcare systems worldwide, finding accurate and efficient diagnosis methods is critical to prevent further disease spread and treat affected patients. In this sense, encouraged by recent successes of machine learning applied to medical imaging analysis [14]—e.g., skin lesion classification [12], brain tumor segmentation [27], cardiac image

© Springer Nature Switzerland AG 2020
J. C. Setubal and W. M. Silva (Eds.): BSB 2020, LNBI 12558, pp. 57–68, 2020.
https://doi.org/10.1007/978-3-030-65775-8_6

analysis [3]—, recent works assessed its performance in COVID-19 diagnostics. They employed convolutional neural networks (CNN) to classify chest X-ray images or CT scans as to belonging to COVID-19 or healthy patients. CNNs are a type of neural network specialized in processing images, which focus on learning, directly from the input, the most discriminative features for the target task. They achieved promising results in pneumonia and lung nodules classification [17,20].

Training a CNN from scratch is challenging, as the optimization process requires a considerable amount of labeled data, which are not always available in medical problems. In this sense, most approaches rely on Transfer Learning [16]—i.e., leveraging from the knowledge learned in a different task to better generalize in the target problem—to fine-tune the network for COVID-19 diagnosis.

Narin et al. [15] evaluated three CNN architectures, pre-trained for object recognition and fine-tuned to a dataset of 100 chest X-ray images (half of them from COVID-19 patients). Apostolopoulos and Bessiana [1] expanded on the previous evaluation, including other networks and employing a dataset of almost 1500 images—224 from patients diagnosed with COVID-19, 700 with pneumonia, and 500 healthy ones. Even though both works achieved good results, the datasets used are very limited in size and patient representativity.

To address such issues, Wang et al. [24] created the COVIDx dataset, containing 14,198 chest X-rays of 14,002 patients—from which 573 were images from 394 COVID-19 patients. Additionally, they proposed COVID-Net, a lightweight CNN architecture, that outperformed traditional networks trained in the same task, such as VGG-19 and ResNet-50. Finally, authors audited the decisions made by their network to validate its reliability.

In this work, we perform an extensive evaluation of CNN architectures published in recent years by the deep learning research community. We use the dataset collected by Wang et al. [24], comparing not only the performance of each network but also their number of parameters, which directly relates to their efficiency. Similar to previous works [1,15], we assess the effectiveness of Transfer Learning, adapting models pre-trained in object recognition to the COVID-19 diagnosis problem. In our analysis, we consider two scenarios: one with three possible diagnostics—COVID-19, pneumonia, or healthy—and a binary scenario in which we are interested in predicting whether the patient has COVID-19 or not. We contrast these experiments with traditional machine learning classifiers, such as Support Vector Machines (SVM) and Random Forest.

The remaining of the text is organized as follows. In Sect. 2, we describe COVIDx dataset, presenting each class and example images. We present details of our methodology in Sect. 3, describing the overall pipeline of our evaluation and the data preparation used. In Sect. 4, we present the experimental evaluation of different classification methods in this problem. Finally, in Sect. 5, we present our final thoughts and draw possible research lines for future work.

2   Dataset

The dataset was obtained from COVIDx [24], which aggregates images from several different datasets. Available data varies in resolution from 400 × 500 to 3520 × 4280 pixels and are grouped in three different classes: COVID-19, Pneumonia, and Normal (cases of no disease). Table 1 shows the patient and chest radiography images distribution in the training and test sets. In the Pneumonia and COVID-19 classes, there are different X-rays of the same patient. Figure 1 illustrates one example of each class.

Table 1. Distribution of patients and chest radiography images, considering Normal, Pneumonia, and COVID-19 diagnostics for the training and test sets.

	Patients		Images	
	Train	Test	Train	Test
Normal	7,966	100	7,966	100
Pneumonia	5,444	98	5,459	100
COVID-19	320	74	473	100
Total	13,730	272	13,898	300

(a)   **(b)**   **(c)**

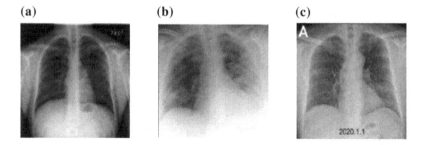

Fig. 1. Examples of X-rays from the COVIDx dataset [24]. (a) Normal, (b) Pneumonia and (c) COVID-19.

Some X-ray images in the dataset present artifacts and noise patterns (Fig. 2), such as medical devices connected to the patient, contour and volume of the breasts, or differences in size and shape of the lungs and rib cage due to the patient being an adult or a child. Even though these patterns might be common to the task, they do not directly relate to the diagnostic outcome and, hence, with enough data a model should learn how to ignore them.

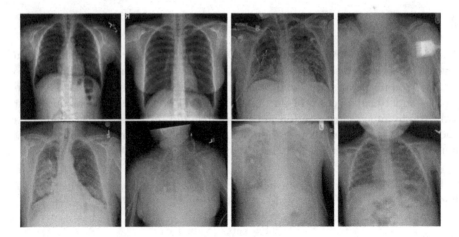

Fig. 2. Examples of images with different patterns, such as medical devices connected to the patient, contour and volume of the breasts and noise.

3   Methodology

Ultimately, our goal is to accurately classify if a chest X-ray image belongs to a patient with COVID-19 or not. In our evaluation, we follow the pipeline depicted in Fig. 3, in which an input image is preprocessed and then analyzed by a classification model.

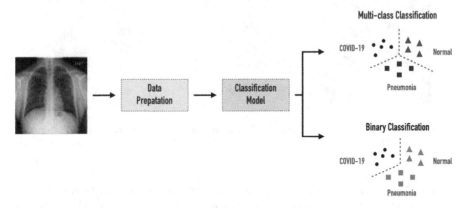

Fig. 3. Overview of our pipeline. An input image is preprocessed and then classified considering two scenarios. In the *Multi-class Classification*, the model assesses if the X-ray image belongs to a COVID-19, a Pneumonia, or a Normal patient. On the *Binary Classification*, the model decides between a COVID-19 or a non-COVID outcome.

We consider two scenarios concerning the output of our models. In the first one, we approach the problem as a multi-class classification with three possible outcomes—i.e., COVID-19, Pneumonia, or Normal. The rationale is that by explicitly differentiating non-COVID cases, the model might better capture subtle differences in each diagnostic. Whereas, in the second scenario, we consider a binary classification between COVID-19 and non-COVID images, in other words, Normal and Pneumonia are grouped into a single class.

In the next subsections, we detail how images are preprocessed before being classified, as well as present the classification models used in our evaluation.

3.1 Data Preparation

Each model evaluated in our analysis expects an input image with particular dimensions and a range of values. In this regard, an input image must be standardized before training and classification.

During training, we resize each image to the network expected input size and normalize its pixel values accordingly. We also apply data augmentation techniques to increase our training set. Considering we have a small dataset that might not realistically represent the variety of images encountered in real scenarios, as seen in Fig. 2, data augmentation is essential to artificially add variability in training and improve model generalization.

We employed the following augmentation techniques: random rotation in the range $[-5°, 5°]$, zoom of at most 10% of the image's dimension, vertical and horizontal shifts up to 10% of the image's height and width, respectively. Considering the set of all possible transformations, each image can generate up to 10,000 slightly altered versions, virtually increasing our training set. We present in Fig. 4 examples of these transformations.

At testing time, before being fed to the classification model, we resize and normalize each input image, without applying any augmentation technique.

(a)   **(b)**   **(c)**   **(d)**   **(e)**

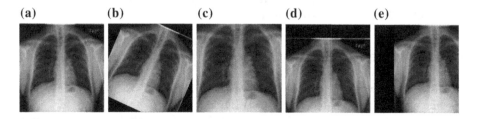

Fig. 4. Data augmentation strategies applied during training. (a) Original image, (b) rotation, (c) zoom, (d) vertical and (e) horizontal shifts.

3.2 Classification Model

In recent years, the research community has proposed several CNN architectures, optimization methods, and regularization techniques. Besides focusing on improving generalization and efficacy in particular tasks, some works also aimed at increasing the efficiency of such networks, allowing them to train faster and run on low-powered devices.

In this work, we evaluate several architectures proposed in the past five years. As training them from scratch would require large datasets, we use pre-trained CNNs optimized on ImageNet [6], an object classification dataset with 14 million images. The rationale behind this is that deep networks tend to learn similar concepts in their initial and intermediate layers common to most visual tasks [26]—from object recognition to medical imaging analysis.

With the knowledge previously obtained, we adapt the networks for the COVID-19 diagnosis task. We remove their last fully-connected layer, responsible for classifying in one of the ImageNet classes, and exchange it for a new fully-connected layer with three or two output neurons—for the *Multi-class* or *Binary* classification scenarios, respectively—activated by softmax operation.

As each architecture imposes different constraints on the learning process, the final characteristics learned by each of them capture distinct and often complementary aspects of the training data. Because of this, we investigate if an ensemble of networks improves the performance over individual models.

We explore three fusion approaches. In the first one, we average the answers of all CNNs, aiming for the mean consensus between them. Secondly, instead of merely averaging their responses, we train a meta-classifier on top of the concatenated answers. This meta-model learns subtle relative patterns between fused classifiers. Finally, we take a step back and extract features from the penultimate layer of each network, optimizing a meta-classifier on their concatenation. This meta-model aggregates the knowledge from intermediate features and learns how to jointly classify them.

In comparison, we also train other machine learning classifiers as baselines for this task. We evaluate SVM, Random Forest, XGBoost [4] and Logistic Regression. In addition to the pre-processing described in Subsect. 3.1, we serialize each image before feeding them to each classifier.

4   Experimental Evaluation

In this section, we present the results of the different CNN architectures and baseline classifiers evaluated in the COVIDx dataset.

To train the predictive models, we organized the original training set into train and validation splits. Initially, we randomly sampled 473 X-ray images of each class (the size of the smaller class), divided into 383 for training and 90 for validation. This was done to mitigate the class unbalance of the dataset, making a balanced sub-sample that allows us to train each architecture efficiently. Using this reduced balanced set, we evaluated all models in terms of accuracy and number of parameters for the *Multi-class* classification scenario.

Considering the top three models, we evaluated their performance applying data augmentation and using the whole unbalanced dataset. To do so, we employed a stratified division of the original training set, using 80% for training and 20% for validation. In this setup, the model has more images to learn from and generalize, even though the training step requires more time. Each model was evaluated in the *Multi-class* and *Binary* classification scenarios, as well as their ensemble, combined under different strategies.

4.1   Model Evaluation

As baselines, we trained SVMs with linear, polynomial, and RBF kernels, Random Forest, Logistic Regression, and XGBoost classifiers. The choice of hyperparameters for each technique was obtained through a grid search, using a 5-fold cross-validation with the balanced training set. Before training, we serialized each input image, transforming them into one-dimensional vectors.

For the evaluation of the CNN architectures, we applied Transfer Learning. For all architectures, we fixed the weights pre-trained on ImageNet dataset and updated only the newly-added classification layer specific to our task. We optimized the networks with cross-entropy loss, using an SGD optimizer with a learning rate of 0.001, momentum of 0.9, and weight decay with a rate of 0.001. Additionally, we employed early stopping, interrupting the training when the validation loss stopped increasing for several epochs. We present in Table 2 the results obtained by the baselines and the CNN architectures in the validation set of the balanced sub-sample of the dataset.

Table 2. Accuracy in the balanced validation set and number of parameters for each evaluated models.

Model	Accuracy in validation (%)	Number of parameters
ResNet50 [8]	87.41	25,636,712
EfficientNetB7 [22]	84.81	66,658,687
MobileNetV2 [9]	81.85	3,538,984
DenseNet121 [11]	80.74	8,062,504
MobileNet [9]	80.37	3,538,984
Random forest	79.25	–
XGBoost [4]	78.52	–
SVM-RBF	78.15	–
Xception [5]	77.78	22,910,480
SVM-Poly	75.93	–
InceptionV3 [21]	74.07	23,851,784
NASNetLarge [29]	72.22	88,949,818
ResNet50V2 [8]	70.37	25,613,800
Logistic regression	68.89	–
SVM-Linear	68.15	–

The best baseline methods were Random Forest, XGBoost, and SVM with RBF kernel, with accuracies ranging from 79.25% to 78.15%. Even though their performances were superior to a few architectures, they are not suitable for dealing with the spatial information of images in the same way as CNNs. By reshaping the image as a vector, the models lose most of the spatial structure between neighboring pixels, which often hinders performance. On the other hand, logistic regression and linear SVM suffer from trying to correctly fit a linear decision plane in the high dimensional space produced by the flattened image.

CNNs that were outperformed by the most baselines have an increased number of parameters, which indicates that network size does not directly relate to the accuracy in the task. This is probably due to the lack of data required to carefully train a network with such complexity for this problem. The exception was EfficientNetB7, which accounts in second both in the number of parameters and accuracy. Despite its size, it outperforms in the ImageNet recognition task several architectures [22]—both with lower and higher number of parameters. This highlights its capability of learning discriminative and rich features, which is essential when adapting it to a task with limited data.

Our experiments show that ResNet50 architecture achieved the highest accuracy in the validation set, with an intermediate number of parameters among the networks evaluated. We select ResNet50, EfficientNetB7 and MobileNetV2 as the base models to perform additional explorations.

4.2   Multi-class and Binary Classification of the Unbalanced Dataset

Once selected, we evaluated the top three models in the complete unbalanced dataset. To do so, we split it into 80% for training and 20% for validation in a stratified division. Due to the class unbalancing, the model would naturally give greater importance to the over-represented classes, i.e., Normal and Pneumonia. To overcome this issue, we assigned weights to each category according to their sizes. The weight of a particular class is equal to the number of images of the largest class divided by the number of samples in it.

Besides that, we also employed the data augmentation techniques from Subsect. 3.1 to increase the diversity in our training. Considering the increased amount of data, we unfroze the initial and intermediate layers of each model, allowing the optimization process to freely update their weights. Each CNN was trained with the same optimizer and hyperparameters in the *Multi-class* and *Binary* classification scenarios. We report the balanced accuracy for each method in the top part of Table 3.

The CNNs obtained high accuracy, with ResNet50 and MobileNetV2 outperforming EfficientNetB7 in both classification scenarios. Similar to the previous experiment, all three CNNs exceeded the baseline Random Forest, which was considerably affected by the class unbalancing, especially in the *Binary* setup.

4.3   Ensemble of CNNs

In addition to evaluating each separate model, we investigated their complementarity and explored how to combine their knowledge in this task. In this analysis, we fused the answers—i.e., the output classification probabilities—and the features from the penultimate layer of each CNN.

Besides combining the answers through average operation, we also trained meta-classifiers on top of the concatenation of probabilities or deep features. We evaluated SVMs (with linear, polynomial, and RBF kernels), Random Forest, and shallow Multi-Layer Perceptrons (MLP) as meta-models. The choice of hyperparameters—such as the number of hidden layers and units of the MLP, or trees in the Random Forest—was done based on the balanced accuracy of the validation set. We present the results in the Table 3.

Table 3. Balanced accuracy in the stratified validation and test sets of *Multi-class* and *Binary* classification for individual CNNs and ensembles. We highlight the best method and ensemble strategies in each scenario and evaluation set.

Method	Multi-class		Binary	
	Accuracy in validation (%)	Accuracy in test (%)	Accuracy in validation (%)	Accuracy in test (%)
Classifiers				
ResNet50	96.81	90.66	96.13	91.99
EfficientNetB7	83.54	79.33	89.21	81.75
MobileNetV2	90.01	82.66	97.87	90.25
Random forest	65.30	62.33	50.55	55.00
Ensemble – Probabilities				
Average	97.22	89.66	98.66	93.25
SVM-Linear	98.37	90.00	98.87	89.75
SVM-Poly	98.70	90.00	99.81	87.25
SVM-RBF	98.47	90.33	98.81	89.50
MLP	99.88	92.00	99.36	93.50
Random forest	98.68	90.33	99.41	84.75
Ensemble – Deep features				
SVM-Linear	97.67	89.66	99.43	89.75
SVM-Poly	98.26	90.00	99.41	87.75
SVM-RBF	98.34	90.00	99.38	89.50
MLP	98.01	83.33	99.43	86.00
Random forest	95.35	88.00	96.82	84.74

Most ensemble strategies outperformed individual CNNs in the validation set, highlighting the importance of combining the characteristics learned by different models. In the *Multi-class* scenario, we improved from 96.81% accuracy with ResNet50 to 99.88% with an MLP over the probabilities. Whereas in the *Binary* classification, a similar gain was obtained, increasing from an accuracy of 97.87% with MobileNetV2 to 99.81% with a polynomial SVM over the probabilities.

Our experiments show that fusing probabilities or deep features achieved similar performance, with slight differences depending on the scenario or meta-classifier. Despite that, working with probabilities proved to be more efficient and more robust to overfitting due to the reduced input dimensionality compared to the concatenated deep features.

4.4   Test Set Evaluation

We evaluated all methods in the test set and present the results in Table 3. The overall performance decreased considerably when comparing validation and test sets for most methods. This was probably due to the severe class unbalance, whose impact could not be entirely prevented by data augmentation and the use of class weights during training.

Considering the individual models, ResNet50 obtained the highest accuracy in both classification scenarios. Whereas, among the ensembles, the MLP optimized over the probabilities outperformed other configurations. We also compare their confusion matrix on Fig. 5. In all cases, the methods are slightly biased towards classifying COVID-19 samples as Normal or Pneumonia, presenting the majority of the false predictions in the bottom row of each matrix. Nonetheless, the ensemble approach is still able to correctly classify 89% of COVID X-rays.

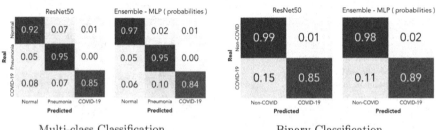

Multi-class Classification.   Binary Classification.

Fig. 5. Confusion matrix for ResNet50 and the best ensemble strategy in both classification scenarios for the test set.

5   Conclusions and Future Work

With the outbreak of COVID-19, healthcare systems worldwide are looking for more accurate and efficient diagnostic methods for the treatment of patients. With the recent advancement of machine learning techniques applied to medical imaging analysis, the automatic classification of X-ray images has become an essential tool for aiding medical diagnoses and a possible approach to diagnosing COVID-19 patients.

In this work, we evaluated several CNN architectures published in the deep learning literature to classify chest X-ray images of COVID-19. The experiments showed that transfer learning techniques can achieve good results by leveraging the generalization of the initial layers in a different domain. To aid model generalization, we employed data augmentation strategies to artificially increase available data. Besides that, we evaluated ensemble techniques applied to the best obtained models, improving the results even further. We achieved an accuracy of 92.00% in the test set considering the classification between COVID-19,

Pneumonia, and Normal classes, while obtaining a 93.50% accuracy when distinguishing between COVID-19 and non-COVID.

As future work, additional preprocessing steps can be investigated, such as segmenting the lung parts of the image, applying filters to reduce noise and highlight particular artifacts, as well as using images with higher resolution. In the post-processing step, explainability techniques, such as Grad-CAM [19], could be used to further audit the network decisions and interpret them. Finally, being a recent problem, datasets for this task are still limited in size and present a considerable class unbalancing. Because of this, a continuous data collection is essential for future research in this problem.

References

1. Apostolopoulos, I.D., Mpesiana, T.A.: Covid-19: automatic detection from X-ray images utilizing transfer learning with convolutional neural networks. Phys. Eng. Sci. Med. **43**(2), 635–640 (2020). https://doi.org/10.1007/s13246-020-00865-4
2. Beck, B.R., Shin, B., Choi, Y., Park, S., Kang, K.: Predicting commercially available antiviral drugs that may act on the novel coronavirus (SARS-CoV-2) through a drug-target interaction deep learning model. Comput. Struct. Biotechnol. J. **18**, 784–790 (2020)
3. Bizopoulos, P., Koutsouris, D.: Deep learning in cardiology. IEEE Rev. Biomed. Eng. **12**, 168–193 (2018)
4. Chen, T., Guestrin, C.: XGBoost: a scalable tree boosting system. In: ACM International Conference on Knowledge Discovery and Data Mining (ACM KDD), pp. 785–794 (2016)
5. Chollet, F.: Xception: deep learning with depthwise separable convolutions. In: IEEE Conference on Computer Vision and Pattern Recognition (CVPR), pp. 1251–1258 (2017)
6. Deng, J., Dong, W., Socher, R., Li, L.J., Li, K., Fei-Fei, L.: ImageNet: a large-scale hierarchical image database. In: IEEE Conference on Computer Vision and Pattern Recognition (CVPR), pp. 248–255 (2009)
7. Gao, Q., Bao, L., Mao, H., Wang, L., Xu, K., Yang, M., Li, Y., Zhu, L., Wang, N., Lv, Z., et al.: Development of an inactivated vaccine candidate for SARS-CoV-2. Science **369**(6499), 77–81 (2020)
8. He, K., Zhang, X., Ren, S., Sun, J.: Deep residual learning for image recognition. In: IEEE Conference on Computer Vision and Pattern Recognition (CVPR), pp. 770–778 (2016)
9. Howard, A.G., et al.: MobileNets: efficient convolutional neural networks for mobile vision applications. arXiv:1704.04861 (2017)
10. Huang, C., Wang, Y., Li, X., Ren, L., Zhao, J., Hu, Y., Zhang, L., Fan, G., Xu, J., Gu, X., et al.: Clinical features of patients infected with 2019 novel coronavirus in wuhan, china. The Lancet **395**(10223), 497–506 (2020)
11. Huang, G., Liu, Z., Van Der Maaten, L., Weinberger, K.Q.: Densely connected convolutional networks. In: IEEE Conference on Computer Vision and Pattern Recognition (CVPR), pp. 4700–4708 (2017)
12. Kawahara, J., Hamarneh, G.: Multi-resolution-tract CNN with hybrid pretrained and skin-lesion trained layers. In: International Workshop on Machine Learning in Medical Imaging, pp. 164–171 (2016)

13. Lalmuanawma, S., Hussain, J., Chhakchhuak, L.: Applications of machine learning and artificial intelligence for Covid-19 (SARS-CoV-2) pandemic: A review. Chaos, Solitons & Fractals **139**, 110059 (2020)

14. Litjens, G., et al.: A survey on deep learning in medical image analysis. Med. Image Anal. **42**, 60–88 (2017)

15. Narin, A., Kaya, C., Pamuk, Z.: Automatic detection of coronavirus disease (COVID-19) using X-ray images and deep convolutional neural networks. arXiv:2003.10849 (2020)

16. Raghu, M., Zhang, C., Kleinberg, J., Bengio, S.: Transfusion: understanding transfer learning for medical imaging. In: Advances in Neural Information Processing Systems (NIPS), pp. 3347–3357 (2019)

17. Rajpurkar, P., et al.: CheXNet: radiologist-level pneumonia detection on chest X-rays with deep learning. arXiv:1711.05225 (2017)

18. Randhawa, G.S., Soltysiak, M.P., El Roz, H., de Souza, C.P., Hill, K.A., Kari, L.: Machine learning using intrinsic genomic signatures for rapid classification of novel pathogens: Covid-19 case study. PLoS ONE **15**(4), e0232391 (2020)

19. Selvaraju, R.R., Cogswell, M., Das, A., Vedantam, R., Parikh, D., Batra, D.: Grad-cam: Visual explanations from deep networks via gradient-based localization. In: IEEE International Conference on Computer Vision (ICCV), pp. 618–626 (2017)

20. Shen, W., Zhou, M., Yang, F., Yang, C., Tian, J.: Multi-scale convolutional neural networks for lung nodule classification. In: International Conference on Information Processing in Medical Imaging (IPMI), pp. 588–599 (2015)

21. Szegedy, C., Vanhoucke, V., Ioffe, S., Shlens, J., Wojna, Z.: Rethinking the inception architecture for computer vision. In: IEEE Conference on Computer Vision and Pattern Recognition (CVPR), pp. 2818–2826 (2016)

22. Tan, M., Le, Q.: EfficientNet: rethinking model scaling for convolutional neural networks. In: IEEE International Conference on Machine Learning (ICML), pp. 6105–6114 (2019)

23. Tuli, S., Tuli, S., Tuli, R., Gill, S.S.: Predicting the growth and trend of COVID-19 pandemic using machine learning and cloud computing. Internet of Things **11**, 100222 (2020)

24. Wang, L., Wong, A.: COVID-Net: a tailored deep convolutional neural network design for detection of COVID-19 cases from chest x-ray images. arXiv:2003.09871 (2020)

25. Yan, L., et al.: An interpretable mortality prediction model for COVID-19 patients. Nat. Mach. Intell. **2**, 1–6 (2020)

26. Yosinski, J., Clune, J., Bengio, Y., Lipson, H.: How transferable are features in deep neural networks? In: Advances in Neural Information Processing Systems (NIPS), pp. 3320–3328 (2014)

27. Zhao, J., Zhang, M., Zhou, Z., Chu, J., Cao, F.: Automatic detection and classification of leukocytes using convolutional neural networks. Med. Biol. Eng. Comput. **55**(8), 1287–1301 (2016). https://doi.org/10.1007/s11517-016-1590-x

28. Zhou, P., Yang, X.L., Wang, X.G., Hu, B., Zhang, L., Zhang, W., Si, H.R., Zhu, Y., Li, B., Huang, C.L., et al.: A pneumonia outbreak associated with a new coronavirus of probable bat origin. Nature **579**(7798), 270–273 (2020)

29. Zoph, B., Vasudevan, V., Shlens, J., Le, Q.V.: Learning transferable architectures for scalable image recognition. In: IEEE Conference on Computer Vision and Pattern Recognition (CVPR), pp. 8697–8710 (2018)

Classification of Musculoskeletal Abnormalities with Convolutional Neural Networks

Guilherme Tiaki Sassai Sato$^{(\boxtimes)}$ [ID], Leodécio Braz da Silva Segundo [ID], and Zanoni Dias [ID]

Institute of Computing, University of Campinas, Campinas, Brazil
guilhermetiaki@me.com, zanoni@ic.unicamp.br

Abstract. Computer-aided diagnosis has the potential to alleviate the burden on medical doctors and decrease misdiagnosis, but building a successful method for automatic classification is challenging due to insufficient labeled data. In this work, we investigate the usage of convolutional neural networks to diagnose musculoskeletal abnormalities using radiographs (X-rays) of the upper limb and measure the impact of several techniques in our model. We achieved the best results by utilizing an ensemble model that employs a support vector machine to combine different models, resulting in an overall AUC ROC of 0.8791 and Kappa of 0.6724 when evaluated using an independent test set.

Keywords: Deep learning · Musculoskeletal abnormalities · X-ray

1 Introduction

Musculoskeletal conditions are extensively present in the population, affecting over 1.3 billion people worldwide [9]. These conditions often cause long-term pain, directly and indirectly reducing the quality of life of those suffering from it and their household [33,34]. In this setting, medical imaging as X-rays plays an essential role as one of the main tools for abnormality detection.

Insufficient medical staff, along with the complexity of diagnosis, creates a system prone to errors. False-negative diagnosis leads to untreated injuries and symptoms such as chronic pain and further complications in the long term, and false-positives diagnosis leads to unnecessary treatment. Computer-Aided Diagnosis (CAD) systems are used to counteract these problems, improving diagnostic accuracy, assist decision-making, and reducing radiologists' workload.

Computer-aided diagnosis has been a topic of research since the 1960s and has significantly evolved due to advances in medicine itself and in computer science. From a task standpoint, CAD has found application in a wide variety of medical disorders. A few of the innumerous works include breast cancer [8,31], lung cancer [19], and Alzheimer's [6]. X-ray classification has been particularly prevalent for the chest area [2,32]. Under the same scope of this work (musculoskeletal abnormalities in the upper limb), 70 works were submitted under a

© Springer Nature Switzerland AG 2020
J. C. Setubal and W. M. Silva (Eds.): BSB 2020, LNBI 12558, pp. 69–80, 2020.
https://doi.org/10.1007/978-3-030-65775-8_7

competition[1] that took place using the same dataset as this work. The reported achieved Cohen's Kappa range from 0.518 to 0.843. However, no further information, such as methodology, is generally available for these works.

Methodology-wise, some of the most successful classifiers include k-nearest neighbors (KNN) [14,20], support vector machines (SVM) [7], random forests [1,15], and neural networks [16,30]. In this work, we will evaluate the use of a neural network classifier due to its promising performance, producing state-of-the-art results in many other applications [4,26].

A wide range of techniques, such as deep learning, image processing, and computer vision, are applied to interpret radiographic images automatically. However, deep learning models' success is highly dependent on the amount of data available, creating a particular challenge for medical images due to privacy concerns and time-consuming labeling requiring experts. In this work, we present a method to classify normal and abnormal X-rays from the upper limb by applying convolutional neural networks and several machine learning techniques aiming to improve the classification and offset the lack of data. The rest of this paper is organized as follows. Section 2 describes the settings of this work and all the experiments executed to reach a final classifier, and Sect. 3 evaluates this classifier using an independent test set and discusses the results obtained, as well as alternative scenarios. Section 4 concludes the paper, including an overview of the results obtained and future work.

2 Methodology

This section goes over the dataset used in this work and details the experiments executed. Over the experiments, we explore different scenarios by applying machine learning and deep learning techniques and measure the impact of the proposed changes when comparing to previously tested scenarios, in a path to maximize the classifier robustness.

2.1 Dataset

In this work, we used the *MURA: Large Dataset for Abnormality Detection in Musculoskeletal Radiographs* [23] dataset, which contains 40,005 radiographic images labeled by radiologists. The MURA dataset is divided into 14,656 studies from the body upper extremities – shoulder, humerus, elbow, forearm, wrist, hand, and finger. Each study is labeled as either normal or abnormal.

The available data is divided into validation and training sets. For testing, a third set was created to be used in a competition and, therefore, not publicly available. To offset this fact and provide a realist measurement of our model's performance in a real-world scenario, we split the training data to create a test set. Our goal was to create a test set the same size as the validation set. As the provided validation set has 8.6% of the size of the training set, the new test set was created by moving, for each body part, 8.6% of the studies from the training to the test set. The number of items in each class is described in Table 1.

---

[1] https://stanfordmlgroup.github.io/competitions/mura/.

Table 1. Distribution of the number of studies (images) contained in each class for the three sets.

	Train		Validation		Test	
	Normal	Abnormal	Normal	Abnormal	Normal	Abnormal
Shoulder	1242 (3838)	1331 (3833)	99 (285)	95 (278)	122 (373)	126 (335)
Humerus	293 (608)	247 (549)	68 (148)	67 (140)	28 (65)	24 (50)
Elbow	997 (2677)	601 (1841)	92 (235)	66 (230)	97 (248)	59 (165)
Forearm	543 (1069)	257 (595)	69 (150)	64 (151)	47 (95)	30 (66)
Wrist	1993 (5237)	1218 (3670)	140 (364)	97 (295)	201 (528)	108 (317)
Hand	1365 (3702)	475 (1354)	101 (271)	66 (189)	132 (357)	46 (130)
Finger	1161 (2834)	600 (1805)	92 (214)	83 (247)	119 (304)	55 (163)

2.2 Experiments

The models used for classification were developed in Python, mainly using the PyTorch framework [21]. Every sample from the dataset has its target class defined among 14 classes (7 body parts, normal or abnormal). Before inputting the images to the model, we normalized each image's pixels values to the mean and standard deviation of the ImageNet dataset [5] and resized to 224×224 pixels.

We trained each model over 40 epochs, with a batch size of 25. The samples are initially shuffled and reshuffled before each epoch. Upon each epoch, the network performance is measured with the scikit-learn library package [22]. To measure the binary classification performance, the 14 class output is condensed into two by ignoring the body part information. To determine the output of a study consisting of several images, the probability distribution output for each image in the study are averaged.

Experiment I: Fit, Pad, or Stretch? Our goal is to use pre-trained networks as a baseline. However, pre-trained networks expect a square input, and our images have a variable aspect ratio, so before testing an assortment of networks, we need to decide how to transform our images: fit, pad, or stretch.

All three transformations come with pros and cons. "Fit" and "Pad" preserve the image aspect ratio, while "Stretch" distorts it. On the other hand, "Fit" loses some amount of information, which might be considerable if the image is very tall or long, "Pad" fills part of the input with irrelevant information, while "Stretch" includes the entire image. Figure 1 illustrates the three transformations.

To decide on the best method, we trained the ResNet-18 network on the three options and compared the performance. Table 2 shows the results obtained. All of them were pretty similar, but "Stretch" achieved the best Kappa. Furthermore, "Stretch" was the fastest operation on our tests. Therefore, we chose the "Stretch" operation to be used in the subsequent experiments.

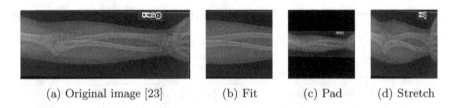

(a) Original image [23]   (b) Fit   (c) Pad   (d) Stretch

Fig. 1. Three transformations to change a rectangular image into square. *Fit* does not deform but loses information. *Pad* also does not deform but adds irrelevant information. *Stretch* includes all pixels but deforms the image.

Table 2. Performance on the validation data for each of the transformations.

	Accuracy	Balanced accuracy	AUC ROC	Kappa
Fit	0.8232	0.8140	0.8720	0.6373
Pad	0.8165	0.8054	**0.8723**	0.6222
Stretch	**0.8274**	**0.8175**	0.8617	**0.6453**

Experiment II: Pre-trained Networks. The use of pre-trained models provides us with a consolidated and validated architecture. These models have demonstrated good results in many similar tasks [10]. In addition, it reduces the training time required compared to the training necessary to achieve similar results without pre-training.

Pre-trained models may, however, have a few caveats. The models were trained using the ImageNet dataset, containing colored images that do not resemble medical images. Therefore, the layer structure of the networks may be suboptimal for this task. Moreover, the ImageNet images contain three channels. Consequently, our inputs must be reshaped to conform to this restriction [35].

We tested several networks among the best-performing ones, including DenseNet [12], EfficientNet [17,29], Inception-v3 [28], Inception-v4 [3,27], Inception-ResNet-v2 [3,27], ResNet [11], and VGG [25]. Table 3 presents the results obtained on each model. All the models performed very similarly, with a Kappa coefficient between 0.6450 and 0.6671.

Experiment III: Data Augmentation. To increase the model capacity to generalize, we tested two data augmentation techniques: cropping and horizontal flip. To verify the effect of these augmentations, we trained the DenseNet-161 network once using only the horizontal flip augmentation, doubling the number of samples (Augmented Dataset A); and once using the original, five random crops and its flips, resulting in a 12-fold increase in the size (Augmented Dataset B), similar to the augmentation proposed by Krizhevsky et al. [13]. Table 4 compares the results of the two experiments with the original dataset.

Table 3. Performance of each network on the validation data.

	Accuracy	Balanced accuracy	AUC ROC	Kappa
DenseNet-121	0.8365	0.8279	0.8874	0.6649
DenseNet-161	**0.8374**	**0.8293**	**0.8892**	**0.6671**
EfficientNet-B7	0.8274	0.8182	0.8873	0.6458
Inception-v3	0.8290	0.8197	0.8769	0.6491
Inception-v4	0.8315	0.8226	0.8783	0.6546
Inception-ResNet-v2	0.8324	0.8229	0.8864	0.6559
ResNet-18	0.8274	0.8175	0.8617	0.6453
ResNet-152	0.8299	0.8220	0.8780	0.6519
VGG-16	0.8357	0.8254	0.8877	0.6621
VGG-19	0.8282	0.8155	0.8840	0.6450

Table 4. Comparison of performance on the validation data using DenseNet-161. *Augmented Dataset A* is the dataset using only horizontal flip, and *Augmented Dataset B* is the dataset using both horizontal flip and crop.

	Accuracy	Balanced accuracy	AUC ROC	Kappa
Original dataset	0.8374	0.8293	0.8892	0.6671
Augmented dataset A	**0.8465**	**0.8370**	**0.8929**	**0.6848**
Augmented dataset B	0.8432	0.8355	0.8922	0.6792

From these results, we could conclude that, despite having more data, Augmented Dataset B performs worse than Augmented Dataset A, but still better than the original. Moreover, since training time is approximately linear in the number of samples, using Augmented Dataset A results in longer training time. Therefore, we trained other networks using Augmented Dataset B (Table 5). These networks were chosen based on the results of Experiment II while avoiding more than one network from the same "family."

Experiment IV: Ensemble Model. An ensemble model combines multiple models to make a final prediction, similar to consulting multiple opinions. This will supposedly improve the model as if a single model performs poorly for a sample, its prediction may be overridden by other models, improving the model stability.

We used the five models from Experiment III and combined its predictions using a set of techniques. First, we made a final prediction based on consensus, i.e., the mode of the five predictions, ignoring the probabilistic distribution. Next, we combined the results using a weighted average, using one weight for each of the five models. Then we tested two techniques using neural networks: a sparsely connected and a fully connected layer to make a final prediction. The architecture of these layers is shown in Fig. 2. While these last two techniques

Table 5. Performance of networks on the validation data for the networks trained using a dataset augmented using only horizontal flip (Augmented dataset A).

	Accuracy	Balanced accuracy	AUC ROC	Kappa
DenseNet-161	**0.8465**	**0.8370**	0.8929	**0.6848**
EfficientNet-B7	0.8399	0.8314	0.8913	0.6719
Inception-ResNet-v2	0.8457	0.8369	0.8892	0.6836
ResNet-152	0.8349	0.8243	0.8931	0.6602
VGG-16	0.8374	0.8302	**0.8967**	0.6676

are also ultimately a weighted average, each element of the output array for each model is now independently weighted. Also, while the mere five weights in the simple weighted were obtained using an exhaustive search, the weights in these two experiments were obtained using gradient descent. Finally, we used a support vector machine (SVM) with a radial basis function (RBF) kernel to ensemble the models. The SVM is implemented by scikit-learn [22].

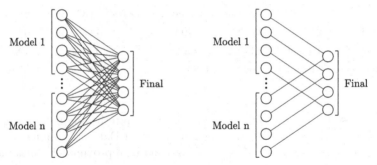

(a) Fully connected neural network architecture

(b) Sparsely connected neural network architecture

Fig. 2. Schematic representation of the two neural network architectures used in this experiment. The number of classes shown here is 4 for simplification, while the real number of classes is 14. The number of models n is 5 in the experiment.

Figure 3 illustrates the final methodology used. The results obtained are presented in Table 6, comparing it with the average results obtained by the models individually. All ensemble models were able to perform better than the average performance of the single models. The single models present a hard to avoid overfitting problem, leading to nearly 100% accuracy in the training set for all the models, therefore making it impossible to find the best way to combine these predictions using the training set since practically any combination will lead to nearly 100% accuracy. We must then tune the parameters in the ensemble layer using the validation set, which should inevitably lead to a higher drop in performance when transitioning to the test set.

Table 6. Comparison of performance on the validation data. The *Consensus* model does not output a probability distribution, only the final prediction, penalizing its AUC ROC score.

	Accuracy	Balanced accuracy	AUC ROC	Kappa
Single model (average)	0.8409	0.8319	0.8926	0.6736
Consensus	0.8449	0.8349	0.8753	0.6811
Weighted average	0.8641	0.8572	0.9086	0.7222
Sparsely connected	0.8590	0.8504	0.9039	0.7110
Fully connected	0.8540	0.8467	0.9102	0.7015
SVM (RBF)	**0.8699**	**0.8638**	**0.9208**	**0.7345**

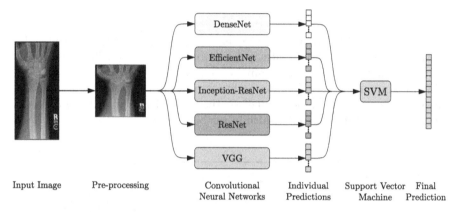

Input Image   Pre-processing   Convolutional   Individual   Support Vector   Final
                                  Neural Networks   Predictions   Machine   Prediction

Fig. 3. Overview of the final classification methodology adopted. The X-ray [23] is squared, processed by five neural networks, and the five independent predictions are combined to reach a final prediction using a SVM.

3   Results and Discussion

The Gradient-weighted Class Activation Mapping (Grad-CAM) [18,24] is a method to visualize areas of the input image that are important to reaching the outputted prediction. Figure 4 shows the heatmaps generated by Grad-CAM for each individual model and the ensemble model, demonstrating the improvement caused by ensembling the models. In this example VGG-16, EfficientNet-B7 and Inception-ResNet-v2 predict "abnormal humerus" while DenseNet-161 and ResNet-152 predict "normal humerus." The Ensemble SVM model prediction is, correctly, "abnormal humerus." The incorrect models focus on a different part of the image but are overridden by other models.

Since the Grad-CAM uses the information from the feature layer to build the visualization and our ensemble models combine the individual models at the output layer level, we have multiple feature layers, which would preclude the use of Grad-CAM. To work around this, we built the ensemble visualization by

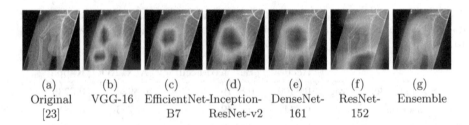

(a)	(b)	(c)	(d)	(e)	(f)	(g)
Original [23]	VGG-16	EfficientNet-B7	Inception-ResNet-v2	DenseNet-161	ResNet-152	Ensemble

Fig. 4. Grad-CAM heatmaps for each individual model and the ensemble model for an abnormal humerus sample. Models VGG-16, EfficientNet-B7 and Inception-ResNet-v2 predict correctly, while DenseNet-161 and ResNet-152 predict incorrectly. The Ensemble SVM model takes all models into consideration and outputs a correct final prediction.

averaging the Grad-CAM outputted matrix (used to create the heatmap) for the five models. Hence, the heatmap generated for the ensemble model is an approximation and not specific to any of the models.

Figure 5 summarizes the Kappa coefficients for all the experiments. Experiment II had an average Kappa of 0.6542; the SVM (RBF) model from Experiment IV improved it to 0.7345, which is our best result.

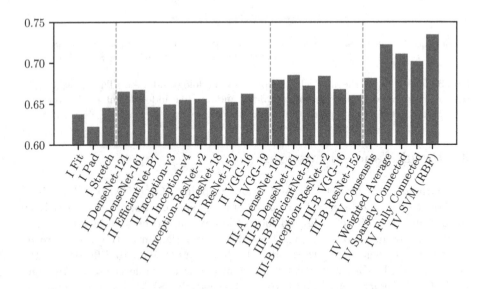

Fig. 5. Summary of the Kappa coefficient for the experiments. The maximum value was 0.7345.

To evaluate our model in a scenario closer to the real world, we will now use our test set, defined in Subsect. 2.1, which has no overlap with the other two sets. Table 7 shows the performance metrics achieved using the Ensemble

SVM model, which was the best performing model on the validation data. The model's performance expectedly decreased due to overfitting.

Our model performed worst on hands, with a Kappa of 0.4717, and best on elbows, with a Kappa of 0.7921. The overall Kappa was 0.6724. In contrast, human radiologists performed worst on fingers and best on wrists [23]. Our model was able to outperform two of the three radiologists evaluated on the elbow classification task, and all of them in the finger classification, but falls behind on other body parts. The MURA competition has ended, so we cannot evaluate our model on the official test set. As a consequence, the human radiologists were evaluated on the official test set, and our model, using our test set, so the performance comparison might not be completely accurate.

Table 7. Performance of the model Ensemble SVM on the test data, broken down by inputted body part.

	Accuracy	Balanced accuracy	AUC ROC	Kappa
Shoulder	0.8266	0.8270	0.8778	0.6535
Humerus	0.8077	0.7976	0.8274	0.6061
Elbow	0.9038	0.8895	0.9023	0.7921
Forearm	0.7922	0.7755	0.8142	0.5578
Wrist	0.8835	0.8633	0.9142	0.7393
Hand	0.8090	0.7225	0.7759	0.4717
Finger	0.8563	0.8558	0.8901	0.6817
Overall	0.8484	0.8319	0.8791	0.6724

The performance difference between the test and validation sets is likely explained due to the fitting of ensemble parameters using the validation set, as described in Experiment IV. We can verify this by running the test set on the Ensemble Consensus model, which does not have any extra parameters. Table 8 shows the results of this test. These results are similar to the results obtained in the validation set (by some metrics, even better), and despite being the worst performer on Experiment IV, it performed better than the Ensemble SVM model, the best ensemble model. Therefore, we could assume that if we were able to train the single models in such a way to reduce overfitting, we could train the ensemble parameters using the train set and reduce the gap between validation and test results. Alternatively, splitting the data into four

Table 8. Performance of the model Ensemble Consensus on the test data. The model has similar performance on the validation set.

Accuracy	Balanced accuracy	AUC ROC	Kappa
0.8593	0.8339	0.8652	0.6899

sets instead of three to fit the ensemble parameters using a separated set could also be beneficial, but would further reduce an already limited dataset.

4   Conclusion

Our work proposed to explore this machine learning classification problem using convolutional neural networks and related methods. The best setting found was to stretch the input to a square, apply horizontal flip data augmentation, and ensemble a variety of architectures using a support vector machine, reaching an AUC ROC of 0.8791 and Kappa of 0.6724.

Although the overall result was lower than that obtained by human radiologists, we were still able to achieve promising results in some scenarios. However, a transition to the clinical setting is challenging. Besides accuracy improvement under a controlled scenario, the algorithm would need, for example, to handle unexpected inputs, provide translation and rotation invariance, and include explainability to mitigate automation bias.

The greatest hindrance in this work was the models' overfitting, to which further experimentation with other methods is needed to adequately address it, such as early stopping and regularization. The development of larger datasets would also provide a big leap on this matter.

References

1. Alickovic, E., Subasi, A.: Medical decision support system for diagnosis of heart Arrhythmia using DWT and random forests classifier. J. Med. Syst. **40**(4), 1–12 (2016). https://doi.org/10.1007/s10916-016-0467-8
2. Arzhaeva, Y., et al.: Development of automated diagnostic tools for pneumoconiosis detection from chest X-ray radiographs (2019)
3. Cadene, R.: Pretrained models for Pytorch (2017). https://github.com/Cadene/pretrained-models.pytorch
4. Chiu, C.C., et al.: State-of-the-art speech recognition with sequence-to-sequence models. In: IEEE International Conference on Acoustics, Speech and Signal Processing (ICASSP), pp. 4774–4778. IEEE (2018). https://doi.org/10.1109/ICASSP.2018.8462105
5. Deng, J., Dong, W., Socher, R., Li, L.J., Li, K., Fei-Fei, L.: ImageNet: a large-scale hierarchical image database. In: IEEE Conference on Computer Vision and Pattern Recognition (CVPR), pp. 248–255. IEEE (2009). https://doi.org/10.1109/CVPR.2009.5206848
6. Ding, Y., et al.: A deep learning model to predict a diagnosis of Alzheimer disease by using 18F-FDG PET of the brain. Radiology **290**(2), 456–464 (2019). https://doi.org/10.1148/radiol.2018180958
7. Dolatabadi, A.D., Khadem, S.E.Z., Asl, B.M.: Automated diagnosis of coronary artery disease (CAD) patients using optimized SVM. Comput. Methods Programs Biomed. **138**, 117–126 (2017). https://doi.org/10.1016/j.cmpb.2016.10.011

8. Garud, H., et al.: High-magnification multi-views based classification of breast fine needle aspiration cytology cell samples using fusion of decisions from deep convolutional networks. In: IEEE Conference on Computer Vision and Pattern Recognition (CVPR), pp. 76–81 (2017). https://doi.org/10.1109/CVPRW.2017.115

9. Global Burden of Disease Collaborative Network: Global burden of disease study 2017 (GBD 2017) results (2018)

10. Habibzadeh, M., Jannesari, M., Rezaei, Z., Baharvand, H., Totonchi, M.: Automatic white blood cell classification using pre-trained deep learning models: ResNet and Inception. In: Proceedings of the 10th International Conference on Machine Vision (ICMV), vol. 10696, p. 1069612. International Society for Optics and Photonics (2017). https://doi.org/10.1117/12.2311282

11. He, K., Zhang, X., Ren, S., Sun, J.: Deep residual learning for image recognition. In: IEEE Conference on Computer Vision and Pattern Recognition (CVPR), pp. 770–778 (2016). https://doi.org/10.1109/CVPR.2016.90

12. Huang, G., Liu, Z., Van Der Maaten, L., Weinberger, K.Q.: Densely connected convolutional networks. In: IEEE Conference on Computer Vision and Pattern Recognition (CVPR), pp. 4700–4708 (2017). https://doi.org/10.1109/CVPR.2017.243

13. Krizhevsky, A., Sutskever, I., Hinton, G.E.: ImageNet classification with deep convolutional neural networks. In: Advances in Neural Information Processing Systems, pp. 1097–1105 (2012)

14. Li, Q., Li, W., Zhang, J., Xu, Z.: An improved k-nearest neighbour method to diagnose breast cancer. Analyst **143**(12), 2807–2811 (2018). https://doi.org/10.1039/C8AN00189H

15. Liu, X., et al.: Automated layer segmentation of retinal optical coherence tomography images using a deep feature enhanced structured random forests classifier. IEEE J. Biomed. Health Inform. **23**(4), 1404–1416 (2018). https://doi.org/10.1109/JBHI.2018.2856276

16. Ma, J., Wu, F., Zhu, J., Xu, D., Kong, D.: A pre-trained convolutional neural network based method for thyroid nodule diagnosis. Ultrasonics **73**, 221–230 (2017). https://doi.org/10.1016/j.ultras.2016.09.011

17. Melas-Kyriazi, L.: EfficientNet PyTorch (2019). https://github.com/lukemelas/EfficientNet-PyTorch

18. Nakashima, K.: Grad-CAM with PyTorch (2017). https://github.com/kazuto1011/grad-cam-pytorch

19. Nishio, M., et al.: Computer-aided diagnosis of lung nodule classification between benign nodule, primary lung cancer, and metastatic lung cancer at different image size using deep convolutional neural network with transfer learning. PLoS ONE **13**(7), e0200721 (2018). https://doi.org/10.1371/journal.pone.0200721

20. Panwar, M., Acharyya, A., Shafik, R.A., Biswas, D.: K-nearest neighbor based methodology for accurate diagnosis of diabetes mellitus. In: Sixth International Symposium on Embedded Computing and System Design (ISED), pp. 132–136. IEEE (2016). https://doi.org/10.1109/ISED.2016.7977069

21. Paszke, A., et al.: PyTorch: an imperative style, high-performance deep learning library. In: Advances in Neural Information Processing Systems, pp. 8026–8037 (2019)

22. Pedregosa, F., et al.: Scikit-learn: machine learning in Python. J. Mach. Learn. Res. **12**, 2825–2830 (2011)

23. Rajpurkar, P., et al.: MURA: large dataset for abnormality detection in musculoskeletal radiographs. arXiv:1712.06957 (2017)

24. Selvaraju, R.R., Cogswell, M., Das, A., Vedantam, R., Parikh, D., Batra, D.: Grad-CAM: visual explanations from deep networks via gradient-based localization. In: IEEE International Conference on Computer Vision (ICCV), pp. 618–626 (2017). https://doi.org/10.1109/ICCV.2017.74

25. Simonyan, K., Zisserman, A.: Very deep convolutional networks for large-scale image recognition. arXiv:1409.1556 (2014)

26. Sun, Y., Liang, D., Wang, X., Tang, X.: DeepID3: face recognition with very deep neural networks. arXiv:1502.00873 (2015)

27. Szegedy, C., Ioffe, S., Vanhoucke, V., Alemi, A.A.: Inception-v4, Inception-ResNet and the impact of residual connections on learning. In: Proceedings of the 31st AAAI Conference on Artificial Intelligence, pp. 4278–4284 (2017)

28. Szegedy, C., Vanhoucke, V., Ioffe, S., Shlens, J., Wojna, Z.: Rethinking the Inception architecture for computer vision. In: IEEE Conference on Computer Vision and Pattern Recognition (CVPR), pp. 2818–2826 (2016). https://doi.org/10.1109/CVPR.2016.308

29. Tan, M., Le, Q.V.: EfficientNet: rethinking model scaling for convolutional neural networks. arXiv:1905.11946 (2019)

30. Tschandl, P., et al.: Expert-level diagnosis of nonpigmented skin cancer by combined convolutional neural networks. JAMA Dermatol. 155(1), 58–65 (2019). https://doi.org/10.1001/jamadermatol.2018.4378

31. Wang, H., Zheng, B., Yoon, S.W., Ko, H.S.: A support vector machine-based ensemble algorithm for breast cancer diagnosis. Eur. J. Oper. Res. 267(2), 687–699 (2018). https://doi.org/10.1016/j.ejor.2017.12.001

32. Wang, X., Peng, Y., Lu, L., Lu, Z., Bagheri, M., Summers, R.M.: ChestX-ray8: hospital-scale chest X-ray database and benchmarks on weakly-supervised classification and localization of common thorax diseases. In: IEEE Conference on Computer Vision and Pattern Recognition (CVPR), pp. 2097–2106 (2017). https://doi.org/10.1109/CVPR.2017.369

33. Woolf, A.D., Pfleger, B.: Burden of major musculoskeletal conditions. Bull. World Health Organ. 81, 646–656 (2003)

34. World Health Organisation: Musculoskeletal conditions fact sheet (2019). https://www.who.int/news-room/fact-sheets/detail/musculoskeletal-conditions

35. Xie, Y., Richmond, D.: Pre-training on grayscale ImageNet improves medical image classification. In: Leal-Taixé, L., Roth, S. (eds.) ECCV 2018. LNCS, vol. 11134, pp. 476–484. Springer, Cham (2019). https://doi.org/10.1007/978-3-030-11024-6_37

Combining Mutation and Gene Network Data in a Machine Learning Approach for False-Positive Cancer Driver Gene Discovery

Jorge Francisco Cutigi[1,2]([⊠]), Renato Feijo Evangelista[2],
Rodrigo Henrique Ramos[1,2], Cynthia de Oliveira Lage Ferreira[2],
Adriane Feijo Evangelista[3], Andre C. P. L. F. de Carvalho[2],
and Adenilso Simao[2]

[1] Federal Institute of Sao Paulo, Sao Carlos, SP, Brazil
cutigi@ifsp.edu.br
[2] University of Sao Paulo, Sao Carlos, SP, Brazil
[3] Barretos Cancer Hospital, Barretos, SP, Brazil

Abstract. An increasing interest in Cancer Genomics research emerged from the advent and widespread use of next-generation sequencing technologies, which have generated a large amount of digital biological data. However, not all of this information in fact contributes to cancer studies. For instance, false-positive-driver genes may contain characteristics of cancer genes but are not actually relevant to the cancer initiation and progression. Including this type of genes in cancer studies may lead to identifying unrealistic trends in the data and mislead biomedical decisions. Therefore, proper screening to detect this specific type of gene among genes considered drivers is of utmost importance. This work is focused on the development of models dedicated to this task. Support Vector Machine (SVM) and Random Forest (RF) machine learning algorithms were selected to induce predictive models to classify supposedly driver genes as real drivers or false-positive drivers based on both mutation data and gene network interactions. The results confirmed that the combination of the two sources of information improves the performance of the models. Moreover, SVM and RF models achieved a classification accuracy of 85.0% and 82.4% over labeled data, respectively. Finally, a literature-based analysis was performed over the classification of a new set of genes to further validate the concept.

Keywords: Cancer bioinformatics · Driver genes · False-positive driver · Complex networks · Machine learning

© Springer Nature Switzerland AG 2020
J. C. Setubal and W. M. Silva (Eds.): BSB 2020, LNBI 12558, pp. 81–92, 2020.
https://doi.org/10.1007/978-3-030-65775-8_8

1   Introduction

Cancer is one of the main cause of death globally, being responsible for around 9.6 million deaths in 2018, according to the World Health Organization[1]. It is considered a complex disease and is caused by the accumulation of genetic alterations in the human body cells, which are called genetic mutations.

The investigation of mutations is crucial for the understanding of cancer initiation and progression. A single cell undergoes a vast number of mutations; nonetheless, not all of them lead to cancer. In this context, mutations can be classified in two types: *Passenger mutations*, which comprehend the majority of mutations in cancer cells and are not significant for the cancer progression, i.e., do not confer a selective advantage to cells; and *Driver mutations*, which are a small group of mutation significant for cancer, i.e., they give cancer cells a growth advantage.

Distinguishing between driver and passenger mutations is a long-standing investigation line in Cancer Genomics. Many computational methods have been developed on this topic [5,9], which are based on various types of data that are currently available. Among them, mutation data analysis has taken a prominent position after the advent of next-sequencing generation technologies (NGS) and thanks to projects such as the TCGA (The Cancer Genome Atlas) [29] that makes large collections of mutation data available. Gene interaction information is also often explored and has an important role in many computational methods [22]. This type of data provides essential information about complex gene interactions among genes and their related proteins. Such interactions are represented by complex networks, in which genes are nodes and edges connect genes that are physically interacting or functionally related [16].

There are also computational methods that benefit from the simultaneous analysis of mutation and gene network data. HotNet [28], HotNet2 [17], Hierarchical HotNet [25], MUFFINN [3], nCOP [13], NetSig [12], and GeNWeMME [6] are methods that employ network algorithms and mutation data analysis to find significantly related mutations and to identify driver genes. Recently, machine learning algorithms, such as DriverML [11], LOTUS [4], and MoProEmbeddings [10] have taken advantage of the massive volume of digital biological data to induce predictive models able to suggest driver genes and find novel biological insights.

Although these methods have been extensively used for driver gene identification, they can misclassify some genes as drivers, thus being necessary expert curation to filter their findings [1]. It occurs because some genes present characteristics of drivers, but are not actually involved in cancer initiation and progression. These genes are referred to as false-positive drivers and may even mislead a biomedical decision or compromise the performance of models that consider them as variables. The avoiding of the misclassification of false-positive-drivers as drivers is still an ongoing challenge. Thus, the development of dedicated tools for further screening these models' findings so that possible misclassified genes can be detected is required. In this work, the terms false-positive drivers and false-drivers are used as synonymous.

---

[1] www.who.int/news-room/fact-sheets/detail/cancer.

In this context, this work presents a machine learning-based approach to induce predictive models able to classify driver gene candidates as real drivers or false-drivers. Random Forest (RF) and Support Vector Machine (SVM) were selected as the supervised learning methods. The proposed approach extracts useful information from both mutation data and gene network interactions as features for the models. The training was performed over a data set composed of 876 labeled genes, created from the combination of somatic mutation data of 33 types of cancer and centrality measures of a union of four gene interactions networks. The models were compared, considering a set of evaluation metrics. Experimental results show that the combination of mutation data and gene interaction data can improve the predictive models' prediction potential. An automated literature-based validation, taken accounted for the number of citations of classified genes, evidenced the potential of models suggesting false-drivers genes.

This paper is organized as follows: Sect. 2 describes the machine learning-based approach employed in this work, from data collection to the induction of predictive models. Next, Sect. 3 presents a thorough evaluation of the models using classification metrics and automated literature-based analysis. Finally, Sect. 4 summarizes the research findings and concludes with final considerations.

2 Method

This section describes the steps in the development of this research. Figure 1 shows a summary of the approach established for the research. In Step 1, cancer mutation data, gene interaction networks, and gene labels are selected from reliable and widely used sources. In Step 2, data is preprocessed, and features are extracted. Finally, in Step 3, a hyper-parameters tuning is performed so that optimized models can be induced and evaluated through stratified k-fold cross-validation. Further assessment of the models' applicability is performed over new genes reported as drivers by other sources.

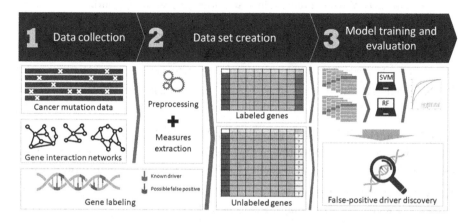

Fig. 1. An overview of the approach.

2.1 Data Collection and Preprocessing

Mutation Data. Data of 33 types of cancer were selected based on a TCGA Pan-Cancer study [1] and downloaded from the cBioPortal[2] [2] using a web API[3]. The collection contains mutation data of two types: (1) Single Nucleotide Variants (SNVs); and (2) Insertions and Deletions (InDels). The mutation data for each type of cancer is structured in a tab-delimited text file referred as MAF (Mutation Annotation Format)[4], with rows listing different somatic mutations and columns containing more than one hundred entries of related information, such as gene name (`Hugo_Symbol`), type of mutation (`Variant_Classification`), and sample/patient id (`Tumor_Sample_Barcode`).

Each MAF file was subjected to a preprocessing routine. This process is crucial in cancer analysis, given that the data contain information that should be suppressed (e.g., a specific type of mutations) for exome analysis. Moreover, hypermutated samples should be removed since they are usually considered outliers and can bias the analyses. The preprocessing was performed in two steps. First, a filter was applied and only nine specific somatic variants were kept in MAF file, namely: `Frame_Shift_Del`, `Frame_Shift_Ins`, `In_Frame_Del`, `In_Frame_Ins`, `Missense_Mutation`, `Nonsense_Mutation`, `Nonstop_Mutation`, `Splice_Site` and `Translation_Start_Site`. These variants were selected because they are non-silent mutations and from coding regions, i.e., they are likely to be mutations that lead to a functional impact. Later, hypermutated samples were removed from the MAF file. Among the possible methods to identify hypermutated samples, the one proposed by Tamborero et al. [27] was applied. According to their criterion, a sample is hypermutated when it contains more than $(Q3 + 4.5 \times IQR)$ somatic mutations, where $Q3$ is the third quartile, and IQR is the interquartile range of the distribution of mutations across all data samples. Finally, the 33 preprocessed MAFs were merged into a single MAF file. Table 1 compares the composition of the mutation data with and without the preprocessing.

Table 1. Mutation data before and after preprocessing routine.

	Non-preprocessed mutation data	Preprocessed mutation data
Patients	10429	9741
Genes	20072	19184
Mutations	2192073	1228126

Gene Interaction Network Data. The following four gene-networks that use protein-protein (PPIs) as the main source of interactions were selected to extract node measures: (1) ReactomeFI (Reactome Functional Interactions) -

---

[2] https://www.cbioportal.org/datasets.

[3] https://www.cbioportal.org/api/swagger-ui.html.

[4] https://docs.gdc.cancer.gov/Data/File_Formats/MAF_Format/.

2019 version; (2) Binary interaction from HINT (High-quality INTeractomes) - March 2020 version; (3) HPRD (Human Protein Reference Database) - Release 9; and (4) HuRI (Human Reference Interactome) - 2020 release. ReactomeFI is an extensive gene network built from curated pathways from many sources, whose data are obtained mainly from reliable PPI and known pathways [14]. HINT is a curated compilation of high-quality PPI, filtered systematically and manually for the removal of low-quality/erroneous interactions [7]. HPRD is a classical and curated human protein interactions, built from PPI, post-translational modifications, enzyme-substrate relationships, and disease associations [15]. HuRI is a protein interaction network built from pairwise combinations of human protein-coding genes involved in binary interactions [19].

The selected gene networks were treated as undirected and unweighted networks. A union operation was applied to these networks, resulting in a single network UGN. Such operation considers the interaction and nodes of all networks, i.e., $UGN = Reactome \cup HINT \cup HPRD \cup HuRI$. For example, if N_1 is a network with node set $\{g_i, g_j, g_k\}$ and interaction set $\{(g_i, g_j), (g_j, g_k)\}$, and N_2 is a network with node set $\{g_i, g_j, g_k, g_l\}$ and interaction set $\{(g_i, g_j), (g_i, g_k), (g_k, g_l)\}$, then the union $N_1 \cup N_2$ has node set $\{g_i, g_j, g_k, g_l\}$ and edge set $\{(g_i, g_j), (g_j, g_k), (g_k, g_l)\}$. Finally, after the union process, a single network was constructed to be used in this research. The resulting network UGN has 18959 genes and 372583 interactions.

Combined Data. In order to train supervised machine learning algorithms, the data set needed to be properly structured. The samples in the unlabeled data set are the genes, while the features are the measures extracted from the mutation and gene network data, as summarized below:

– ten features were extracted for each gene from the MAF file to create a mutation data set DS_{MUT}. One feature is related to the gene's coverage, i.e., the number of patients in which the gene is mutated. Nine features were extracted for each somatic variant, representing each specific somatic variant's number of mutations. Thus, the data set DS_{MUT} is composed of 19184 samples and ten features.
– ten features were extracted for each gene (node) in the network UGN to create a data set DS_{GN}. The features are the following centrality measures: degree, betweenness, closeness, eigenvector, coreness, clustering coefficient, average of neighbors' degree, leverage, information, and bridging. Such measures consider distinct aspects of the network structure and topology to characterize the importance of a node, thus highlighting its central role according to each of measures [21]. The data set DS_{GN} is composed of 18959 samples and ten features.

A combined data set was obtained by merging DS_{MUT} and DS_{GN}. Some of the genes were not contained in both data sets, therefore only their intersection was taken. The merging leads to the data set DS_{COMB}, composed of 16281 samples and 20 features.

Gene Labels. A total of 250 genes listed as possible-false-positive drivers[5] by the Network of Cancer Genes (NCG) [24] were used as a reference to label genes in the dataset as false-drivers (FD) for the induction of predictive models by the supervised machine learning algorithms. For the drivers class D, two sets D_{NCG} and D_{CGC} of known driver genes were extracted from the NCG[6] and Cancer Gene Census (CGC)[7] [26], respectively. The D_{NCG} set contains 711 known cancer drivers, while D_{CGC} 723. A union of both sets was performed, resulting in a known driver gene set $D = D_{NCG} \cup D_{CGC}$ of 729 genes. However, a total of 49 genes in this set were also present in FD, therefore, they were removed from D. The remaining 680 genes were then used as a reference to label the drivers.

Finally, the genes in the data set DS_{COMB} contained in the FD and D lists were extracted and properly labeled. Considering only the labeled samples, the resulting data set DS_{COMB_L} is composed of 876 samples, 20 numeric features, and one class label (647 drivers and 229 false-drivers). The unlabeled data set DS_{COMB_U} is composed of 15405 samples. A z-score standardization was applied to all features DS_{COMB_L}, and the class imbalance was also addressed prior to the training process, as described in Sect. 2.2.

2.2 Machine Learning Training Process

Supervised machine learning algorithms were trained with the data set DS_{COMB_L} to induce predictive models that classify genes as drivers or false-drivers. Scikit-learn [23], a Python module for machine learning, was used in all processes described in this section.

Predictive Models. Two machine learning algorithms were selected to induce the predictive models: Support Vector Machine (SVM) and Random Forest (RF). The models were induced using a stratified 5-fold cross-validation scheme with under-sampling applied to every training portion of folds to avoid over-fitting and address class-imbalance. The under-sampling was performed by randomly removing samples from the majority class, i.e., the driver class. It is important to note that both the under-sampling and the folds split procedures were repeated, taking different random states on every new training process.

Hyper-parameter Selection. Different hyper-parameters sets were assessed by training multiple models in a grid-search process, always using 5-fold cross-validation with under-sampling and repeating 10 times in order to account for the possible influence of randomness. SVM hyper-parameters C and gamma were both varied from 1 to 10, in addition to the auto and scale standard options. The linear, sigmoid and rbf kernels were considered. RF was tested with a

---

[5] Version 6.0 – http://ncg.kcl.ac.uk/false_positives.php.

[6] Version 6.0 – http://ncg.kcl.ac.uk/download.php.

[7] Version 91 – https://cancer.sanger.ac.uk/census.

total number of estimators in the range from 20 to 500. The maximum depth was tested with limits from 3 to 20 layers, including the no restriction scenario. Both `gini` and `entropy` criteria were evaluated. The possibilities for maximum numbers of features considered to perform a split varied according to the following functions: `None`, `auto`, `sqrt`, `log2`. The mean prediction accuracy was used as a metric for the comparison and was calculated over all folds and repetition for each model. The best model for each algorithm was selected for further evaluation.

Evaluation and Validation Criteria. Accuracy, Precision, Recall, and F1-score were also quantified to assess the selected models. These metrics are important because they can help to identify possible systematic trends on the miss-classifications. The receiver operating characteristic (ROC) curves were generated for the models trained, considering both mutation and gene network features. They were then compared to the ROC curves obtained from models trained using just a single source of features (i.e., either mutation or gene network data). The models with a single source of features were also trained using the same methodology and hyper-parameters selected through new grid-searches. The areas under the ROC curves (AUC) were also calculated for comparison.

Further evaluation was conducted over the models' classifications of unlabeled genes that have been recently reported as drivers [20]. The findings were compared to the literature through a systematic citation frequency analysis using a literature-mined database of cancer genes [18]. Such analysis was designed to bring additional evidence that the induced models are, in fact, capable of identifying potential false-positive-driver genes, thus validating the concept.

3   Results

3.1   Model Evaluation

The optimal hyper-parameters obtained from the grid-searches performed for the three labeled data sets, with different features, are provided in Table 2. Because some of the optimal parameters obtained for the SVM were in the limits of the tested ranges, these were then expanded, but no significant improvement in performance was observed. The data sets are referred to as follow. DS_{COMB_L}: labeled data set with features from mutation data and gene network data; DS_{MUT_L}: labeled data set with features only from mutation data; and DS_{GN_L}: labeled data set with features only from gene network data.

The models induced using these hyper-parameters were further investigated using the metrics described in Sect. 2.2. Once again, the mean of each metric was calculated over 30 repetitions of the whole training process. Table 3 shows the calculated averages and standard deviations for each selected evaluation metric.

Both models trained with combined features presented satisfactory results, without significant differences in performance based on the selected metrics. Moreover, the construction of models using combined mutation and network

Table 2. Optimal hyper-parameters for data sets containing different features.

	SVM				RF		
	DS_{COMB_L}	DS_{MUT_L}	DS_{GN_L}		DS_{COMB_L}	DS_{MUT_L}	DS_{GN_L}
Kernel	linear	linear	rbf	Estimators	300	100	150
C	1	1	1	max. depth	10	12	5
gamma	1	1	10	max. features	None	auto	auto
				Criterion	gini	gini	gini

features outperforms the other models induced with a single source of features. This trend is also observed when analyzing ROC curves, presented in Fig. 2. Additionally, DeLong test [8] was performed over the 30 repetitions to compare ROC curves of DS_{COMB_L} and DS_{MUT_L}. A p-value < 0.05 was observed in 24 and 27 tests for SVM and RF, respectively.

Table 3. Comparison between models induced by different features: DS_{COMB_L}, DS_{MUT_L}, and DS_{GN_L}.

	SVM			RF		
	DS_{COMB_L}	DS_{MUT_L}	DS_{GN_L}	DS_{COMB_L}	DS_{MUT_L}	DS_{GN_L}
Accuracy	0.850 ± 0.007	0.828 ± 0.006	0.704 ± 0.011	0.824 ± 0.007	0.785 ± 0.007	0.659 ± 0.009
Precision	0.916 ± 0.005	0.898 ± 0.004	0.784 ± 0.006	0.931 ± 0.004	0.923 ± 0.006	0.853 ± 0.009
Recall	0.877 ± 0.007	0.865 ± 0.009	0.827 ± 0.014	0.823 ± 0.009	0.774 ± 0.009	0.651 ± 0.018
F1	0.896 ± 0.005	0.881 ± 0.004	0.805 ± 0.008	0.874 ± 0.005	0.842 ± 0.006	0.738 ± 0.010

3.2   Concept Application and Validation

The results have shown that combining features extracted from mutation and gene network data improves the predictive potential of the models induced by machine learning algorithms. However, these results were obtained using the labels defined based on the NCG and the CGC and, therefore, there could still be open questions regarding the application of the models on the prediction for currently unlabeled genes. Thus, the models were applied to a new set of driver genes recently reported by the IntOGen[8] (Integrative Onco Genomics) [20].

A subset of 131 driver genes from IntOGen that was also contained in the unlabeled portion of DS_{COMB} consisted of a new data set for the validation. The previously trained SVM and RF models were applied to classify this new set of genes. Data were processed following the same steps applied to create the set DS_{COMB_L}. The SVM model detected 12 possible false-drivers (FD) genes and 119 drivers (D), while the RF identified 15 FD and 116 D.

The consistency of these findings was further investigated based on current literature reports using CarcerMine[9], a literature-mined database of drivers [18].

---

[8] www.intogen.org/.
[9] bionlp.bcgsc.ca/cancermine/ – *query performed in October, 2020.*

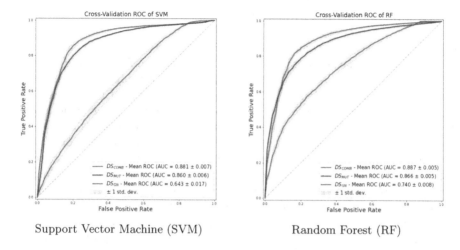

Support Vector Machine (SVM)   Random Forest (RF)

Fig. 2. ROC curve comparison

CancerMine extracts literature evidence of cancer genes, classifying them as drivers, oncogenes, and tumor suppressors genes. Among the 131 genes, 78 genes (59.5%) have been reported as cancer genes in at least one research paper.

Figure 3 shows the frequency of citations found in CancerMine for genes classified as D or FD by both models. For each model, the figure presents the frequencies taking a threshold of at least one, three, and ten citations as drivers, oncogenes, and tumor suppressors genes in research papers. Genes classified as D were reported as cancer genes more often than the ones detected by the models as FD. At higher thresholds, there are no longer citations of FD genes, while the D genes are still fairly cited. This difference between the frequency in citations for genes classified as D and FD is literature-based evidence that the models, in fact, identified genes that are likely to be false-positive-drivers. Finally, when comparing the two models' results, the SVM findings were in higher agreement with the literature.

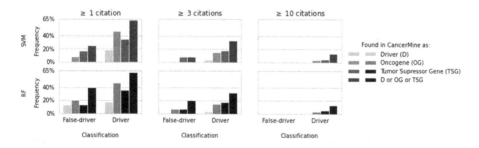

Fig. 3. Frequency of citations of the classified genes and cancer genes.

4   Discussion and Conclusion

This work proposed a machine learning approach to detect potential false-drivers cancer genes. The natural application of this discovery tool is to avoid the misclassification of false-positive-drivers as drivers and possibly eliminate unnecessary further analysis. Detecting false-drivers is also crucial to prevent their inclusion in data analyses or on the development of models, which could lead to the identification of unrealistic patterns. For this task, ten measures from mutation data and ten from gene interactions were extracted, and Support Vector Machines and Random Forest models were induced using the combined source of features. Data were properly preprocessed, and stratified k-fold cross-validation was applied to the models' training. Hyper-parameters-optimization was also conducted. In general, both models achieved satisfactory classification performance, benefited by the combination of mutation and gene interaction features. Furthermore, the discovery over a new set of genes recently reported as drivers were in agreement with the literature, based on a citation frequency analysis. However, it is important to note that this concept has been implemented considering the currently available data, which is scarce and still under continuous investigation. Therefore, it is expected that the proposed tool can be eventually revisited and improved as new information becomes available.

Supplementary Information. Source codes, scripts of experiments and the complete list of libraries and versions used in this work are available on the following link: https://github.com/jcutigi/FalseDriverDiscovery_BSB2020.

References

1. Bailey, M.H., et al.: Comprehensive characterization of cancer driver genes and mutations. Cell **173**(2), 371–385.e18 (2018). https://doi.org/10.1016/j.cell.2018.02.060
2. Cerami, E., et al.: The cBio cancer genomics portal: an open platform for exploring multidimensional cancer genomics data. Cancer Discov. **2**(5), 401–404 (2012). https://doi.org/10.1158/2159-8290.CD-12-0095
3. Cho, A., Shim, J.E., Kim, E., Supek, F., Lehner, B., Lee, I.: MUFFINN: cancer gene discovery via network analysis of somatic mutation data. Genome Biol. **17**(1), 129 (2016). https://doi.org/10.1186/s13059-016-0989-x
4. Collier, O., Stoven, V., Vert, J.P.: LOTUS: a single- and multitask machine learning algorithm for the prediction of cancer driver genes. PLoS Comput. Biol. **15**(9), 1–27 (2019). https://doi.org/10.1371/journal.pcbi.1007381
5. Cutigi, J.F., Evangelista, A.F., Simao, A.: Approaches for the identification of driver mutations in cancer: a tutorial from a computational perspective. J. Bioinform. Comput. Biol. **18**(03), 2050016 (2020). https://doi.org/10.1142/S021972002050016X. pMID: 32698724
6. Cutigi, J.F., Evangelista, A.F., Simao, A.: GeNWeMME: a network-based computational method for prioritizing groups of significant related genes in cancer. In: Kowada, L., de Oliveira, D. (eds.) BSB 2019. LNCS, vol. 11347, pp. 29–40. Springer, Cham (2020). https://doi.org/10.1007/978-3-030-46417-2_3

7. Das, J., Yu, H.: HINT: high-quality protein interactomes and their applications in understanding human disease. BMC Syst. Biol. **6**, 92 (2012). https://doi.org/10.1186/1752-0509-6-92

8. DeLong, E.R., DeLong, D.M., Clarke-Pearson, D.L.: Comparing the areas under two or more correlated receiver operating characteristic curves: a nonparametric approach. Biometrics **44**, 837–845 (1988)

9. Dimitrakopoulos, C.M., Beerenwinkel, N.: Computational approaches for the identification of cancer genes and pathways. Wiley Interdisc. Rev.: Syst. Biol. Med. **9**(1), e1364 (2017). https://doi.org/10.1002/wsbm.1364

10. Gumpinger, A.C., Lage, K., Horn, H., Borgwardt, K.: Prediction of cancer driver genes through network-based moment propagation of mutation scores. Bioinformatics **36**(Supplement_1), i508–i515 (2020). https://doi.org/10.1093/bioinformatics/btaa452

11. Han, Y., et al.: DriverML: a machine learning algorithm for identifying driver genes in cancer sequencing studies. Nucleic Acids Res. **47**(8), e45–e45 (2019)

12. Horn, H., et al.: NetSig: network-based discovery from cancer genomes. Nat. Methods **15**, 61–66 (2018). https://doi.org/10.1038/nmeth.4514

13. Hristov, B.H., Singh, M.: Network-based coverage of mutational profiles reveals cancer genes. Cell Syst. **5**(3), 221–229 (2017)

14. Jassal, B., et al.: The reactome pathway knowledgebase. Nucleic Acids Res. **48**(D1), D498–D503 (2020)

15. Keshava Prasad, T.S., et al.: Human protein reference database-2009 update. Nucleic Acids Res. **37**(Database issue), D767–D772 (2009). https://doi.org/10.1093/nar/gkn892

16. Kim, Y., Cho, D., Przytycka, T.M.: Understanding genotype-phenotype effects in cancer via network approaches. PLoS Comput. Biol. **12**(3), e1004747 (2016). https://doi.org/10.1371/journal.pcbi.1004747

17. Leiserson, M.D.M., et al.: Pan-cancer network analysis identifies combinations of rare somatic mutations across pathways and protein complexes. Nat. Genet. **47**(2), 106–114 (2015). https://doi.org/10.1038/ng.3168

18. Lever, J., Zhao, E.Y., Grewal, J., Jones, M.R., Jones, S.J.: CancerMine: a literature-mined resource for drivers, oncogenes and tumor suppressors in cancer. Nat. Methods **16**(6), 505–507 (2019)

19. Luck, K., et al.: A reference map of the human binary protein interactome. Nature **580**, 1–7 (2020)

20. Martínez-Jiménez, F., et al.: A compendium of mutational cancer driver genes. Nat. Rev. Cancer **20**, 1–18 (2020)

21. Oldham, S., Fulcher, B., Parkes, L., Arnatkeviciute, A., Suo, C., Fornito, A.: Consistency and differences between centrality measures across distinct classes of networks. PLoS ONE **14**(7), 1–23 (2019). https://doi.org/10.1371/journal.pone.0220061

22. Ozturk, K., Dow, M., Carlin, D.E., Bejar, R., Carter, H.: The emerging potential for network analysis to inform precision cancer medicine. J. Mol. Biol. **430**(18), 2875–2899 (2018)

23. Pedregosa, F., et al.: Scikit-learn: machine learning in Python. J. Mach. Learn. Res. **12**, 2825–2830 (2011)

24. Repana, D., et al.: The Network of Cancer Genes (NCG): a comprehensive catalogue of known and candidate cancer genes from cancer sequencing screens. Genome Biol. **20**(1), 1 (2019). https://doi.org/10.1186/s13059-018-1612-0

25. Reyna, M.A., Leiserson, M.D.M., Raphael, B.J.: Hierarchical HotNet: identifying hierarchies of altered subnetworks. Bioinformatics **34**(17), i972–i980 (2018). https://doi.org/10.1093/bioinformatics/bty613
26. Sondka, Z., Bamford, S., Cole, C.G., Ward, S.A., Dunham, I., Forbes, S.A.: The COSMIC cancer gene census: describing genetic dysfunction across all human cancers. Nat. Rev. Cancer **18**(11), 696–705 (2018)
27. Tamborero, D., et al.: Comprehensive identification of mutational cancer driver genes across 12 tumor types. Sci. Rep. **3**, 2650 (2013)
28. Vandin, F., Upfal, E., Raphael, B.J.: Algorithms for detecting significantly mutated pathways in cancer. J. Comput. Biol. **18**(3), 507–522 (2011). https://doi.org/10.1089/cmb.2010.0265. pMID: 21385051
29. Weinstein, J.N., et al.: The cancer genome atlas pan-cancer analysis project. Nat. Genet. **45**(10), 1113 (2013)

Unraveling the Role of Nanobodies Tetrad on Their Folding and Stability Assisted by Machine and Deep Learning Algorithms

Matheus Vitor Ferreira Ferraz[1,2] (ID), Wenny Camilla dos Santos Adan[2] (ID), and Roberto Dias Lins[1,2(✉)] (ID)

[1] Department of Fundamental Chemistry, Federal University of Pernambuco, Recife, PE 50670-560, Brazil
[2] Department of Virology, Aggeu Magalhães Institute, Oswaldo Cruz Foundation, Recife, PE 50670-420, Brazil
roberto.lins@cpqam.fiocruz.br

Abstract. Nanobodies (Nbs) achieve high solubility and stability due to four conserved residues referred to as the Nb tetrad. While several studies have highlighted the importance of the Nbs tetrad to their stability, a detailed molecular picture of their role has not been provided. In this work, we have used the Rosetta package to engineer synthetic Nbs lacking the Nb tetrad and used the Rosetta Energy Function to assess the structural features of the native and designed Nbs concerning the presence of the Nb tetrad. To develop a classification model, we have benchmarked three different machine learning (ML) and deep learning (DL) algorithms and concluded that more complex models led to better binary classification for our dataset. Our results show that these two classes of Nbs differ significantly in features related to solvation energy and native-like structural properties. Notably, the loss of stability due to the tetrad's absence is chiefly driven by the entropic contribution.

Keywords: Camelid antibodies · Rosetta Energy Function · Machine learning

1 Introduction

Ever since their discovery, single-domain binding fragment of heavy-chain camelid antibodies [1], referred to as nanobodies (Nbs), have gained considerable attention in translational research as therapeutic and diagnostic tools against human diseases and pathogens [2]. Along with its small size (15 kDa) and favorable physical-chemical properties (e.g., thermal and environmental stabilities), Nbs display binding affinities equivalent to conventional antibodies (cAbs) [1, 3]. Moreover, its heterologous expression in bacteria allows overcoming cAbs production pitfalls, such as high production cost and need of animal facility [4, 5]. Hence, Nbs are considered as a promising tool against numerous diseases. A variety of Nbs is currently being investigated under pre-clinical and clinical stages against a wide range of viral infections [6, 7].

© Springer Nature Switzerland AG 2020
J. C. Setubal and W. M. Silva (Eds.): BSB 2020, LNBI 12558, pp. 93–104, 2020.
https://doi.org/10.1007/978-3-030-65775-8_9

The general structural topology of the Nbs is depicted in Fig. 1. It is characterized by a core structure composed of a pair of β-sheets, built from 4 and 5 antiparallel β-strands linked by loops and a disulfide bridge. In contrast to cAbs, which contains six variable loops, Nbs display three highly variable loops H1, H2, and H3. These loops correspond to the Complementary Determining Region (CDR), which is responsible for antigenic binding and recognition, hence providing the target specificity of the Nbs. The overall structure of the Nbs is maintained by four conserved portions, termed as the framework. A significant difference regarding Nbs and cAbs arises from the lack of the variable light chain, and as a consequence, the light-heavy domains interface. To compensate for this loss, four highly conserved residues referred to as the Nb Tetrad are found to replace the nonpolar side chains with polar ones [8, 9]. The Nb tetrad comprises the residues Y/P37, E44, R/C45 and G47. Presumably, these substitutions increase hydrophilicity and solubility of the Nbs, being crucial for their stability [10].

Fig. 1. Cartoon representation of the overall topology of an Nb (PDB ID: 3DWT) [11]. The Nb domain consists of 9 β-sheets linked by loop regions, 3 of these constitute the CDR region and are colored in green, blue, and red. The framework region separated by the hypervariable loops are colored in silver. The Nb tetrad residues are highlighted in yellow. (Color figure online)

These four residues' presence is a hallmark characteristic of Nbs as it has been shown by several sequence alignments studies [12, 13]. The high conservation of these residues indicates an evolutionary-driven constraint, and it highlights their pivotal role in Nbs structure. To ascertain that changes in the Nb tetrad would negatively impact the Nb folding, we have previously designed a Nb by altering the tetrad residues. The obtained chimera presented low expression yields and the absence of a well-defined globular three-dimensional structure due to aggregation (unpublished data). On the contrary, attempts to "camelize" human/murine Abs by grafting the Nb tetrad to the Ab heavy chain's corresponding position has resulted in structural deformations of the framework β-sheet, leading to scarce stability and aggregation [14]. Although it has been described a phage-display library derived from llamas that has produced a set of stable and soluble Nbs devoid of the Nb tetrad [15], these Nbs are unusual, and their stability should be explained in the light of an alternative mechanism.

Given the Nb's tetrad importance in maintaining its folded structure and stability, these residues can be considered key to engineer novel Nbs. It has been shown that molecular dynamics simulations are not sufficient to capture the lower stability of aggregating Nbs, and it does not elucidate the structural and thermodynamic features role of the Nb tetrad to the structures of the Nbs [16]. In this study, we seek to identify the impact of the Nb tetrad from a molecular perspective. To gain insight into the thermodynamic contributions to the folded Nbs, we have used the Rosetta Energy Function components combined with machine learning to identify whether there are differences in the structural pattern of natural Nbs and the corresponding Nbs without the presence of the tetrad sidechains, by replacing them for methyl groups (Alanine). We benchmarked two machine learning (ML) models (Support Vector Machine [17] and Random Forest [18]) and one deep learning (DP) model (Artificial neural network by Multilayer Perceptron [19]) to evaluate the performance of these algorithms in effectively capture the differences among the classes from the multivariate nature of the data.

2   Computational Details

2.1   Dataset Preparation

A total of 30 non-redundant X-ray derived Nb structures, with resolution lower than 3 Å, were retrieved from the Protein Data Bank (PDB). To alleviate bad atomic contacts, the nearest local minimum in the energy function was achieved by geometry-minimizing their initial coordinates using the Rosetta package v. 3.10 [20] and the linear-Broyden-Fletcher-Goldfarb-Shanno minimization flavor conditioned to the Armijo-Goldstein rule. The minimization protocol was carried out in a stepwise fashion, where the sidechain angles were initially geometry-minimized, followed by full rotamer packing and minimization of the orientation of sidechain, backbone, and rigid body. To enhance sampling, $\chi 1$ and $\chi 2$ rotamers angles were used for all residues that pass an extra-chi cutoff of 1. Hydrogen placement was optimized during the protocol. For each of the minimized structures, the Nb tetrad residues were identified and replaced by alanine using the *RosettaScripts* in the four positions, and the obtained structures were geometry-minimized accordingly to the previously described protocol. Thus, the final dataset consisted of 60 instances.

To evaluate folding propensity, the Nb structures were scored using the all-atom Rosetta Energy Function 2015 (REF2015) [21] to calculate the energy of all atomic interactions within the proteins. The REF2015 possesses 20 terms and these were used as the features. The terms can be found in the GitHub (https://github.com/mvfferraz/NanobodiesTetrad), and a detailed description of each term can be found in reference [21]. The score function is a model parametrized to approximate the energy for a given protein conformation. Thus, it consists of a weighted sum of energy terms expressed as mathematical functions based on fundamental physical theories, statistical-mechanical models, and protein structures observations. The Rosetta package is a state-of-art prime-tool to the modeling and design of proteins, and its empirical energy function successfully allows for a valid assessment of the relative thermodynamic stability of folded proteins. The weights for each energy term were kept as default. The parsed command lines, PDB codes, and dataset are available in the GitHub.

2.2 Classification Methods

All the algorithms were written using Python v. 3, and the Scikit Learn Library [22] was employed in conjunction with the Pandas [23] and Numpy [24] packages. In addition, Tensorflow [25] and Keras [26] were used for the NN algorithm. The dataset was split as training (70%) and test (30%) sets. The data features vector was standardized using preprocessing tools to normally distribute the data by scaling the data to a zero mean and unit variance. Details of the code can be found in the GitHub.

Linear Discriminant Analysis (LDA). To identify whether the folding propensity of the Nbs containing the Nb tetrad and those that do not, are linearly separable, a one-component LDA was carried out [27]. LDA projects the input data to a linear subspace constituted of directions to maximize the separation between the classes and minimize the separation among a class. Bayes' statistics are applied to fit conditional class densities for each sample of the data. To select the significant variables, the ensemble learning method of extremely randomized trees (Extra Tree) classifier [28] was used. The number of estimators was kept as 100. The number of features to consider when searching for the best split was assigned as 2, and the quality of a split was measured using the entropy criterion. The LDA was solved using eigenvalue decomposition and was performed by fitting the data and then transforming it without additional parameters. The weights of the LD were used to detect which features are responsible for separating the classes explicitly. To compare if two means were statically different, two-tailed paired t-test was used (GraphPad Prism 8 [29]). Differences were considered statistically significant for a p-value such that $p < 0.05$, at the 95% confidence level.

Support Vector Machine (SVM). SVM consists of a non-probabilistic binary linear classifier, and wherein classification is performed by the construction of a set of hyperplanes in a high-dimensional space. SVM seeks to find a line of separation between the hyperplanes from each class. This line is optimally drawn for maximizing the distance between the closest points regarding each class. C-Support Vector Classification (SVC) was used with a linear Kernel with $C = 1$ hyperparameter, identified with a grid-search over pre-defined values for C (0.001, 0.01, 0.1, 1, 10, 100) and different types of Kernel (Linear and Radial basis function). The linear Kernel, K, is defined as a function of the vectors in the input space, x and y, as $K(x, y) = x^T y$, for $x, y \in \mathbb{R}^d$.

Random Forest (RF). RF is a meta estimator that builds a number of decision trees on bootstrapped training samples and uses averaging from random samples for each split in a tree. All parameters were implemented as the default, save by the criterion to measure the split's quality, set as entropy envisioning information gain.

Neural Network (NN). TensorFlow library was used in conjunction with the Keras high-level application programming interface. The classification was performed using the Multi-layer Perceptron (MLP) Classifier with 100 hidden layers. An MLP is a feed-forward artificial NN class, which learns a function $f(\cdot) : R^m \rightarrow R^n$ by training on a dataset with m input dimensions and n output dimensions, and it contains hidden layers in between the input and output layer. Each hidden layer contains a weight propagated for each posterior layer as a weighted linear summation and followed by a non-linear activation function $g(\cdot) : R \rightarrow R$. The weight optimization was conducted by stochastic gradient descent, and the step-size for updating the weights was defined as 0.01. Maximum iterations number was set as 500, or until it reaches convergence by considering the default tolerance.

Diagnostic Performance Evaluation. Four performance measures were assessed. The accuracy of the models was computed using the 10-fold cross-validation. To verify the model's performance, the confusion matrix, and the Receiver Operating Characteristic (ROC) curve were evaluated along with the models' learning curve. For a binary classification task, precision, recall, and f1-score are defined according to the assigned classification (true positive (tp), true negative (tn), false positive (fp) and false negative (fn)) as described by Eqs. 1–3. For a detailed description of each metric, see [30].

$$\text{Precision} = \frac{\text{tp}}{\text{tp} + \text{fp}} \tag{1}$$

$$\text{Recall} = \frac{\text{tp}}{\text{tp} + \text{fn}} \tag{2}$$

$$\text{f1} = \frac{2}{\text{recall}^{-1} + \text{precision}^{-1}} = \frac{\text{tp}}{\text{tp} + \frac{1}{2}(\text{fp} + \text{fn})} \tag{3}$$

3   Results and Discussion

3.1   Features Selection

The Rosetta energy terms are convenient mathematical approximations to the physics that governs protein structure and stability. The Rosetta Energy Function (REF) ranks the relative fitness of several amino acid sequences for a given protein structure, and it is capable of predicting the threshold for protein stability by discriminating native-like from non-native structures in a decoy [31]. The functional form relies upon pairwise decomposability of energy terms. The decomposition limits the number of energetic contributions to $1/2N(N - 1)$, where N is the atom's number in the system.

When using Rosetta energy function to calculate the score of a protein, *i.e.*, the relative energy for a given conformation reasoned by specific parameters of the Hamiltonian, it yields a total of 20 energetic terms [21]. A feature selection was performed to reduce the effects of noise or irrelevant variables to construct the models. A feature was considered relevant and non-redundant if it presented a feature importance score higher than 0.05 (Fig. 2). From the obtained split, a total of 7 features were filtered:

- *fa_dun*: the probability of a given rotamer is a native-like state based on Dunbrack's statistics for a given ϕ and ψ angles;
- *hbond_sc*: energy for the sidechain-sidechain hydrogen bond;
- *lk_ball_wtd*: asymmetric solvation energy;

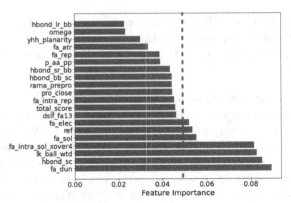

Fig. 2. Feature selection based on REF terms using Extra trees classifier. Feature importance greater than 0.05 was regarded as a relevant feature. The red dashed line represents the threshold for a given feature to be filtered. (Color figure online)

- *fa_intra_sol_xover4*: Lazaridis-Karplus solvation energy for intra-residue interactions;
- *fa_sol*: Lazaridis-Karplus solvation energy model based on Gaussian exclusion;
- *ref*: An approximation to the relative energies of the unfolded-state ensembles;
- *fa_elec*: Coulombic electrostatic potential.

The short descriptions of the terms were retrieved from [21]. As can be seen, almost half of the selected terms are related to the system's solvation properties. Since the replacement of hydrophilic residues for alanine increases the hydrophobic content of the Nbs lacking Nb tetrad, these structural differences have potentially been captured by the REF. These observations corroborate the well-described importance of the Nbs tetrad for solubility.

3.2 Linear Separability of the Data

LDA was used to evaluate whether the filtered features' combination can discriminate natural Nbs from Nbs lacking Nb tetrad. Since we have two classes, the LDA was performed in a one-dimensional fashion. LDA is a supervised dimensionality reductor that identifies the attributes that mostly account for the classes' variance. From fitting a Gaussian density to each class, a single LD was able to separate the class linearly. Figure 3A shows the one-dimensional separability for the classes. In general, natural Nbs lead to a negative value for the LD, and the contrary is observed for the Nbs that lack the Nb tetrad.

To investigate the features that account for the most separation between the classes, the LD loadings were assessed. The loadings indicate the contribution of each feature in predicting class assignment and are shown in Fig. 3B. A higher weight (relatively to their modulus) are *fa_dun*, *lk_ball_wtd*, and *ref*. These results highlight the importance of the Nbs tetrad to solubility and stability of Nbs. The *lk_ball_wtd* consists of the orientation-dependent solvation of polar atoms when assuming the ideal water geometry. As already stated, the REF was able to capture the solvation contribution to the Nb tetrad presence.

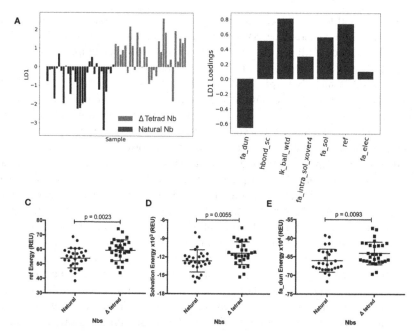

Fig. 3. Discriminant features assessed by LDA. (A) One dimensional LDA. The bars in red are the LD values for the Nbs lacking the Nb tetrad whereas the bars in blue consist of the natural Nbs; (B) Loading of each feature used to calculate de LDs; (C–E) *ref* energy, *lk_ball_wdt*, *fa_dun* energy terms, respectively, for each class. The p-value < 0.05 indicates significant differences in the means. (Color figure online)

Moreover, the other two features are related to native-like conformations properties. Thus, these results show that the Nb tetrad potentially impacts Nbs' backbone ϕ and ψ angles distribution as it is found in the Dunbrack's library of rotamers. Given the importance of the torsion angles for protein folding, a putative explanation for the Nbs tetrad's role in maintaining the structure arises from a geometrical issue. It must be noted that regarding the geometric features, this effect is unlikely to be an artifact from the modeling, since the replacement by alanine residues are not expected to cause significant structural changes, due to the small size of its sidechain, it can be positioned and matched for in any part of the protein (except when replacing tightly buried glycine residues).

To ensure the average of these features were statistically different between the classes, a two-tailed t-test was used. Figure 3(C–E) shows the distribution of the data for each class, along with averages and standard deviations. All three features presented a p-value < 0.05, indicating that these averages are statistically different.

3.3 Classification Algorithms

Given that the selected features can discriminate between the natural Nbs and Nbs that lack the Nb tetrad, classification algorithms were employed to compare the different performance in capturing the classes' structural differences. In this benchmark, machine learning (SVM and RF) and deep learning (ANN-MLP) were assessed regarding their binary classification performance. SVM is an instance-based learning model, and RF is an ensemble method. MLP is a class of NN, and here it has been employed more than three hidden-layers, and therefore, consists of a DL approach.

All the models have been prepared with the same data and training set. All the 20 features were taken into account to carry out the classifications, since using the selected features from extra trees classifier resulted in poor performances (Data not shown for conciseness). Since it is a small dataset, it is prone to suffer from overfitting the data (high variance). Thus, we performed several performance evaluations. Initially, the models were compared regarding their threshold metrics. Threshold metrics are useful for diagnosing classification prediction errors. Initially, the scores (Fig. 4A), which are directly associated with a combination of the precision and the recall values, were calculated using two approaches: 1) Evaluation was performed considering the initial training/test set; 2) A 10-fold cross-validation was employed. In the latter flavor of evaluating the estimator performance, the training set is split into k sets, and the metrics are calculated in a loop for the different generated sets. The performance is then measured by the average of each k-fold cross-validation. The SVM model presented a remarkable performance in properly assigning the classes, with an accuracy of 0.94 for the initial test set, and an accuracy average of 0.80 when considering ten different subsets. Followed by SVM, MLP also presented good metrics, even though with a slightly lower value. From the three models, the one with the poorest metrics was the RF algorithm. The two formers are more complex and robust models, so that the classification task is likely not trivial, in such a way, a simpler algorithm will not capture the main differences between the classes. The algorithms were compared using the confusion matrix (Fig. 4B–D). The diagonal elements of the matrixes consist of the number of true label classification, whereas, off-diagonal elements represent the mislabeled classifications. The SVM and MLP algorithms outperformed the RF model. The performance metrics are summarized in Table 1 and demonstrate the SVM and MLP algorithms' efficacy for our dataset.

To identify how much the models can benefit from adding more data, learning curves were plotted. Two learning curves were constructed: 1) Train learning curve: calculated based on the training set and diagnosis how well the model is learning, and 2) Validation learning curve: calculated based on a hold-out validation set and diagnosis how well the model is generalizing. Figure 5D–F shows the learning curve for the models. For SVM, the training curve modestly decreases as more samples are added, and the learning curve

Table 1. Threshold performance metrics for each binary classification model

Model	Nb	Precision	Recall	F1-score	Instances	Accuracy
SVM	Natural	0.90	1.00	0.95	9	0.94
	Δ Tetrad	1.00	0.89	0.94	9	
RF	Natural	0.83	0.56	0.67	9	0.72
	Δ Tetrad	0.67	0.89	0.76	9	
MLP	Natural	0.82	1.00	0.90	9	0.89
	Δ Tetrad	1.00	0.78	0.88	9	

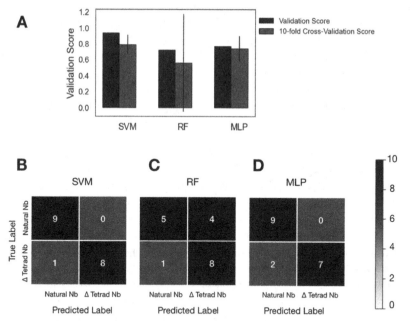

Fig. 4. Threshold metrics for the models' performance (A) Validation scores for the SMV, RF and MLP models considering the accuracy for the train/test split and for the 10-fold cross-validation; (B–D) Confusion matrix for SVM, RF, and MLP

increases until reaching a plateau at a score of nearly 0.80. As can be seen, the model fits the data well, but its generalization has a slightly lower value for the score. Thus, the SVM model might be slightly overfitted. However, its learning capability is progressively increased as more samples are added, indicating that the set number is small. For MLP, a similar trend is observed. However, for the same number of samples, SVM acquires a higher score for the learning curve, suggesting a better model's performance. These results indicate that one source of difficulty for classifying using this dataset resides in the small number of samples.

Furthermore, it shows that the algorithm's training and learning process is not straightforward, given that MLP presents a higher score for the training, proposing that the more complex fitting to the data is required. The RF model did not reflect sensitivity to increasing the number of samples, and a decrease in the learning curve is observed. Thus, the RF model does not benefit from increasing the dataset, and its overfitting cannot be attributed solely to the small size of the dataset, but rather to the simplicity of the algorithm over a complex classificatory task.

These information show that SVM and MLP have the potential to classify between the classes. Such a model is of fundamental relevance for a myriad of protein design algorithms that rely on Monte Carlo sampling. Since a large number of decoys are usually generated, identifying the Nbs that possess native-like characteristics is of enormous advantage to time and resources saving for experimental characterization. From our benchmarking, the RF model is not a proper model to learn from the data. Besides SVM

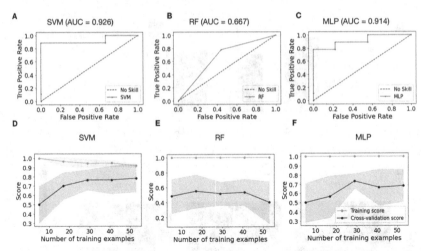

Fig. 5. Assessment of the models' performance through their characteristic curves. (A–C) ROC curves for SVM, RF, and MLP models, respectively; (D–F) Learning curves and the validation score as function of the number of training examples for SVM, RF, and MLP, respectively

having a slight advantage over MLP, the latter is a promising alternative since its training curve perfectly fits the training data, and its increasing learning curve is a promising indicator of its potential. The SVM presented a satisfactory performance, and from searching for different parameters combination, a considerable gain in the predictivity capacity might be observed. The ML and DL algorithms' performance confirms that there are traits that allow for the discrimination of Nbs containing the tetrad or not. Our results show that abolishing the tetrad associates with loss of folding stability in agreement with literature data. It is captured by the *ref* term, which in turn is shown to have significant contributions from the solvation energy and torsional dihedral motion terms. Therefore, the loss of Nbs stability due to eliminating the tetrad is a mostly entropic-driven phenomenon.

4   Conclusions

We have compared the structural features, calculated by the REF's energy term, of natural Nbs containing the Nbs tetrad and a synthetic set of Nbs lacking the tetrad. Data mining analyses revealed that the two classes of nanobodies differ mainly by folding and solvation features, corroborating with previous studies suggesting the tetrad's importance for stability and solubility. This work's findings expand the knowledge on the impact of the Nbs tetrad from a molecular-level perspective by highlighting the importance of entropic contributions to their stability.

Acknowledgements. This work has been funded by FACEPE, CAPES, CNPq, and FIOCRUZ. We acknowledge the LNCC for the availability of resources and support.

References

1. Muyldermans, S.: Nanobodies: natural single-domain antibodies. Annu. Rev. Biochem. **82**, 775–797 (2013)
2. Mir, M.A., Mehraj, U., Sheikh, B.A., Hamdani, S.S.: Nanobodies: The "Magic Bullets" in therapeutics, drug delivery and diagnostics. Hum. Antib. **28**, 29–51 (2020)
3. Vincke, C., Muyldermans, S.: Introduction to heavy chain antibodies and derived Nanobodies. Methods Mol. Biol. **911**, 15–26 (2012)
4. Morrison, C.: Nanobody approval gives domain antibodies a boost. Nat. Rev. Drug. Discov. **18**, 485–487 (2019)
5. Jovčevska, I., Muyldermans, S.: The Therapeutic potential of Nanobodies. BioDrugs **34**(1), 11–26 (2019). https://doi.org/10.1007/s40259-019-00392-z
6. Beghein, E., Gettemans, J.: Nanobody technology: A versatile toolkit for microscopic imaging, Protein–Protein interaction analysis, and protein function exploration. Front. Immunol. **8**, 771 (2017)
7. Konwarh, R.: Nanobodies: Prospects of expanding the Gamut of neutralizing antibodies against the novel coronavirus, SARS-CoV-2. Front. Immunol. **11**, 1531 (2020)
8. Revets, H., De Baetselier, P., Muyldermans, S.: Nanobodies as novel agents for cancer therapy. Expert. Opin. Biol. Ther. **5**, 111–124 (2005)
9. Muyldermans, S.: Single domain camel antibodies: Current status. J. Biotechnol. **74**, 277–302 (2001)
10. Barthelemy, P.A., et al.: Comprehensive analysis of the factors contributing to the stability and solubility of autonomous human VH domains. J. Biol. Chem. **283**, 3639–3654 (2008)
11. Vincke, C., Loris, R., Saerens, D., Martinez-Rodriguez, S., Muyldermans, S., Conrath, K.: General strategy to humanize a camelid single-domain antibody and identification of a universal humanized nanobody scaffold. J. Biol. Chem. **284**, 3273–3284 (2009)
12. Mitchell, L.S., Colwell, L.J.: Comparative analysis of nanobody sequence and structure data. Proteins **86**, 697–706 (2018)
13. Kunz, P., et al.: Exploiting sequence and stability information for directing nanobody stability engineering. Biochim. Biophys. Acta Gen. Subj. **1861**, 2196–2205 (2017)
14. Rouet, R., Dudgeon, K., Christie, M., Langley, D., Christ, D.: Fully human VH single domains that rival the stability and cleft recognition of camelid antibodies. J. Biol. Chem. **290**, 11905–11917 (2015)
15. Tanha, J., Dubuc, G., Hirama, T., Narang, S.A., MacKenzie, C.R.: Selection by phage display of llama conventional V(H) fragments with heavy chain antibody V(H)H properties. J. Immunol. Methods **263**, 97–109 (2002)
16. Soler, M.A., de Marco, A., Fortuna, S.: Molecular dynamics simulations and docking enable to explore the biophysical factors controlling the yields of engineered nanobodies. Sci. Rep. **6**, 34869 (2016)
17. Hearst, M.A., Dumais, S.T., Osuna, E., Platt, J., Scholkopf, B.: Support vector machines. IEEE Intell. Syst. Appl. **13**, 18–28 (1998)
18. Breiman, L.: Random forests. Mach. Learn. **45**, 5–32 (2001)
19. Pal, S.K., Mitra, S.: Multilayer perceptron, fuzzy sets, classifiaction. IEEE. Trans. Newural. Netw. **3**(5), 683–697 (1992)
20. Leaver-Fay, A., et al.: ROSETTA3: An object-oriented software suite for the simulation and design of macromolecules. Methods Enzymol. **487**, 545–574 (2011)
21. Alford, R.F., et al.: The Rosetta all-atom energy function for macromolecular modeling and design. J. Chem. Theory Comput. **13**, 3031–3048 (2017)
22. Pedregosa, F., et al.: Scikit-learn: Machine learning in Python. J. Mach. Learn. Res. **12**, 2825–2830 (2011)

23. McKinney, W.: Data structures for statistical computing in python. In: Proceedings of the 9th Python in Science Conference, pp. 56–61. Austin (2010)
24. Harris, C.R., et al.: Array programming with NumPy. Nature **585**, 357–362 (2020)
25. Abadi, M., et al.: Tensorflow: Large-scale machine learning on heterogeneous distributed systems. arXiv preprint arXiv:1603.04467 (2016)
26. Gulli, A., Pal, S.: Deep learning with Keras. Packt Publishing Ltd, Birmingham (2017)
27. Fisher, R.A.: The use of multiple measurements in taxonomic problems. Annals Eugen. **7**, 179–188 (1936)
28. Geurts, P., Ernst, D., Wehenkel, L.: Extremely randomized trees. Mach. Learn. **63**, 3–42 (2006)
29. Prism, G.: Graphpad software. San Diego, CA, USA (1994)
30. Powers, D.M.: Evaluation: From precision, recall and F-measure to ROC, informedness, markedness and correlation. J. Mach. Learn. Technol. **2**, 37–63 (2011)
31. Cunha, K.C., Rusu, V.H., Viana, I.F., Marques, E.T., Dhalia, R., Lins, R.D.: Assessing protein conformational sampling and structural stability via de novo design and molecular dynamics simulations. Biopolymers **103**, 351–361 (2015)

Experiencing DfAnalyzer for Runtime Analysis of Phylogenomic Dataflows

Luiz Gustavo Dias[1]([✉])[ID], Marta Mattoso[2][ID], Bruno Lopes[1][ID], and Daniel de Oliveira[1][ID]

[1] Fluminense Federal University, IC/UFF, Niterói, Brazil
lgdias@id.uff.br, {bruno,danielcmo}@ic.uff.br
[2] Federal University of Rio de Janeiro, COPPE/UFRJ, Rio de Janeiro, Brazil
marta@cos.ufrj.br

Abstract. Phylogenomic experiments provide the basis for evolutionary biology inferences. They are data- and CPU-intensive by nature and aim at producing phylogenomic trees based on an input dataset of protein sequences of genomes. These experiments can be modeled as scientific workflows. Although workflows can be efficiently managed by Workflow Management Systems (WfMS), they are not often used by bioinformaticians, which traditionally use scripts to implement their workflows. However, collecting provenance from scripts is a challenging task. In this paper, we specialize the DfAnalyzer tool for the phylogenomics domain. DfAnalyzer enables capturing, monitoring, debugging, and analysing dataflows while being generated by the script. Additionally, it can be invoked from scripts, in the same way bioinformaticians already import libraries in their code. The proposed approach captures strategic domain data, registering provenance and telemetry (performance) data to enable queries at runtime. Another advantage of specializing DfAnalyzer in the context of Phylogenomic experiments is the capability of capturing data from experiments that execute either locally or in HPC environments. We evaluated the proposed specialization of DfAnalyzer using the SciPhylomics workflow and the proposed approach showed relevant telemetry scenarios and rich data analyses.

Keywords: Scientific workflow · Provenance · Dataflow analysis

1 Introduction

Over the past years, several categories of experiments in the bioinformatics domain became more and more dependent on complex computational simulations. One example are the Phylogenomic analyses that provide the basis for evolutionary biology inferences, and they have been fostered by several technologies (*e.g.*, DNA sequencing methods [10] and novel mathematical and computational algorithms). This leads to a high submission rate of protein sequences

This work was partially supported by CNPq, CAPES, and FAPERJ.

J. C. Setubal and W. M. Silva (Eds.): BSB 2020, LNBI 12558, pp. 105–116, 2020.
https://doi.org/10.1007/978-3-030-65775-8_10

to databases such as UniProt, which now contain millions of sequences [18] that can be used in such analyses. Due to the need of processing such volume of data, the execution of this type of experiment became data- and CPU-intensive, thus requiring High-Performance Computing (HPC) environments to process data and analyze results in a timely manner.

A phylogenomic analysis experiment may be modeled as a workflow [4]. A workflow is an abstraction that allows for the user (*e.g.*, a bioinformatician) to compose a series of activities connected by data dependencies, thus creating a dataflow. These activities are typically legacy programs (*e.g.*, MAFFT, BLAST, *etc*). Bioinformaticians often fall back on Workflow Management Systems (WfMSs), such as Galaxy [2], Pegasus [5] and SciCumulus [12], to model and manage the execution of these workflows, but they have to rewrite the workflow into their languages and restrictions to their execution environments. Thus, despite collecting provenance data [6] (the derivation history of a data product, starting from its original sources - *e.g.*, a dataset containing many DNA and RNA sequences), which is a key issue to analyze and reproduce results (and allied with domain-specific data and telemetry data, it provides an important framework for data analytics), WfMSs are often not used by bioinformaticians.

This way, several bioinformaticians prefer to implement their experiments using scripts (*e.g.*, Shell, Perl, Python) [9]. More recently, bioinformaticians also started to explore efficient Data-Intensive Scalable Computing (DISC) systems and migrate their data- and CPU-intensive experiments to frameworks like Apache Spark (https://spark.apache.org/), *e.g.*, SparkBWA [1] and ADAM (https://adam.readthedocs.io/en/latest/).

Collecting provenance data (and domain-specific data) from scripts and DISC frameworks is challenging. There are several alternatives to WfMSs focused on capturing and analyzing provenance from scripts and DISC frameworks that can be applied in phylogenomic analyses [7,15]. However, they also present some limitations. The first one is that some approaches require specific languages (*e.g.*, noWorkflow [15] works only with Python) and specific versions of the framework (*e.g.*, SAMbA-RaP [7] requires a specific version of Apache Spark). Flexibility to define the level of granularity is also an issue. In general, automatic provenance data capturing generates fine-grained provenance, which commonly overwhelms bioinformaticians with a large volume of undesired data to analyze (*e.g.*, access to files and databases). On the other hand, automatic capturing coarse-grained provenance may not provide enough data for analysis (even existing WfMS provide non-flexible level of granularity). In addition, capturing domain-specific and telemetry data is also an issue in these approaches. We consider that If this integrated database (provenance, domain-specific data and telemetry data) is available, bioinformaticians can focus on analyzing just relevant data, reproduce the results and also observe a specific pattern to infer that something is not going well in the script at runtime, deciding to stop it or change parameters.

To address these issues, DfAnalyzer [17] was recently proposed to provide an agnostic way for scientists to define the granularity of the provenance data capture. The DfAnalyzer provenance database can be queried at runtime (*i.e.*, during the execution of the experiment), using a W3C PROV compliant data model. This portable and easy-to-query provenance database is also a step towards the reproducibility of bioinformatics experiments, regardless how they are implemented, since DfAnalyzer can be coupled to scripts, Spark applications and existing WfMSs. In this paper, we specialize DfAnalyzer to the context of phylogenomic experiments and present the benefits for bioinformatics data analyses and debugging. Thus, the main contribution of this paper is an extension of DfAnalyzer to collect telemetry (performance) data and its customization for the bioinformatics domain.

The remainder of the paper is organized as follows. Section 2 provides background concepts, describing the DfAnalyzer tool and discusses related work. Section 3 introduces the extensions and customizations in DfAnalyzer tool and presents the evaluations in a case study with a phylogenomic experiment. Finally, Sect. 4 concludes the paper and points future directions.

2 Background

2.1 DfAnalyzer: Runtime Dataflow Analysis of Scientific Applications Using Provenance

DfAnalyzer [17] is a W3C provenance compliant system that captures and stores provenance during the execution of experiments, regardless of how they are implemented. DfAnalyzer is based on a formal dataflow representation to register the flow of datasets and data elements. It allows for analyzing and debugging dataflows at runtime. One important characteristic of DfAnalyzer is that it captures only relevant data (as defined by the user), thus avoiding overloading users with a large volume of low level data. DfAnalyzer captures "traditional" provenance data (*e.g.*, data derivation path) but also domain-specific data through raw data extraction, *e.g.*, a DNA sequence or the e-value. These characteristics are essential since experiments may generate massive datasets, while only a small sub-set of provenance and domain data is relevant for analysis [14]. The original architecture of DfAnalyzer has five components: (i) Provenance Data Extractor (PDE); (ii) Raw Data Extractor (RDE); (iii) Dataflow Viewer (DfViewer); (iv) Query Interface (QI); and (v) Provenance Database. The first two components are invoked by plugging calls on the script, while the other three have independent interfaces for the user to submit data analyses at runtime.

After deploying the DfAnalyzer, users are required to identify relevant data in their own script, model these data and instrument the code (add DfAnalyzer calls) in order that DfAnalyzer automatically captures data and populates the database. It is worth noticing that the database tables are automatically created based on the instrumentation performed in the code. This data identification is based on the following: Dataflow: a tag to identify the dataflow that is being captured; Transformations: The data transformations that are part of the dataflow; Tasks: The smaller unit of processing, a transformation may be executed by several tasks; Datasets: Group of data elements consumed by tasks and transformations. A transformation consumes an output produced by other transformation. <u>Data elements:</u> The attributes that compose datasets. They represent either domain-specific data or input parameters of the experiment. Such information is inserted at the beginning of the script or program. Before inserting the tags on the script, it is needed to map these concepts to the script's dataflow. Identifiying the dataflow on the script is essential to represent the data transformations, dependencies, and data elements that need to be stored in the provenance database.

Listing 1 shows an example of code instrumentation in DfAnalyzer. To instrument the code, we have a 3-step process: (i) import DfAnalyzer packages in the script; (ii) define the prospective provenance (*i.e.*, the definitions of the dataflow and transformations—lines 2 to 17); and (iii) define the retrospective provenance (*i.e.*, activities and data elements to capture—lines 20 to 27). When the scripts start running each call sends to DfAnalyzer the prospective and retrospective data to be stored in the database. Figure 1 presents a fragment of the provenance database schema. Each dataset identified in the script has an associated table in the database. It is worth noticing that this instrumentation is performed only once and the script may be executed several times after that. There are datasets, data elements, and data transformations that are typically used in phylogenomic workflows. To avoid this repetitive step for bioinformaticians and allow for a consistent data representation, this work provides specialized services for DfAnalyzer users in this domain.

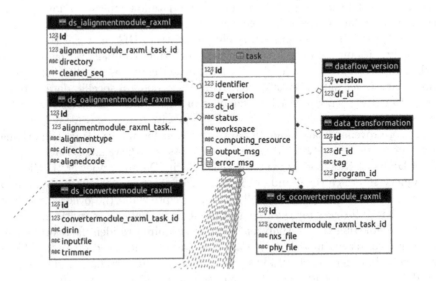

Fig. 1. A fragment of the provenance database schema

Listing 1 DfAnalyzer Prospective and Retrospective example code.

```
1   #define prospective model
2   dataflow_tag = "experiment_name"
3   df = Dataflow(dataflow_tag)
4   tf3 = Transformation("Alignment_"+dataflow_tag)
5   tf3_input = Set("iAlignment_"+dataflow_tag, SetType.INPUT,
6       [Attribute("path", AttributeType.TEXT),
7       Attribute("alignment", AttributeType.TEXT),
8       Attribute("sequence", AttributeType.TEXT)])
9   tf3_output = Set("oAlignmentModule_"+dataflow_tag, SetType.OUTPUT,
10          [Attribute("alignment", AttributeType.TEXT),
11          Attribute("path", AttributeType.TEXT),
12          Attribute("sequence", AttributeType.TEXT)])
13  tf3.set_sets([tf3_input, tf3_output])
14  tf2_output.set_type(SetType.INPUT)
15  tf2_output.dependency=tf2._tag
16  tf2.set_sets([tf2_output, tf3_input, tf3_output])
17  df.add_transformation(tf3)
18
19  #capturing retrospective data
20  t3 = Task(3, dataflow_tag, "Alignment_"+dataflow_tag, "t3")
21  t3_input = DataSet("iAlignment_"+dataflow_tag, [Element([dirin, alignment, corrected_file])])
22  t3.add_dataset(t3_input)
23  t3.begin()
24          oalignment = execute_alignment(dirin, corrected_file, alignment, trimmer)
25  t3_output = DataSet("oAlignment_"+dataflow_tag, [Element([alignment, dirin, oalignment])])
26  t3.add_dataset(t3_output)
27  t3.end()
```

2.2   Related Work

Hondo *et al.* [8] use provenance to the analysis of the transcriptome of the fungus Schizosaccharomyces pombe in four different conditions. Like in DfAnalyzer, they also adopt the PROV-DM model [11], but use different noSQL database systems to represent provenance data. Despite the flexibility of PROV-DM, the data granularity level chosen in [8] is coarse-grain, just capturing the transformation name, the program used to execute it, and the transformation execution time. Afgan *et al.* [2] focus on providing support in RNA sequencing experiments in Galaxy WfMS for non-specialists. They encapsulate the complexity of the environment configuration and data analysis. Although the tool supports the design and execution of workflows, provenance capture has a fixed granularity level and domain-specific data is not captured.

Carvalho *et al.* [3] propose an approach that converts code from interactive notebooks into workflows to capture provenance and identify the dataflow. The approach identifies transformations and tasks automatically, *e.g.*, functions are transformations. Although this reduces the instrumentation effort, it is dependent on the organization of the code, *i.e.*, if the programmer does not use functions, the identification of the transformations may be compromised. In addition, it does not capture domain-specific data neither supports parallel scripts (which is very common in the bioinformatics domain). Pimentel *et al.* [15] propose the noWorkflow tool to automatically collect provenance from the execution of Python scripts. noWorkflow is easy to deploy, but it is specific for Python scripts and does not support parallel executions. Pina *et al.* [16] also specialized DfAnalyzer, but they have focused on fine-tuning parameters in scripts of

Convolutional Neural Networks. It does not help capturing provenance in the genomics domain neither telemetry data.

3   Evaluating DfAnalyzer for Phylogenomics Experiments

This section presents the extensions implemented in DfAnalyzer to be specialized for the phylogenomics domain and the evaluation with the SciPhylomics workflow [13].

3.1   Specializing DfAnalyzer

DfAnalyzer already provides ways to capture provenance and domain-specific data, but based on previous experiments, performance data are also very important for data analytics. Bioinformaticians often have to analyze the domain-specific data together with the performance data to evaluate the trade-off between producing results with high quality and the time needed to produce such results.

Therefore, in this subsection we explain the specialization of DfAnalyzer to capture performance data. We have added a new component to the DfAnalyzer architecture named *Telemetry Data Extractor*. This component is build on top of the *psutil* library (version 5.7.2). Psutil is cross-platform and leads to retrieve performance data of running processes. It also leads to capture system resources usage (*e.g.*, CPU, memory, *etc.*). Before the execution of each task or transformation, the TDE component is invoked to monitor the resource usage. The monitoring process ends when the task or transformation finishes execution. All performance data are stored in a specific table that is associated to the *Task* table presented in Fig. 1.

3.2   SciPhylomics Workflow

SciPhylomics [13] is a phylogenomic analysis workflow that aims to construct phylogenetic trees comparing hundreds of different genomes. SciPhylomics is composed by nine transformations. The first four transformations are associated to phylogenetic analysis (or gene phylogeny): (i) multiple sequence alignment (MSA), (ii) MSA conversion, (iii) search for the best evolutionary model, and (iv) construction of phylogenetic trees. After the execution of the gene phylogeny activities, the data quality analysis activity is executed. This data quality analysis allows for filtering results that do not meet a quality criteria. The last four activities represent the phylogenomic analysis (or genome phylogeny): (vi) concatenation of MSA to obtain a "superalignment", (vii) election of an evolutionary model for the "phylogenomic tree" construction based on the "superalignment", (viii) construction of phylogenomic trees and (ix) the phylogenomic tree election.

SciPhylomics is data- and CPU-intensive, so it requires HPC environments to produce results in a feasible time. In this evaluation, we implemented SciPhylomics in a Python script that is build on top of Parallel Python library. Parallel

Python is a python module which provides mechanisms for parallel execution of python scripts on multiple processors/cores and clusters.

As aforementioned, the definition of the dataflow is essential for identifying capturing relevant provenance, domain-specific and telemetry data. This way, in this evaluation we considered a fragment of Sciphylomics composed of five transformations associated to gene phylogeny, as presented in Fig. 2: (i) Sequence Cleaning; (ii) Sequence Alignment; (iii) Sequence Conversion (iv) Model Generation; (v) Tree Generation. It is worth noticing that the transformations of the dataflow were defined specifically for the context of this paper, being possible to split these transformations into smaller ones (or merging them into a bigger one).

After defining the transformations in the script, we need to list the programs that execute them. In this fragment of SciPhylomics, each transformation may be implemented by several programs (*i.e.*, a variability). The gray transformations presented in Fig. 2 are the variant transformations, *e.g.*, data transformation (2) can be executed by the applications MAFFT, ClustalW, and Muscle. Similarly, in transformation (5) the programs RAxML and MrBayes can be used.

This characteristic is interesting for the analysis phase since we can compare the results and performance of different programs that implement the same transformation. After identifying SciPhylomics structure and the variabilities, we instrumented the SciPhylomics python script according to Listing 1 to specify the data transformations, tasks, data dependencies, datasets, and data elements.

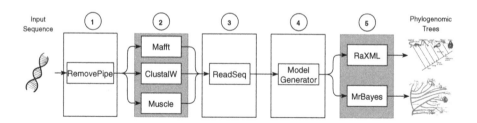

Fig. 2. A fragment of the SciPhylomics dataflow

3.3   Setup of the Experiment

The fragment of SciPhylomics presented in Fig. 2 was executed varying the programs in transformations 2 and 5. We defined two different executions of SciPhylomics (called Experiments A and B) depending on the chosen tree generation program. In addition, for each experiment, we can vary the chosen MSA program. Thus, six variants were performed with a dataset composed of 98 different multifasta files (each variant was executed 10 times).

The machine configuration used to execute SciPhylomics was an AMD FX(tm)-8150 8-Core Processor, 32 GB RAM, and 2 TB hard disk. The overview of the parameter values used in the executions is in Table 1. The parameters values were defined following an specialist. In special, *Ngen* is related to the number of cycles that the Markov Chain Monte Carlo (MCMC) algorithm is executed, the main algorithm goal is to make small random changes in some parameters to accept or reject according to the probability. Also, *Nchains* is the parameter related to the number of different parallel MCMCMC chains within a single execution run, *printfreq* is related to the frequency that the information is shown on the screen, *nruns* how many independent analyses are started simultaneously. Concerning the MSA programs, it was not necessary to set parameters and the programs were executed in default mode. More information are find on Muscle (www.ebi.ac.uk/Tools/msa/muscle/), Mafft (mafft.cbrc.jp/alignment/software/), Clustalw (www.genome.jp/tools-bin/clustalw), RAxML (http://cme.h-its.org/exelixis/web/software/raxml/) and MrBayes (http://mrbayes.sourceforge.net/commref_mb3.2.pdf) documentation.

Table 1. Parameters of the variants of SciPhylomics fragment

Experiment A			Experiment B		
Wf. step	Program	Parameter	Wf. step	Program	Parameter
Clean	RemovePipe	Total	Clean	RemovePipe	Total
Alignment	Mafft	Default	Alignment	Mafft	Default
	ClustalW	Default		ClustalW	Default
	Muscle	Default		Muscle	Default
Converter	ReadSeq	Default	Converter	ReadSeq	Default
Evolutive Model generator	Model generator	Default	Evolutive model generator	Model generator	Default
Tree generator	RaXML	Default	Tree generator	MrBayes	ngen	100000
					nchains	4
					printfreq	1000
					burnin	0
					nruns	2
					rates_mrbayes	4

3.4   Data Analysis in Practice

In this Section, we aim at presenting the advantages of analyzing provenance from phylogenomic analyses in practice using a series of analytical queries. The experimental process was executed in 75.08h. For the analysis presented in this section, we used six DfAnalyzer tables: *dataflow* (contains information about the dataflow), *data_set* (contains details of the datasets), *data_transformation* (contains details of the transformations), *task* (contains details of the tasks), *dataflow_execution* (contains statistics of the execution of the dataflow), *data_transformation_execution* (contains statistics of the execution of each transformation). For more details about the DfAnalyzer database schema please refer to [17].

The first analysis is related the evolutionary model generated by the fourth transformation, implemented by ModelGenerator program. Let us assume that the user needs to analyze what are the models chosen for the input data. If the bioinformatician executed SciPhylomics without DfAnalyzer, one has to open each file to check this information, which is time-consuming, tedious, and error-prone. By using DfAnalyzer, one has only to submit the query shown on Listing 2. In this query, the user wants to discover the number of times a specific evolutionary model was used, but just when the length of the input sequence is larger than 20.

Listing 2 Number of occurrences of a specific evolutionary model
---
```
1  SELECT count(*) as occurence, model
2              FROM ds_omodelgeneratormodule_mrb
3              GROUP BY model
4              ORDER BY occurence DESC;
```

The result of the query presented in Listing 2 is shown in Fig. 3. One can state that WAG and RtREV models are the most common ones. WAG presented higher likelihoods than any of the other models. This type of query can be adapted to other attributes of the database, such as quality of the generated tree, e-value, etc. Another performed analysis was related to the execution time of each transformation and resource usage. Capturing this type of data can impact the performance of the experiment since it is usually captured in a short time interval and may introduce overhead. Thus, in this analysis we defined an interval for capturing performance data (30 s window). The box-plots of the execution time behaviour for the six variations of SciPhylomics are presented in Fig. 4.

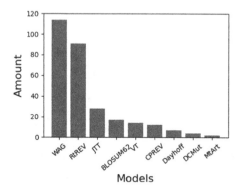

Fig. 3. Number of occurrences of each evolutionary model.

As presented in Fig. 4, there is a non-negligible variation in execution time according to the tree generator program used in the experiments. The experiments executed with RaXML (Fig. 4b) finished faster than those performed with MrBayes (Fig. 4a). The execution time difference can be explained based on the technique used for each program. MrBayes performs a Bayesian inference and

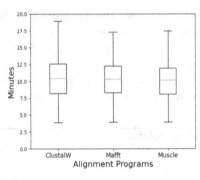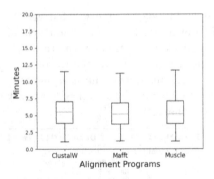

(a) SciPhylomics with MrBayes   (b) SciPhylomics with RaXML

Fig. 4. Execution time of SciPhylomics varying the MSA programs (ClustalW, Mafft and Muscle) and tree generator programs (MrBayes and RaXML).

model choice across a wide range of phylogenetic and evolutionary models and is more costly than RAxML. The programs execute different operations and generate different datasets that are composed of different attributes as well. While RAxML does not execute the operation of "search for evolutionary model" (since the ModelGenerator program is executed), MrBayes execute this operation, and in this experiment variability the operation "search for evolutionary model" is executed twice. These characteristics explain the difference in time execution in both cases, and the resource usage as well, as shown on Fig. 5.

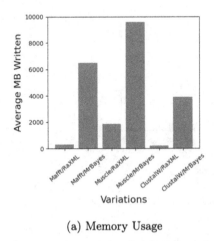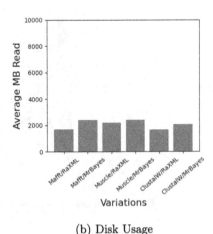

(a) Memory Usage   (b) Disk Usage

Fig. 5. SciPhylomics resource usage: memory and disk

4   Conclusion

This paper presents an approach to capture provenance data from phylogenomic dataflows without the need to rewrite the workflow to a specific programing language or Workflow Management System engine. We specialized the novel DfAnalyzer tool in the context of Phylogenomic experiments so that in an existing workflow it is possible to achieve a flexible granularity-level data capture and create an integrated database composed of domain-specific, provenance and telemetry data. DfAnalyzer is prospective provenance based. By modelling prospective provenance data, retrospective provenance is automatically captured while the workflow executes, and the data created by its transformations are stored in a relational database for further querying. Differently from other approaches, the provenance is captured with flexible granularity, and the bioinformaticians can specify what is important for their analysis and reduce the experiment cost in different spheres.

Another advantage of applying DfAnalyzer in the context of Phylogenomic experiments is the capability of capturing data from experiments that execute either locally or HPC environments, due to the fact that DfAnalyzer is asynchronous and request-based, it can execute in different environments. This asynchronous characteristic contributes that the instrumentation does not cause delays in the workflow execution. In addition, we extended DfAnalyzer to capture telemetry (performance data). This way, users are allowed to perform analyses based on both the provenance data and performance metrics. We evaluated the proposed specialization of DfAnalyzer using the previously defined SciPhylomics workflow and the proposed approach showed relevant telemetry scenarios and rich data analyses. In future work, we intend to evaluate reproducibility in experiments based on the analysis of the provenance database.

References

1. Abuín, J.M., Pichel, J.C., Pena, T.F., Amigo, J.: SparkBWA: speeding up the alignment of high-throughput DNA sequencing data. PLoS ONE **11**(5), e0155461 (2016)
2. Afgan, E., et al.: The galaxy platform for accessible, reproducible and collaborative biomedical analyses: 2016 update. Nucleic Acids Res. **44**(W1), W3–W10 (2016)
3. Carvalho, L.A.M.C., Wang, R., Gil, Y., Garijo, D.: NIW: converting notebooks into workflows to capture dataflow and provenance. In: Tiddi, I., Rizzo, G., Corcho, Ó. (eds.) Proceedings of Workshops and Tutorials of the 9th International Conference on Knowledge Capture (K-CAP 2017), Austin, Texas, USA, 4 December 2017. CEUR Workshop Proceedings, vol. 2065, pp. 12–16. CEUR-WS.org (2017)
4. de Oliveira, D.C.M., Liu, J., Pacitti, E.: Data-intensive workflow management: for clouds and data-intensive and scalable computing environments (2019)
5. Deelman, E., et al.: Pegasus, a workflow management system for science automation. FGCS **46**, 17–35 (2015)
6. Freire, J., Koop, D., Santos, E., Silva, C.T.: Provenance for computational tasks: a survey. CS&E **10**(3), 11–21 (2008)

7. Guedes, T., et al.: Capturing and analyzing provenance from spark-based scientific workflows with SAMbA-RaP. Future Gener. Comput. Syst. **112**, 658–669 (2020)
8. Hondo, F., et al.: Data provenance management for bioinformatics workflows using NoSQL database systems in a cloud computing environment. In: 2017 IEEE International Conference on Bioinformatics and Biomedicine (BIBM), pp. 1929–1934. IEEE (2017)
9. Marozzo, F., Talia, D., Trunfio, P.: Scalable script-based data analysis workflows on clouds. In: WORKS, pp. 124–133 (2013)
10. Masulli, F.: Comput. Methods Programs Biomed. **91**(2), 182 (2008)
11. Moreau, L., et al.: PROV-DM: the PROV data model. W3C Recommendation **30**, 1–38 (2013)
12. Oliveira, D., Ocaña, K.A.C.S., Baião, F.A., Mattoso, M.: A provenance-based adaptive scheduling heuristic for parallel scientific workflows in clouds. JGC **10**(3), 521–552 (2012)
13. de Oliveira, D., et al.: Performance evaluation of parallel strategies in public clouds: a study with phylogenomic workflows. Future Gener. Comput. Syst. **29**(7), 1816–1825 (2013)
14. Olma, M., Karpathiotakis, M., Alagiannis, I., Athanassoulis, M., Ailamaki, A.: Slalom: coasting through raw data via adaptive partitioning and indexing. Proc. VLDB Endow. **10**(10), 1106–1117 (2017)
15. Pimentel, J.F., Murta, L., Braganholo, V., Freire, J.: noWorkflow: a tool for collecting, analyzing, and managing provenance from python scripts. Proc. VLDB Endow. **10**(12), 1841–1844 (2017)
16. Pina, D.B., Neves, L., Paes, A., de Oliveira, D., Mattoso, M.: Análise de hiperparâmetros em aplicações de aprendizado profundo por meio de dados de proveniência. In: Anais Principais do XXXIV Simpósio Brasileiro de Banco de Dados, pp. 223–228. SBC (2019)
17. Silva, V., de Oliveira, D., Valduriez, P., Mattoso, M.: Dfanalyzer: runtime dataflow analysis of scientific applications using provenance. Proc. VLDB Endow. **11**(12), 2082–2085 (2018)
18. The UniProt Consortium: UniProt: the universal protein knowledgebase. Nucleic Acids Res. **45**(D1), D158–D169 (2016)

Sorting by Reversals and Transpositions with Proportion Restriction

Klairton Lima Brito[1]([envelope]) [ID], Alexsandro Oliveira Alexandrino[1] [ID],
Andre Rodrigues Oliveira[1] [ID], Ulisses Dias[2] [ID], and Zanoni Dias[1] [ID]

[1] Institute of Computing, University of Campinas, Campinas, Brazil
{klairton,alexsandro,andrero,zanoni}@ic.unicamp.br
[2] School of Technology ,University of Campinas, Limeira, Brazil
ulisses@ft.unicamp.br

Abstract. In the field of comparative genomics, one way of comparing two genomes is through the analysis of how they distinguish themselves based on a set of mutations called rearrangement events. When considering that genomes undergo different types of rearrangements, it can be assumed that some events are more common than others. To model this assumption one can assign different weights to different events, where more common events tend to cost less than others. However, this approach, called weighted, does not guarantee that the rearrangement assumed to be the most frequent will be also the most frequently returned by proposed algorithms. To overcome this issue, we investigate a new problem where we seek the shortest sequence of rearrangement events able to transform one genome into the other, with a restriction regarding the proportion between the events returned. Here we consider two rearrangement events: reversal, that inverts the order and the orientation of the genes inside a segment of the genome, and transposition, that moves a segment of the genome to another position. We present an approximation algorithm applicable to any desired proportion, for both scenarios where the orientation of the genes is known or unknown. We also show an improved (asymptotic) approximation algorithm for the case where the gene orientation is known.

Keywords: Rearrangement events · Proportion restriction · Approximation algorithm

1 Introduction

When comparing two genomes, one of the main goals is to determine the sequence of mutations that occurred during the evolutionary process capable of transforming a genome into another. In comparative genomics, we estimate this sequence through genome rearrangements, evolutionary events (mutations) affecting a large sequence of the genome.

Two genomes G_1 and G_2 can be computationally represented as the sequence of labels assigned to their shared genes (or shared blocks of genes). Labels are

© Springer Nature Switzerland AG 2020
J. C. Setubal and W. M. Silva (Eds.): BSB 2020, LNBI 12558, pp. 117–128, 2020.
https://doi.org/10.1007/978-3-030-65775-8_11

usually integer numbers. In addition, we associate a positive or negative sign in each of the numbers, reflecting the orientation of that gene (or block) inside of the genomes. Assuming that the genomes do not contain duplicated genes, this representation results in a signed permutation, when the orientation of the genes is known, and in an unsigned permutation otherwise. One of the genomes can be seen as the identity permutation, in which the elements are in ascending order, so problems dealing with genome rearrangements are usually treated as sorting problems, in which the goal is to transform a given permutation into the identity.

Two of the most studied genome rearrangements in the literature are the reversal, that inverts the order and the orientation of the genes inside a segment of the genome, and transposition, that moves a segment of the genome to another position. The SORTING BY REVERSALS problem has an exact polynomial algorithm for signed permutations [5] but it is NP-hard for unsigned permutations [4].

The SORTING BY TRANSPOSITIONS problem is NP-hard [3]. When we allow the use of reversals and transpositions, and assuming that both events occur with the same frequency (unweighted approach), we have the SORTING BY REVERSALS AND TRANSPOSITIONS (SBRT) problem that is NP-hard on signed and unsigned permutations [7].

In the weighted approach each type of event has an associated cost, and the goal is to find a sequence of rearrangement events that transforms one genome into another minimizing the sum of the costs. Oliveira *et al.* [7] showed that SORTING BY WEIGHTED REVERSALS AND TRANSPOSITIONS (SBWRT) problem is NP-hard on signed and unsigned permutations when the ratio between the cost of a transposition and the cost of a reversal is less than or equal to 1.5. Oliveira *et al.* [8] developed a 1.5-approximation algorithm for SBWRT on signed permutations considering costs 2 and 3 for reversals and transpositions, respectively.

The problem with weighted approaches is that they do not guarantee that lower cost rearrangements, i.e., assumed to be most frequent, will be the most frequently used by the algorithms. To overcome this issue we propose and investigate the Sorting by Reversals and Transpositions with Proportion Restriction problem on signed and unsigned permutations. In this problem, we seek a sorting sequence with an additional constraint in which the ratio between the number of reversals and the size of the sequence must be greater than or equal to a given parameter $k \in [0..1]$. We provide an algorithm that guarantees an approximation for any value of k on signed and unsigned permutations. We also show an asymptotic algorithm for the signed case with an improved approximation factor.

This manuscript is organized as follows. Section 2 provides definitions used throughout the paper. Section 3 presents an approximation algorithm for the signed and unsigned cases. Section 4 presents an asymptotic approximation algorithm for the signed case with an improved approximation factor. Section 5 concludes the paper.

2   Basic Definitions

This section formally presents the definitions used in the genome rearrangement problems. Given two genomes $\mathcal{G}_1$ and $\mathcal{G}_2$, each synteny block (common block of genes between the two genomes) is represented by an integer that also has a positive or negative sign to indicate its orientation, if known. Therefore, each genome is a permutation of integers. We assume that one of them is represented by the identity permutation $\iota_n = (+1\ +2\ \ldots\ +n)$ and the other is represented by a signed (or unsigned) permutation $\pi = (\pi_1\ \pi_2\ \ldots\ \pi_n)$.

We define a rearrangement model $\mathcal{M}$ as the set of rearrangement events allowed to compute the distance. Given a rearrangement model $\mathcal{M}$ and a permutation π, the rearrangement distance $d(\pi)$ is the minimum number of rearrangements of $\mathcal{M}$ that sorts π (i.e., that transforms π into ι). The goal of the Sorting by Genome Rearrangements problems consists in finding such distance and the sequence that reflects it.

In this work, we will assume that $\mathcal{M}$ contains both reversals and transpositions. Let us formally define these events.

Definition 1. *Given a signed permutation $\pi = (\pi_1\ \ldots\ \pi_n)$, a reversal $\rho(i,j)$, with $1 \leq i \leq j \leq n$, transforms π in the permutation $\pi \cdot \rho(i,j) = (\pi_1\ \ldots\ \pi_{i-1}$ $\underline{-\pi_j\ \ldots\ -\pi_i}\ \pi_{j+1}\ \ldots\ \pi_n)$.*

Definition 2. *Given an unsigned permutation $\pi = (\pi_1\ \ldots\ \pi_n)$, a reversal $\rho(i,j)$, with $1 \leq i < j \leq n$, transforms π in the permutation $\pi \cdot \rho(i,j) = (\pi_1\ \ldots\ \pi_{i-1}\ \underline{\pi_j\ \ldots\ \pi_i}\ \pi_{j+1}\ \ldots\ \pi_n)$.*

Definition 3. *Given a permutation $\pi = (\pi_1\ \ldots\ \pi_n)$, a transposition $\tau(i,j,k)$, with $1 \leq i < j < k \leq n+1$, applied to π transforms it in the permutation $\pi \cdot \tau(i,j,k) = (\pi_1\ \ldots\ \pi_{i-1}\ \underline{\pi_j\ \ldots\ \pi_{k-1}\ \pi_i\ \ldots\ \pi_{j-1}}\ \pi_k\ \ldots\ \pi_n)$. The effect of a transposition is the same on signed and unsigned permutations.*

The following definition helps us to formally define the problem of sorting by reversals and transpositions with a constraint on the number of reversals used in the sorting sequence.

Definition 4. *Given a sequence of reversals and transpositions S, let $|S|$ denote the number of events in S and let $|S_\rho|$ denote the number of reversals in S.*

SORTING BY REVERSALS AND TRANSPOSITIONS WITH PROPORTION RESTRICTION (SBRTWPR)

    Input: A permutation π, that can be signed or unsigned, and a rational number $k \in [0..1]$.

    Task: Find the shortest sequence S of reversals and transpositions that turns π into ι, such that $\frac{|S_\rho|}{|S|} \geq k$.

Note that when $k = 1$ the SBRTwPR problem becomes the Sorting by Reversals problem on signed [5] and unsigned [4] permutations. Moreover, when $k = 0$ we have the Sorting by Reversals and Transpositions problem on signed [10] and unsigned [9] permutations.

Example 1 shows an optimal solution S for $\pi = (-1 +4 -8 +3 +5 +2 -7 -6)$ considering the SBRT and the SBWRT problems (SBWRT using costs 2 for reversals and 3 for transpositions). Note that half of the operations in S are reversals and half are transpositions, even using a higher cost for transpositions.

Example 1.

$$
\begin{array}{rrl}
\pi = & & (-1 +4 -8 +3 +5 +2 -7 -6) \\
\pi^1 = & \pi \cdot \rho(1,5) = & (\underline{-5 -3 +8 -4 +1} +2 -7 -6) \\
\pi^2 = & \pi^1 \cdot \tau(2,4,9) = & (-5 \underline{-4 +1 +2} \, \underline{-7 -6 -3 +8}) \\
\pi^3 = & \pi^2 \cdot \tau(1,3,7) = & (\underline{+1 +2} \, \underline{-7 -6 -5 -4} -3 +8) \\
\pi^4 = & \pi^3 \cdot \rho(3,7) = & (+1 +2 \underline{+3 +4 +5 +6 +7} +8) \\
S = & & \{\rho(1,5), \tau(2,4,9), \tau(1,3,7), \rho(3,7)\}
\end{array}
$$

Example 2 shows an optimal solution S' for the same signed permutation π considering the SBRTwPR problem, adopting $k = 0.6$ (i.e., at least 60% of the operations in S must be reversals). Compared with Example 1, the sequence S' has only one more operation than S, while ensuring the minimum proportion of reversals and using both reversals and transpositions.

Example 2.

$$
\begin{array}{rrl}
\pi = & & (-1 +4 -8 +3 +5 +2 -7 -6) \\
\pi^1 = & \pi \cdot \rho(2,8) = & (-1 \underline{+6 +7 -2 -5 -3 +8} -4) \\
\pi^2 = & \pi^1 \cdot \rho(2,4) = & (-1 \underline{+2 -7 -6} -5 -3 +8 -4) \\
\pi^3 = & \pi^2 \cdot \tau(6,8,9) = & (-1 +2 -7 -6 -5 \, \underline{-4} \, \underline{-3 +8}) \\
\pi^4 = & \pi^3 \cdot \rho(1,1) = & (\underline{+1} +2 -7 -6 -5 -4 -3 +8) \\
\pi^5 = & \pi^4 \cdot \rho(3,7) = & (+1 +2 \underline{+3 +4 +5 +6 +7} +8) \\
S' = & & \{\rho(2,8), \rho(2,4), \tau(6,8,9), \rho(1,1), \rho(3,7)\}
\end{array}
$$

In the following, we present breakpoints and the cycle graph, both widely used to obtain bounds for the distance and to develop algorithms.

2.1 Breakpoints

Given a permutation $\pi = (\pi_1 \ \dots \ \pi_n)$, we extend π by adding the elements $\pi_0 = 0$ and $\pi_{n+1} = n + 1$, with these elements having positive signs when considering signed permutations. We observe that these elements are not affected by rearrangement events. From now on, we work on extended permutations.

Definition 5. *For an unsigned permutation π, a pair of elements π_i and π_{i+1}, with $0 \le i \le n$, is a breakpoint if $|\pi_{i+1} - \pi_i| \ne 1$.*

The number of breakpoints in a permutation π is denoted by $b(\pi)$. Given an operation γ, let $\Delta b(\pi, \gamma) = b(\pi) - b(\pi \cdot \gamma)$, that is, $\Delta b(\pi, \gamma)$ denotes the change in the number of breakpoints after applying γ to π.

Remark 1. The identity permutation ι is the only permutation with $b(\pi) = 0$.

2.2   Cycle Graph

For a signed permutation π, we define the cycle graph $G(\pi) = (V, E)$, such that $V = \{+\pi_0, -\pi_1, +\pi_1, -\pi_2, +\pi_2, \ldots, -\pi_n, +\pi_n, -\pi_{n+1}\}$ and $E = E_b \cup E_g$, where $E_b = \{(-\pi_i, +\pi_{i-1}) \mid 1 \le i \le n+1\}$ and $E_g = \{(+(i-1), -i) \mid 1 \le i \le n+1\}$. We say that E_b is the set of black edges and E_g is the set of gray edges.

Note that each vertex is incident to two edges (a gray edge and a black edge) and, so, there exists a unique decomposition of edges in cycles. The size of a cycle $C \in G(\pi)$ is the number of black edges in C. A cycle C is trivial if it has size 1. If C has size less than or equal to 3, then C is called short and, otherwise, C is called long. The identity permutation ι_n is the only one with a cycle graph containing $n+1$ cycles, which are all trivial.

The number of cycles in $G(\pi)$ is denoted by $c(\pi)$. Given an operation γ, let $\Delta c(\pi, \gamma) = c(\pi \cdot \gamma) - c(\pi)$, that is, $\Delta c(\pi, \gamma)$ denotes the change in the number of cycles after applying γ to π.

The cycle graph $G(\pi)$ is drawn in a way to highlight characteristics of the permutation, as shown in Fig. 1. In this representation, we draw the vertices in a horizontal line, from left to right, following the order $+\pi_0, -\pi_1, +\pi_1, \ldots, -\pi_n$, $+\pi_n, -\pi_{n+1}$. The black edges are horizontal lines and the gray edges are arcs.

For $1 \le i \le n+1$, the black edge $(-\pi_i, +\pi_{i-1})$ is labeled as i. We represent a cycle C by the sequence of labels of its black edges following the order they are traversed, assuming that the first black edge is the one with highest label (rightmost black edge of C) and it is traversed from right to left. Assuming this representation, if a black edge is traversed from left to right we add a minus sign to its label (the first black is always positive since it is traversed from right to left by convention).

Two black edges of a cycle C are divergent if their labels have different signs, and convergent otherwise. A cycle C is divergent if at least one pair of black edges of C are divergent, and it is convergent otherwise.

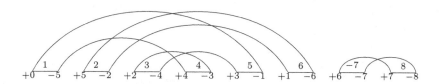

Fig. 1. Cycle Graph for $\pi = (+5 \ +2 \ +4 \ +3 \ +1 \ +6 \ -7)$. In this cycle graph, we have the cycles $C_1 = (5, 3, 4, 1)$, $C_2 = (6, 2)$, and $C_3 = (8, -7)$.

We also classify convergent cycles as oriented and non-oriented. A cycle $C = (c_1, c_2, \ldots, c_k)$ is non-oriented if $c_i > c_{i+1}$, for all $1 \leq i < k$. Otherwise, we say that C is oriented.

Two cycles $C = (c_1, c_2, \ldots, c_k)$ and $D = (d_1, d_2, \ldots, d_k)$ are interleaving if either $|c_1| > |d_1| > |c_2| > |d_2| > \ldots > |c_k| > |d_k|$ or $|d_1| > |c_1| > |d_2| > |c_2| > \ldots > |d_k| > |c_k|$.

Let g_1 be a gray edge adjacent to black edges with labels x_1 and y_1, such that $|x_1| < |y_1|$ and let g_2 be a gray edge adjacent to black edges with labels x_2 and y_2, such that $|x_2| < |y_2|$. We say that two gray edges g_1 and g_2 intersect if $|x_1| < |x_2| \leq |y_1| < |y_2|$. Two cycles C and D intersect if an edge from C intersect with an edge from D.

An open gate is a gray edge from a cycle C that does not intersect with any other gray edge from C. An open gate g_1 from C is closed if another gray edge (which is not from C) intersects with g_1. All open gates of $G(\pi)$ must be closed [8].

In the example of Fig. 1, the cycle $C_1 = (5, 3, 4, 1)$ is convergent and oriented, the cycle $C_2 = (6, 2)$ is convergent and non-oriented, and the cycle $C_3 = (8, -7)$ is divergent. The gray edge from C_1 adjacent to black edges 1 and 4 intersects with the gray edge from C_2 adjacent to black edges 2 and 6, so the cycles C_1 and C_2 intersect.

3   Approximation Algorithms

In this section, we present approximation algorithms considering both unsigned and signed permutations.

3.1   Unsigned Case

Here we present an approximation algorithm with a factor of $3 - k$ based on breakpoints for SBRTwPR on unsigned permutations.

Lemma 1 (Kececioglu and Sankoff [6]). *For any reversal ρ, $\Delta b(\pi, \rho) \leq 2$.*

Lemma 2 (Walter et al. [10]). *For any transposition τ, $\Delta b(\pi, \tau) \leq 3$.*

Lemma 3. *Given an instance (π, k) for SBRTwPR on unsigned permutations, and an optimal sequence of events S, the average number of breakpoints decreased by an operation in S is less than or equal to $3 - k$.*

Proof. Since $|S|$ is an optimal sequence for the instance (π, k), we have that at least $|S|k$ operations present in S are reversals. By Lemmas 1 and 2, we have that a reversal can remove up to two breakpoints while a transposition can remove up to three. Let $\phi b(S)$ denote the average number of breakpoints decreased by an operation in S, we have that

$$\phi b(S) \leq \frac{(2|S|k) + (3|S|(1-k))}{|S|} = 2k + 3(1-k) = 3 - k. \qquad \square$$

Theorem 1. *Given an instance* (π, k) *for* SBRTWPR *on unsigned permutations, we have that* $d_k(\pi) \geq \frac{b(\pi)}{3-k}$.

Proof. Since $b(\pi)$ breakpoints must be removed in order to turn the permutation π into ι and, by Lemma 3, up to $3-k$ breakpoints are removed per operation on average, the theorem follows.   □

Theorem 2 (Kececioglu and Sankoff [6]). *It is possible to turn an unsigned permutation* π *into* ι *using at most* $b(\pi)$ *reversals.*

Theorem 3. SBRTWPR *is approximable by a factor of* $3 - k$ *on unsigned permutations.*

Proof. By Theorem 2, we can turn any unsigned permutation π into ι using at most $b(\pi)$ reversals. Since we use only reversals, the constraint $\frac{|S_\rho|}{|S|} \geq k$ is not violated. By the lower bound showed in Theorem 1, we have $\frac{b(\pi)}{\frac{b(\pi)}{3-k}} = 3 - k$.   □

In order to avoid solutions for the problem consisting exclusively of reversals, we propose the Algorithm 1. This algorithm guarantees the same approximation factor for the problem and tends to provide solutions in which the ratio between the number of reversals and the size of the sorting sequence is close to k.

Algorithm 1: An approximation algorithm for SBRTWPR on unsigned permutations.

Input: An unsigned permutation π and a value k
Output: A sequence of reversals and transpositions that sorts π
1 Let $S \leftarrow \{\}$
2 **while** $\pi \neq \iota$ **do**
3   **if** $\frac{|S_\rho|}{|S|+1} \geq k$ *and there is a transposition* τ *such that* $\Delta b(\pi, \tau) \geq 1$ **then**
4   Apply τ in π
5   Append τ to S
6   **else**
7   Let S' be a sequence of reversals that decreases, on average, one or more breakpoints per operation [6]
8   Apply S' in π
9   Append S' to S
10 **return** S

Note that a transposition τ is only applied if two constraints are fulfilled: (i) $\frac{|S_\rho|}{|S|+1} \geq k$, this ensures that the sorting sequence will comply with the main restriction of the problem that $\frac{|S_\rho|}{|S|} \geq k$. (ii) $\Delta b(\pi, \tau) \geq 1$, this constraint ensures that the sorting sequence will contain a maximum of $b(\pi)$ operations, since every reversal sequence removes, on average, one or more breakpoints per operation. Since Algorithm 1 removes one or more breakpoints by iteration, it guarantees that the permutation π will be sorted. In addition, no more than $b(\pi)$ operations will be used to sort π, maintaining the approximation factor of $3 - k$. Since each operation (reversal or transposition) can be found in linear time and $|S| \leq b(\pi) \leq n + 1$, the running time of Algorithm 1 is $\mathcal{O}(n^2)$.

3.2   Signed Case

Here we present an approximation algorithm with a factor of $3 - \frac{3k}{2}$ based on the cycle graph for SBRTwPR on signed permutations.

Lemma 4 (Hannenhalli and Pevzner [5]). *For any reversal ρ, $\Delta c(\pi, \rho) \leq 1$.*

Lemma 5 (Bafna and Pevzner [1]). *For any transposition τ, $\Delta c(\pi, \tau) \leq 2$.*

Lemma 6. *Given an instance (π, k) for SBRTwPR on signed permutations, and an optimal sequence of events S, the average number of cycles increased by an operation in S is less than or equal to $2 - k$.*

Proof. Since $|S|$ is an optimal sequence for the instance (π, k), we have that at least $|S|k$ operations in S sequence are reversals. By Lemmas 4 and 5, we have that a reversal creates at most one new cycle, while a transposition creates at most two new cycles. Let $\phi c(S)$ denote the average number of cycles increased by an operation in S, we have that:

$$\phi c(S) \leq \frac{(1|S|k) + (2|S|(1 - k))}{|S|} = 1k + 2(1 - k) = 2 - k. \qquad \square$$

Theorem 4. *Given an instance (π, k) for SBRTwPR on signed permutations, we have that $d_k(\pi) \geq \frac{n + 1 - c(\pi)}{2 - k}$.*

Proof. Since $(n + 1) - c(\pi)$ new cycles must be created in order to turn the permutation π into ι and, by Lemma 6, up to $2 - k$ new cycles are created per operation on average, the theorem follows. $\qquad \square$

Theorem 5. *Given a signed permutation π, there exists a sequence of reversals S that transforms π into ι such that the average number of cycles increased by any reversal in S is greater than or equal to $2/3$.*

Proof. If at any stage $G(\pi)$ has a divergent cycle C, then there exists a reversal applied to C that increases the number of cycles by one unit [10]. Otherwise, $G(\pi)$ has only convergent cycles, and one of the following is true [8]:

- there exists a long oriented cycle (Fig. 2, Case 1);
- there exists a short cycle C whose open gates are closed by another non-trivial cycle D (Fig. 2, Case 2);
- there exists a long non-oriented cycle C whose open gates are closed by one or more non-trivial cycles (Fig. 2, Case 3).

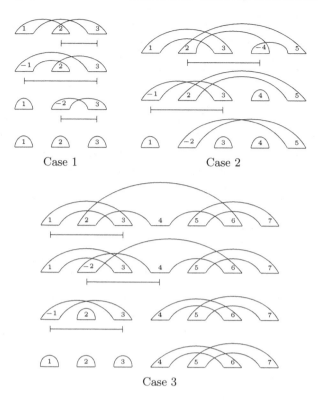

Fig. 2. Operations applied in each case of Theorem 5.

If $G(\pi)$ has an oriented long cycle C, then we can apply a reversal on its black edges in such a way that it turns C into a divergent cycle C'. Since C' is long, we can apply at least two reversals on C' that increase the number of cycles by one unit each (Fig. 2, Case 1).

In the other two cases we can turn the cycle C into an oriented cycle C' by applying one reversal to a cycle D that closes an open gate from C. If C' is short, we can break it into two trivial cycles with a reversal, and this second reversal turns D into a divergent cycle D', which guarantees that we can apply a third reversal to D' that increases the number of cycles by one (Fig. 2, Case 2). If C' is long, then we can apply at least two reversals that increase the number of cycles by one unit each (Fig. 2, Case 3).

In the three cases above we applied three reversals that increased the number of cycles by two, and the theorem follows.   □

Theorem 6. SBRTwPR *is approximable by a factor of* $3 - \frac{3k}{2}$ *on signed permutations.*

Proof. By Theorem 5, we can turn any signed permutation π into ι using at most $\frac{3(n+1-c(\pi))}{2}$ reversals. Since we use only reversals, the constraint $\frac{|S_\rho|}{|S|} \geq k$ is not violated. By the lower bound showed in Theorem 4, we have:

$$\frac{\frac{3(n+1-c(\pi))}{2}}{\frac{n+1-c(\pi)}{2-k}} = 3 - \frac{3k}{2}.$$

$\square$

Note that in order to avoid a solution composed exclusively of reversals, the approach used in Algorithm 1 can be adapted to be applied in this case as well. In Sect. 4, we will present an asymptotic approximation algorithm with an improved approximation factor for the signed case.

4   Asymptotic Approximation for the Signed Case

In this section we show an asymptotic algorithm for SBRTwPR on signed permutations, where $k \in [0, 1]$ with an approximation factor of $\left(\frac{2-k}{1-\frac{k}{3}}\right)$.

Definition 6. *Let $\mathcal{A}_\rho$ be an algorithm that sorts a permutation using only signed reversals and guarantees a ratio of 2/3 of cycles increased by applied reversals (Theorem 5), and let $\mathcal{A}_\rho(\pi)$ represents the sequence of reversals returned by the algorithm that sorts π.*

Now consider Algorithm 2.

Algorithm 2: An approximation algorithm for SBRTwPR on signed permutations.

Input: A signed permutation π and a value k
Output: A sequence of reversals and transpositions that sorts π

1 Let $S \leftarrow \{\}$
2 **while** $|\mathcal{A}_\rho(\pi)| > k(|S| + |\mathcal{A}_\rho(\pi)|)$ **do**
3   **if** $G(\pi)$ *has a divergent cycle* **then**
4   Let ρ be a reversal that increases one cycle in $G(\pi)$
5   Apply ρ in π
6   Append ρ to S
7   **else**
8   Let S' be a sequence of at most two transpositions that increases two cycles in $G(\pi)$ [2, Theorem 3.4]
9   Apply S' in π
10   Append S' to S
11 Apply $\mathcal{A}_\rho(\pi)$ in π
12 Append $\mathcal{A}_\rho(\pi)$ to S
13 **if** $|S_\rho| < k|S|$ **then**
14   Replace the last two transpositions of S with six reversals [8]
15 **return** S

Lemma 7. *Given a signed permutation π, Algorithm 2 sorts π using at most $(n + 1 - c(\pi))/(1 - k/3) + 4$ operations.*

Proof. Let $S = (S_1, \ldots, S_{|S|})$ be the sorting sequence generated by the algorithm without considering the substitution of transpositions by reversals applied in line 14. Let S' be the subsequence of operations applied in the while loop of lines 2 to 10. Each operation in S' increases the number of cycles by at least one unit, and each operation in $S \setminus S'$ (that is, the operations applied outside the while loop) increases on average in $2/3$ the number of cycles. By the condition of line 2, we have that $|S'| \geq (1 - k)|S|$ and, therefore, the average increase in the number of cycles in S is at least $\frac{(1-k)|S| + k|S|2/3}{|S|} = 1 - k/3$. Since these operations increase at most $n + 1 - c(\pi)$ cycles, we have that $|S| \leq \frac{n+1-c(\pi)}{1-k/3}$. In the final sequence, we may increase four operations by replacing the last two transpositions with six reversals (only if necessary). Therefore, the size of this sequence is at most $\frac{n+1-c(\pi)}{1-k/3} + 4$.

Theorem 7. *Algorithm 2 is a $\frac{2-k}{1-k/3}$-asymptotic approximation algorithm for* SBRTwPR.

Proof. Since the algorithm only adds transpositions while the condition of line 2 is satisfied and at most two transpositions are added in the sorting sequence in one iteration, we guarantee that $|S_\rho| \geq k$ by replacing the last two transpositions by reversals. By Lemma 7 and Theorem 4, the sequence S returned by Algorithm 2 satisfies $|S| \leq \frac{n+1-c(\pi)}{1-k/3} + 4 \leq \frac{2-k}{1-k/3}d_k(\pi) + 4$. Therefore, it is a $\frac{2-k}{1-k/3}$-asymptotic approximation algorithm for SBRTwPR. □

5   Conclusion

We investigated the Sorting by Reversals and Transpositions with Proportion Restriction problem and presented an approximation algorithm with a factor of $3 - k$ for unsigned permutations, and an approximation and an asymptotic approximation algorithm with factors $3 - \frac{3k}{2}$ and $\frac{2-k}{1-\frac{k}{3}}$ for signed permutations, respectively.

As future work, we intend to test the proposed algorithms and develop heuristics for the problems. Another interesting research line would be to investigate the complexity of the problems when $0 < k < 1$.

Acknowledgments. This work was supported by the National Council of Technological and Scientific Development, CNPq (grants 400487/2016-0, 140272/2020-8, and 425340/2016-3), the Coordenação de Aperfeiçoamento de Pessoal de Nível Superior - Brasil (CAPES) - Finance Code 001, and the São Paulo Research Foundation, FAPESP (grants 2013/08293-7, 2015/11937-9, 2017/12646-3, and 2019/27331-3).

References

1. Bafna, V., Pevzner, P.A.: Sorting permutations by transpositions. In: Proceedings of the Sixth Annual ACM-SIAM Symposium on Discrete Algorithms (SODA 1995), pp. 614–623. Society for Industrial and Applied Mathematics, Philadelphia (1995)
2. Bafna, V., Pevzner, P.A.: Sorting by transpositions. SIAM J. Discret. Math. **11**(2), 224–240 (1998)
3. Bulteau, L., Fertin, G., Rusu, I.: Sorting by transpositions is difficult. SIAM J. Discret. Math. **26**(3), 1148–1180 (2012)
4. Caprara, A.: Sorting permutations by reversals and Eulerian cycle decompositions. SIAM J. Discret. Math. **12**(1), 91–110 (1999)
5. Hannenhalli, S., Pevzner, P.A.: Transforming cabbage into turnip: polynomial algorithm for sorting signed permutations by reversals. J. ACM **46**(1), 1–27 (1999)
6. Kececioglu, J.D., Sankoff, D.: Exact and approximation algorithms for sorting by reversals, with application to genome rearrangement. Algorithmica **13**, 180–210 (1995)
7. Oliveira, A.R., Brito, K.L., Dias, U., Dias, Z.: On the complexity of sorting by reversals and transpositions problems. J. Comput. Biol. **26**, 1223–1229 (2019)
8. Oliveira, A.R., Brito, K.L., Dias, Z., Dias, U.: Sorting by weighted reversals and transpositions. J. Comput. Biol. **26**, 420–431 (2019)
9. Rahman, A., Shatabda, S., Hasan, M.: An approximation algorithm for sorting by reversals and transpositions. J. Discret. Algorithms **6**(3), 449–457 (2008)
10. Walter, M.E.M.T., Dias, Z., Meidanis, J.: Reversal and transposition distance of linear chromosomes. In: Proceedings of the 5th International Symposium on String Processing and Information Retrieval (SPIRE 1998), pp. 96–102. IEEE Computer Society, Los Alamitos (1998)

Heuristics for Breakpoint Graph Decomposition with Applications in Genome Rearrangement Problems

Pedro Olímpio Pinheiro$^{(\boxtimes)}$, Alexsandro Oliveira Alexandrino ,
Andre Rodrigues Oliveira , Cid Carvalho de Souza , and Zanoni Dias

Institute of Computing, University of Campinas, Campinas, Brazil
pedro.pinheiro@students.ic.unicamp.br,
{alexsandro,andrero,cid,zanoni}@ic.unicamp.br

Abstract. The breakpoint graph of a permutation is a well-known structure used in genome rearrangement problems. Most studies use the decomposition of such graph into edge-colored disjoint alternating cycles to develop algorithms for these problems. The goal of the Breakpoint Graph Decomposition (BGD) problem is to find a decomposition of the breakpoint graph with maximum number of cycles. For unsigned permutations, which model genomes without information about gene orientation, the BGD problem is NP-hard. In this work, we developed a greedy and a Tabu Search algorithm which are compared experimentally with the approximation algorithm presented by Lin and Jiang [10]. The experiments revealed that our algorithms find significantly better solutions. Finally, we used our algorithms as part of algorithms for genome rearrangement problems and the distances calculated in this way have largely improved.

Keywords: Breakpoint graph · Genome rearrangements · Maximum cycle decomposition

1 Introduction

The rearrangement distance between two genomes is a problem in comparative genomics that aims to find a minimum sequence of rearrangements required to transform one genome into the other.

Genomes in this problem are generally represented as permutations, where each element of the permutation corresponds to a gene. When the orientation of the genes is known, we use a plus or minus sign in each element to indicate the orientation, and we say that the permutation is *signed*. Otherwise, elements do not have signs, and we say that the permutation is *unsigned*.

Since we can model one of the genomes as the *identity permutation* (i.e., permutation $(1\ 2\ \dots\ n)$ or $(+1\ +2\ \dots\ +n)$), the problem of transforming one genome into another by rearrangements is equivalent to that of sorting permutations by rearrangements. We often assume that these permutations have two extra elements 0 and $n+1$ at the beginning and at the end, respectively.

© Springer Nature Switzerland AG 2020
J. C. Setubal and W. M. Silva (Eds.): BSB 2020, LNBI 12558, pp. 129–140, 2020.
https://doi.org/10.1007/978-3-030-65775-8_12

The most studied rearrangements are the reversal, transposition, and Double cut-and-join (DCJ) operations. The problems of Sorting by Reversals and Sorting by DCJs are solvable in polynomial time on signed permutations [8,13], while they are NP-hard on unsigned permutations [3]. The problems of Sorting by Transpositions [2] and Sorting by Reversals and Transpositions [11] are also NP-hard.

The first bounds for these problems used the concept of *breakpoints*, which are pairs of elements that are adjacent in the given permutation but not in the identity permutation. Later, improved bounds and new algorithms were developed using the *breakpoint graph* of a permutation [1]. We formally define the breakpoint graph of a permutation in Sect. 2.

These bounds for the rearrangement distance are based on the number of cycles in a maximum cardinality decomposition of the breakpoint graph into edge-colored disjoint cycles. This decomposition is unique on signed permutations or when the model allows only transpositions on unsigned permutations. When considering reversals on unsigned permutations, Caprara [3] showed that finding a maximum cardinality decomposition of a breakpoint graph of an unsigned permutation is NP-hard, and the same is valid for DCJs.

Using the Breakpoint Graph Decomposition as a subproblem, Bafna and Pevzner [1] presented a 7/4-approximation algorithm for the Sorting by Reversals. This factor was improved by Christie [5] to 1.5 and by Lin and Jiang [10] to $1.4193 + \epsilon$. Based on a similar strategy, Chen [4] presented a $(1.4167 + \epsilon)$-approximation algorithm for the Sorting by DCJs problem. More recently, Jiang *et al.* [9] presented a randomized FPT algorithm for the Sorting by DCJs with an approximation factor of $(4/3 + \epsilon)$.

In this paper, we propose a greedy algorithm and an algorithm based on the Tabu Search metaheuristic for the Breakpoint Graph Decomposition. We analyze the performance of these algorithms in practice and compare them with the algorithm of Lin and Jiang [10]. Furthermore, we present experimental results of these algorithms applied to the genome rearrangement distance considering a model with DCJs and a model with reversals and transpositions.

This paper is organized as follows. Section 2 introduces the concepts used in the algorithms and formalizes the problem. Section 3 presents the heuristics created for the Breakpoint Graph Decomposition problem. In Sect. 4, we show the experimental results for the heuristics of Sect. 3. At last, in Sect. 5, we give our final remarks and discuss directions of future work.

2   Preliminaries

Let π be a permutation $(\pi_1 \ \pi_2 \ \ldots \ \pi_n)$, where $\pi_i \in \{1, \ldots, n\}$ and $\pi_i = \pi_j$ iff $i = j$, for $1 \leq i, j \leq n$. The *identity permutation* ι_n is the permutation $(1 \ 2 \ \ldots \ n)$, which is the target of the sorting by rearrangements problems. We *extend* π by adding the elements $\pi_0 = 0$ and $\pi_{n+1} = n + 1$. In the next definitions, we assume that permutations are in its extended form.

We say that (π_i, π_{i+1}) is a *breakpoint* if $|\pi_{i+1} - \pi_i| \neq 1$, for $0 \leq i \leq n$. The number of breakpoints in a permutation π is denoted by $b(\pi)$.

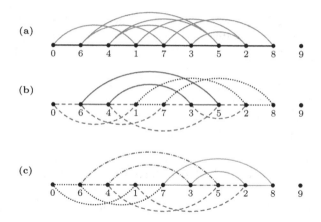

Fig. 1. (a) Breakpoint graph $G(\pi)$ for $\pi = (0\ 6\ 4\ 1\ 7\ 3\ 5\ 2\ 8\ 9)$, where $b(\pi) = 8$. **(b)** Decomposition of $G(\pi)$ into three cycles $C_1 = (8, 2, 1, 7)$, $C_2 = (5, 3, 4, 6)$, and $C_3 = (2, 5, 4, 1, 0, 6, 7, 3)$. **(c)** Decomposition of $G(\pi)$ into four cycles $C_1 = (8, 2, 3, 7)$, $C_2 = (2, 5, 4, 1)$, $C_3 = (5, 3, 4, 6)$, and $C_4 = (7, 1, 0, 6)$.

Example 1. The permutation $\pi = (0\ 4\ 3\ 5\ 1\ 2\ 6)$ has breakpoints $(0, 4)$, $(3, 5)$, $(5, 1)$, and $(2, 6)$. Therefore, $b(\pi) = 4$.

The *breakpoint graph* $G(\pi) = (V, E)$ of a permutation π is an edge-colored undirected graph with vertices $V = \{\pi_0, \pi_1, \ldots, \pi_n, \pi_{n+1}\}$ and edges $E = E_b \cup E_g$, where E_b is the set of black edges and E_g is the set of gray edges. For all $0 \leq i \leq n$, there exists a *black edge* (π_i, π_{i+1}) and a *gray edge* $(\pi_i, \pi_i + 1)$ if the pair (π_i, π_{i+1}) is a breakpoint.

The set of black edges E_b connects elements adjacent in π and the set of gray edges E_g connects elements adjacent in ι_n. Elements adjacent in both π and ι_n are not connected in $G(\pi)$. By convention, we draw the breakpoint graph by placing each vertex on a horizontal line in the order they appear in the permutation. Black edges are represented as horizontal lines and gray edges are represented as arcs. An example is shown in Fig. 1(a).

An *alternating cycle* C is a cycle such that every pair of consecutive edges have distinct colors. An alternating cycle C of $G(\pi)$ is a *k-cycle* if it has k black edges. By convention, we list the vertices in an alternating cycle $C = (\pi_{x_1}, \pi_{x_2}, \ldots, \pi_{x_m})$ assuming that the first vertex is the rightmost element from π (i.e. $x_1 \geq x_i$, for $2 \leq i \leq m$) and that (π_{x_1}, π_{x_2}) is a black edge.

Since each vertex is incident to the same number of black and gray edges, there exists at least one decomposition of $G(\pi)$ into edge-disjoint alternating cycles [3]. In the Breakpoint Graph Decomposition problem, we are interested in finding a set of edge-disjoint alternating cycles with maximum cardinality. For a permutation π, we denote by $c(\pi)$ the number of alternating cycles in an optimal solution for the Breakpoint Graph Decomposition problem.

In Figs. 1(b) and 1(c), we show two distinct examples of decompositions of a breakpoint graph into alternating cycles. The decomposition in Fig. 1(c) is maximum for the breakpoint graph of the permutation $\pi = (0\ 6\ 4\ 1\ 7\ 3\ 5\ 2\ 8\ 9)$.

BREAKPOINT GRAPH DECOMPOSITION (BGD)
  Input: A permutation π.
  Goal: Find a set of edge-disjoint alternating cycles $\mathcal{H} = \{C_1, C_2, \ldots\}$, such that $|\mathcal{H}|$ is maximum.

---

3   Heuristics for Breakpoint Graph Decomposition

In the following sections, we present two general heuristics for the BGD problem.

3.1   Greedy Algorithm

As shown in Fig. 1(a), each vertex in the breakpoint graph has a maximum of two black edges and two gray edges. Our greedy algorithm has a subroutine called bfs_cycle that, starting from a given vertex with incident edges, performs a breadth first search for an alternating cycle, as we explain in detail later in this section. Depending on which vertex the algorithm chooses to start the search, different variations could be applied:

- using the leftmost vertex with incident edges, which we call the FIRST approach;
- choosing a vertex at random among those still having incident edges, which we call the RANDOM approach;
- executing for all vertices still having incident edges, creating a list with the returned cycles, and, at the end, picking the smallest cycle from this list, which we call the ALL approach.

The selected cycle is added to the list of cycles and all edges in that cycle are removed from the breakpoint graph. These steps are repeated until there are no edges left in the graph, in which case the greedy algorithm stops. Variations FIRST and RANDOM take $\mathcal{O}(n^2)$ while the variation ALL takes $\mathcal{O}(n^3)$.

As the RANDOM variation is non-deterministic, we also developed a variation called MAX, that runs the RANDOM k times and returns the decomposition with the largest number of cycles. The best trade-off between execution time and solution quality was achieved using $k = n\sqrt{n}$, so the MAX variation has a time complexity of $\mathcal{O}(n^3\sqrt{n})$.

The bfs_cycle subroutine: given a vertex v, we perform a breadth first search starting at v with the constraint that we explore a new vertex u only if the explored path from v to u is an alternating path (i.e., consecutive edges have distinct colors).

When exploring the edges of a vertex u, if there exists an edge (u, v) such that the path from v to u plus the edge (u, v) forms an alternating cycle, then we stop the search returning this cycle. Since we do this using a breadth first search, this subroutine finds the smallest alternating cycle starting with v.

As an example consider the breakpoint graph of Fig. 1(a). The FIRST approach returns the decomposition of Fig. 1(c). It starts at vertex 0, finds the smallest alternating cycle starting at 0, which is the cycle with vertices $(0, 6, 7, 1)$, and removes its edges from the graph. Then, it proceeds to find the smallest cycle starting at vertex 6, which still has black edges. The algorithm finds the cycle with vertices $(6, 4, 3, 5)$ and removes its edges from the graph. The algorithm continues until there are no edges in the graph.

Consider the same example but executing the RANDOM approach. Suppose that the vertices 6, 1, and 0 were chosen in each iteration, in this order. Then, the algorithm returns the decomposition of Fig. 1(b). In the first iteration, the algorithm finds the smallest cycle starting at vertex 6, which is the cycle with vertices $(6, 4, 3, 5)$. In the next iteration, the algorithm finds the smallest cycle starting at 1, which is the cycle with vertices $(1, 7, 8, 2)$. At this point, the graph has only one cycle that is chosen in the last iteration.

As a heuristic, our greedy algorithm described above cannot guarantee an approximation factor. However, it is possible to adapt it in such a way that it guarantees the same approximation as the Lin and Jiang algorithm (L&J) [10]. Let C be the cycle decomposition returned by L&J. Append to our cycle list every cycle from C with less than 4 black edges (i.e., the list of cycles that is enough to guarantee the approximation factor of $1.4193 + \epsilon$). Remove their edges from the breakpoint graph and start the breadth first search over this modified breakpoint graph until no more edges exist. The L&J algorithm requires $\mathcal{O}(n^3)$ time to generate all possible short cycles (i.e., cycles of size 4 and 6) plus $\mathcal{O}(3^m)$ to calculate the greatest subset of disjoint short cycles, where m is the number of short cycles, with $\epsilon = 0$.

3.2 Tabu Search

Tabu Search is a strategy to solve combinatorial optimization problems capable of using many other methods, such as linear programming and specialized heuristics, to overcome local optima [6, 7].

Glover [6, 7] describes the following generic optimization problem to introduce the Tabu Search metaheuristic: given the set X of feasible solutions to an optimization problem and a objective function f, find an element $x \in X$ with maximum (or minimum) value of $f(x)$.

Given a feasible solution $x \in X$, a *movement* is an operation s that generates a new feasible solution $s(x) = x'$. The set of possible movements that can be applied to a solution x is denoted by $S(x)$. Usually, we are interested in movements that create better solutions in terms of the objective function.

To avoid local optima, movements that do not increase the objective value are also executed. However, allowing these movements can trap the search into a set of solutions without improvements in the objective value. To prevent this scenario, we create a tabu list T with the movements performed in recent iterations of the search that should not be done again or undone.

Algorithm 1. Tabu Search Heuristic

1: Choose an initial solution $x \in X$ and make $x^* \leftarrow x$, where x^* is the best solution
   found so far. Start the iteration counter $k \leftarrow 0$ and the tabu list $T \leftarrow \emptyset$.
2: If $S(X) \setminus T = \emptyset$, then go to step 4. Otherwise, make $k \leftarrow k + 1$ and choose
   $s_k(x) \in S(x) \setminus T$ such that $s_k(x) = \arg\max_{s \in S(x) \setminus T}\{f(s(x))\}$.
3: Make $x \leftarrow s_k(x)$. If $f(x) > f(x^*)$, then $x^* \leftarrow x$.
4: If the maximum number of iterations was reached or $S(x) \setminus T = \emptyset$, stop the algorithm and return the solution x^*. Otherwise, update T and go to step 2.

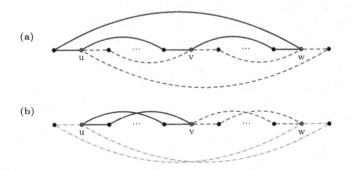

Fig. 2. (a) Example of a solution for BGD with two cycles C_1 (straight edges) and
C_2 (dashed edges). These cycles have three vertices (u, v, and w) in common and
they satisfy the conditions to apply the first movement. (b) Three cycles generated by
applying the first movement in the cycles C_1 and C_2.

Algorithm 1 describes the Tabu Search heuristic presented by Glover [6].

For the Breakpoint Graph Decomposition problem, the objective function is
to maximize the number of cycles in the solution. The initial solution is given
by one of the algorithms of Sect. 3.1 and the two movements developed are
described next.

The first movement is applied in two cycles C_1 and C_2, such that they have
at least three common vertices u, v, and w, transforming them into three cycles.

Given two vertices x and y from the three common vertices u, v, and w, let
P^1_{xy} and P^2_{xy} denote the paths in the cycles C_1 and C_2, respectively, that go from
x to y and do not include the third common vertex. To apply the first movement,
the paths P^1_{uv} and P^2_{uv} must have lengths with same parity and the first edge
of these two paths must have different colors. The same condition must hold for
paths P^1_{vw} and P^2_{vw} and for the paths P^1_{wu} and P^2_{wu}.

When these conditions are satisfied, a solution with one more cycle is created
by replacing cycles C_1 and C_2 with three cycles C'_1, C'_2, and C'_3, where C'_1 is the
concatenation of P^1_{uv} and P^2_{uv}, C'_2 is the concatenation of P^1_{vw} and P^2_{vw}, and C'_3
is the concatenation of P^1_{wu} and P^2_{wu}.

Figure 2 shows an example of the first movement being applied to two cycles.
Since this movement always creates a solution with one more cycle, the Tabu
Search algorithms execute it whenever possible.

The second movement is applied in two cycles C_1 and C_2, such that they have at least two common vertices u and v, transforming them into two distinct cycles.

Let P_{uv}^1 and Q_{vu}^1 be the two distinct paths from u to v and from v to u in cycle C_1, respectively, such that P_{uv}^1 begins with a black edge. Let P_{uv}^2 and Q_{vu}^2 be the two distinct paths from u to v and from v to u in cycle C_2, respectively, such that P_{uv}^2 begins with a black edge. Note that both Q_{vu}^1 and Q_{vu}^2 end with a gray edge.

The second movement can be applied if the length of P_{uv}^1 has the same parity of the length of P_{uv}^2. This movement creates two cycles C_1' and C_2', such that C_1' is the path P_{uv}^1 concatenated with Q_{vu}^2 and C_2' is the path P_{uv}^2 concatenated with Q_{vu}^1.

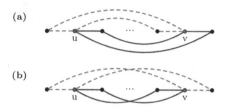

(a)

(b)

Fig. 3. (a) Example of a solution for BGD with two cycles C_1 (dashed edges) and C_2 (straight edges). These cycles have two vertices (u and v) in common and they satisfy the conditions to apply the second movement. **(b)** Two new cycles generated by applying the second movement in the cycles C_1 and C_2.

Figure 3 shows an example of the second movement applied to two cycles. When this movement is executed, the pair (u, v) is included to the tabu list T. Note that this movement creates a new solution with the same number of cycles. This operation is applied only if the pair (u, v) is not in the tabu list T.

The second movement is useful to continue searching for new solutions after the algorithm reaches a local optimum. Since, in this case, there is no pair of cycles that satisfies the conditions of the first movement, the second movement alters the cycles of the solution possibly enabling the use of the first movement in the new solution.

We limit the second movement with the tabu list, because, otherwise, the Tabu Search could enter into an infinite loop, applying the second movement replacing two cycles C_1 and C_2 by two new cycles C_1' and C_2' and, in the next iteration, reverting the previous operation by applying the second movement to the new cycles C_1' and C_2' transforming them back to C_1 and C_2.

The complexity of each iteration is $\mathcal{O}(n^3)$. The best trade-off between execution time and solution quality was achieved using the maximum number of iterations equals to n. Then the complete search becomes $\mathcal{O}(n^4)$.

4   Experimental Results

In order to check the efficiency of our heuristics, we implemented Lin and Jiang's algorithm (L&J), the greedy heuristic, and the Tabu Search. To compare them, we generated sets of permutations such that the number of breakpoints is as large as possible (i.e., $n+1$ for a permutation of size n). Thus, every breakpoint graph has $n+1$ black edges and every vertex has two black edges and two gray edges, with the exception of the first and the last vertices which have only one edge of each color. Our sets are separated by permutation size, from 10 to 100 in intervals of 10, with 100 permutations each.

Four different experiments were performed: the first two consist of (i) comparing the different versions of the greedy algorithm against the L&J; and (ii) indicating to the greedy algorithm that it must contain the same short cycles as those of L&J (so that our greedy algorithm also guarantees the same approximation factor). The other two experiments consist of running the Tabu Search on top of the first two tests to check its improvement. Results for variant RANDOM of the greedy algorithm are the average of 100 executions for each permutation.

Table 1 shows the average number of cycles returned by L&J and the four variants of the greedy algorithm, namely FIRST, ALL, RANDOM and MAX. We can see that, except for permutations of size up to 20, all variations of the greedy algorithm returned decompositions with a greater number of cycles on average compared to those returned by L&J. Besides, the decompositions obtained by MAX have on average more cycles than L&J for all permutations of size greater than or equal to 20. For permutations of size greater than or equal to 60, the MAX heuristic returned cycle decompositions whose number of cycles are at least 50% greater on average than the cycle decompositions returned by L&J. Results of Experiment 2 show that by using the same set of short cycles from L&J in the cycle decomposition, the variations FIRST, ALL, and RANDOM returned cycle decompositions with more cycles compared to their results on Experiment 1. However, the MAX variation, which produces the best results on average, returned cycle decompositions with a smaller number of cycles on average compared to the results obtained on Experiment 1, for permutations sizes between 20 and 80.

Table 2 shows the average number of cycles returned by Tabu Search using the output of L&J and the four variants of the greedy algorithm. We can see that Tabu Search was able to improve the results of all algorithms, especially L&J that had a great improvement (probably because it returned the lowest average number of cycles, with a large room for improvement). The MAX variation remains as the one that returns the greatest number of cycles on average, and the same behavior of Experiment 2 happened here: on average, the results of Experiment 4 are slightly worse than the results of Experiment 3, for permutations sizes between 20 and 90.

Table 1. Average number of cycles returned in the Experiments 1 and 2. Experiment 1 consists of the **L&J** algorithm and all variations of the greedy heuristic. Experiment 2 consists of using the same short cycles returned by **L&J** in the solution of our greedy heuristic. Each value represents the average for 100 permutations of size n, which is indicated in the first column. For the RANDOM approach, we also did the average of 100 executions for each permutation.

n	Experiment 1					Experiment 2			
	L&J	FIRST	ALL	RANDOM	MAX	FIRST	ALL	RANDOM	MAX
10	4.14	3.99	4.09	3.98	4.14	4.14	4.14	4.14	4.14
20	6.58	6.23	6.63	6.35	6.89	6.75	6.75	6.75	6.75
30	8.01	8.04	8.75	8.27	9.22	8.89	8.90	8.88	8.91
40	9.12	9.98	10.70	10.06	11.39	10.81	10.89	10.84	10.97
50	9.70	11.39	12.75	11.67	13.45	12.61	12.74	12.62	12.97
60	10.20	13.06	14.38	13.20	15.38	14.25	14.47	14.27	14.88
70	10.97	14.45	16.36	14.73	17.13	16.09	16.33	16.02	16.84
80	11.52	15.69	18.16	16.30	19.01	17.77	18.24	17.77	18.95
90	11.73	17.27	19.51	17.60	20.44	19.14	19.63	19.13	20.72
100	12.31	18.35	21.35	19.03	22.17	20.67	21.27	20.71	22.54

4.1   Applications in Genome Rearrangement Distance

Algorithms for the Breakpoint Graph Decomposition problem are often used as a part of approximation algorithms for genome rearrangement problems considering unsigned permutations [4,10,12].

Due to the relation between these problems, we can also use the value $b(\pi) - |\mathcal{H}|$ to evaluate a decomposition $\mathcal{H}$. We recall that $b(\pi)$ is the number of breakpoints in a permutation π and $c(\pi)$ is the number of cycles in an optimal solution for the BGD problem. Since $b(\pi)$ is constant for a given permutation π, searching for a decomposition $\mathcal{H}$ of $G(\pi)$ with maximum cardinality is equivalent to find a decomposition $\mathcal{H}$ of $G(\pi)$ such that $b(\pi) - |\mathcal{H}|$ is minimum.

The approximation algorithms mentioned in Sect. 1 use the value $b(\pi) - c(\pi)$ to calculate the approximation factor, which is also a lower bound for the reversal and DCJ distances.

Theorem 1 (Bafna and Pevzner [1]). *For any permutation π, the reversal distance $d_r(\pi) \geq b(\pi) - c(\pi)$.*

Theorem 2 (Chen [4] and Yancopoulos et al. [13]). *For any permutation π, the DCJ distance $d_{DCJ}(\pi) \geq b(\pi) - c(\pi)$. Also, given a decomposition $\mathcal{H}$ for $G(\pi)$, the DCJ distance $d_{DCJ}(\pi) \leq b(\pi) - |\mathcal{H}|$.*

Theorem 3 (Rahman et al. [12]). *For any permutation π, the reversal and transposition distance $d_{rt}(\pi) \geq (b(\pi) - c(\pi))/2$. Also, given a decomposition $\mathcal{H}$ for $G(\pi)$, the reversal and transposition distance $d_{rt}(\pi) \leq b(\pi) - |\mathcal{H}|$.*

Table 2. Average number of cycles returned by the Tabu Search heuristic using the cycle decomposition returned by each algorithm as input on Experiments 3 and 4. Experiment 3 consists of using the Tabu Search heuristic in the solutions of Experiment 1 and Experiment 4 consists of using the Tabu Search heuristic in the solutions of Experiment 2. Each value represents the average for 100 permutations of size n, which is indicated in the first column. For the RANDOM approach, we also did the average of 100 executions for each permutation.

n	Experiment 3					Experiment 4			
	L&J	FIRST	ALL	RANDOM	MAX	FIRST	ALL	RANDOM	MAX
10	4.14	4.09	4.12	4.14	4.14	4.14	4.14	4.14	4.14
20	6.76	6.50	6.80	6.89	6.89	6.77	6.77	6.81	6.81
30	8.87	8.54	9.00	9.22	9.22	9.03	9.02	9.13	9.13
40	10.85	10.50	11.12	11.39	11.40	11.08	11.14	11.33	11.33
50	12.55	12.28	13.06	13.45	13.58	12.99	13.08	13.40	13.42
60	14.39	13.97	14.90	15.40	15.61	14.75	14.85	15.34	15.41
70	16.16	15.72	16.78	17.15	17.48	16.60	16.79	17.35	17.41
80	17.90	17.18	18.67	19.11	19.63	18.39	18.77	19.42	19.54
90	19.24	18.65	20.17	20.58	21.33	19.87	20.24	21.14	21.26
100	20.72	20.29	21.96	22.33	23.27	21.56	21.99	22.94	23.14

We used our algorithms as part of the approximation algorithms for the genome rearrangement distance to evaluate how the distinct decompositions affect the resulting sorting sequences. The general idea is that a decomposition of $G(\pi)$ in alternating cycles is associated with a signed permutation, and we can use algorithms for signed permutations once we have found a decomposition of $G(\pi)$.

We recall that in sorting problems the goal is to find a sequence that sorts a permutation π with minimum length. Note that the identity permutation ι is the only one without breakpoints and, consequently, the breakpoint graph $G(\iota)$ has no edges. In this way, we can interpret sorting a permutation π as removing all breakpoints and cycles of $G(\pi)$.

Next, we describe the results of our experiments using algorithms for the DCJ distance and the reversals and transpositions distance of unsigned permutations. Table 3 presents the results for the DCJ distance and the reversals and transpositions distance.

For a permutation π and a decomposition $\mathcal{H}$, the DCJ distance is equal to $b(\pi) - |\mathcal{H}|$ [4,13] and, consequently, the results of this table are similar to the ones of Experiment 4 of Table 2, since all permutations of size n used in the experiment have $n + 1$ breakpoints.

For the experiment with reversals and transpositions distance, we used the $2k$-approximation algorithm presented by Rahman et al. [12], where k is the approximation factor of the algorithm used for the BGD problem. For a permutation π and a decomposition $\mathcal{H}$, this algorithm returns a sorting sequence S such that $(b(\pi) - |\mathcal{H}|)/2 \leq |S| \leq b(\pi) - |\mathcal{H}|$, where $|S|$ is the length of the sequence. When considering reversals and transpositions, some characteristics of the cycles define how easy or difficult it is to sort the permutation. As an example, when considering reversals, a sequence can remove the two cycles from Fig. 3(b) using two reversals, but at least three reversals are needed to remove the two cycles from Fig. 3(a) [8]. For this reason, a decomposition $\mathcal{H}_1$ can yield a sorting sequence with more operations than a sorting sequence using a decomposition $\mathcal{H}_2$ that has fewer cycles than $\mathcal{H}_1$.

All the proposed algorithms of Sect. 3 had significantly better results than Lin and Jiang's algorithm. Although the ALL approach results in solutions with more cycles than the FIRST approach, we can see that, for some values of n, the FIRST approach yields better results for the distance. Overall, the MAX approach also had better results.

Table 3. Results for the DCJ distance and the reversals and transpositions (RT) distance using the original L&J algorithm (Table 1) and the algorithms of the fourth experiment (Table 2) as a subroutine for the genome rearrangements algorithms. Each value represents the average for 100 permutations of size n, which is indicated in the first column. For the RANDOM approach, we also did the average of 100 executions for each permutation.

n	L&J		FIRST		ALL		RANDOM		MAX	
	DCJ	RT	DCJ	RT	DCJ	RT	DCJ	RT	DCJ	RT
10	6.86	6.02	6.86	6.23	6.86	6.23	6.86	6.20	6.86	6.21
20	14.42	12.16	14.23	12.17	14.23	12.15	14.19	12.06	14.19	11.91
30	22.99	18.64	21.97	17.70	21.98	17.79	21.87	17.83	21.87	17.70
40	31.88	25.11	29.92	23.17	29.86	23.22	29.67	23.26	29.67	23.24
50	41.30	32.86	38.01	29.03	37.92	28.89	37.60	28.83	37.58	28.79
60	50.80	39.91	46.25	33.98	46.15	34.26	45.66	34.24	45.59	33.68
70	60.03	46.83	54.40	39.48	54.21	39.44	53.65	39.48	53.59	39.04
80	69.48	54.69	62.61	44.45	62.23	44.56	61.58	44.74	61.46	44.29
90	79.27	61.12	71.13	49.55	70.76	50.10	69.86	50.11	69.74	49.56
100	88.69	68.59	79.44	55.04	79.01	54.91	78.06	55.20	77.86	54.13

5   Conclusion

In this paper, we studied the problem of Breakpoint Graph Decomposition, which is associated with the genome rearrangement distance on unsigned permutations. We developed a greedy algorithm and an algorithm based on the Tabu Search metaheuristic for this problem. In our experiments, the proposed algorithms yielded better results than the algorithm proposed by Lin and Jiang [10]. The Tabu Search algorithm, which is given a feasible solution and looks for better ones through local search, was able to improve the solutions returned by all the algorithms.

We also evaluated the solutions in our experiments using them in algorithms for the rearrangement distance that receive a decomposition of the breakpoint graph as input. As before, the performance of our algorithms was significantly better than that of the approximation algorithm of Lin and Jiang.

As future works, we intend to develop algorithms based on the GRASP and Genetic Algorithms metaheuristics.

Acknowledgments. This work was supported by the National Council of Technological and Scientific Development, CNPq (425340/2016-3), the Coordenação de Aperfeiçoamento de Pessoal de Nível Superior - Brasil (CAPES) - Finance Code 001 , and the São Paulo Research Foundation, FAPESP (grants 2013/08293-7 , 2015/11937-9 , 2017/12646-3 , 2019/25410-3 , and 2019/27331-3).

References

1. Bafna, V., Pevzner, P.A.: Genome rearrangements and sorting by reversals. SIAM J. Comput. **25**(2), 272–289 (1996)
2. Bulteau, L., Fertin, G., Rusu, I.: Sorting by transpositions is difficult. SIAM J. Discret. Math. **26**(3), 1148–1180 (2012)
3. Caprara, A.: Sorting permutations by reversals and Eulerian cycle decompositions. SIAM J. Discret. Math. **12**(1), 91–110 (1999)
4. Chen, X.: On sorting unsigned permutations by double-cut-and-joins. J. Comb. Optim. **25**(3), 339–351 (2013). https://doi.org/10.1007/s10878-010-9369-8
5. Christie, D.A.: A 3/2-approximation algorithm for sorting by reversals. In: Proceedings of the 9th Annual ACM-SIAM Symposium on Discrete Algorithms (SODA 1998), pp. 244–252. Society for Industrial and Applied Mathematics, Philadelphia (1998)
6. Glover, F.W.: Tabu search - Part I. INFORMS J. Comput. **1**(3), 190–206 (1989)
7. Glover, F.W.: Tabu search - Part II. INFORMS J. Comput. **2**(1), 4–32 (1990)
8. Hannenhalli, S., Pevzner, P.A.: Transforming cabbage into turnip: polynomial algorithm for sorting signed permutations by reversals. J. ACM **46**(1), 1–27 (1999)
9. Jiang, H., Pu, L., Qingge, L., Sankoff, D., Zhu, B.: A randomized FPT approximation algorithm for maximum alternating-cycle decomposition with applications. In: Wang, L., Zhu, D. (eds.) COCOON 2018. LNCS, vol. 10976, pp. 26–38. Springer, Cham (2018). https://doi.org/10.1007/978-3-319-94776-1_3
10. Lin, G., Jiang, T.: A further improved approximation algorithm for breakpoint graph decomposition. J. Comb. Optim. **8**(2), 183–194 (2004). https://doi.org/10.1023/B:JOCO.0000031419.12290.2b
11. Oliveira, A.R., Brito, K.L., Dias, U., Dias, Z.: On the complexity of sorting by reversals and transpositions problems. J. Comput. Biol. **26**, 1223–1229 (2019)
12. Rahman, A., Shatabda, S., Hasan, M.: An approximation algorithm for sorting by reversals and transpositions. J. Discret. Algorithms **6**(3), 449–457 (2008)
13. Yancopoulos, S., Attie, O., Friedberg, R.: Efficient sorting of genomic permutations by translocation inversion and block interchange. Bioinformatics **21**(16), 3340–3346 (2005)

Center Genome with Respect
to the Rank Distance

Priscila Biller[1,2] (iD), João Paulo Pereira Zanetti[4] (iD), and João Meidanis[1,3(✉)] (iD)

[1] Institute of Computing, University of Campinas, Campinas, Brazil
meidanis@unicamp.br
[2] Department of Mathematics, Simon Fraser University, Burnaby, Canada
[3] Scylla Bioinformatics, Campinas, Brazil
[4] R&D Seismology and Acoustics, Royal Netherlands Meteorological Institute
(KNMI), De Bilt, Netherlands

Abstract. The rank distance between matrices has been applied to genome evolution, specifically in the area of genome rearrangements. It corresponds to looking for the optimal way of transforming one genome into another by cuts and joins with weight 1 and double-swaps with weight 2. In this context, the genome median problem, which takes three genomes A, B, and C and aims to find a genome M such that $d(A, M) + d(B, M) + d(C, M)$ is minimized, is relevant. This problem can be stated for any genomic distance, not just the rank distance. In many cases, the genome median problem is NP-hard, but a number of approximate methods have been developed.

Here we examine a related problem, the so-called center genome problem, where we aim to minimize the maximum (instead of the sum) of pairwise distances between the center genome and the inputs. We show that, for the rank distance, and for two genomic inputs A and B, it is not possible to always attain the well-known lower bound $\lceil d(A, B)/2 \rceil$. The issue arises when A and B are co-tailed genomes (i.e., genomes with the same telomeres) with $d(A, B)$ equal to twice an odd number, when the optimal attainable score is 1 unit larger than the lower bound. In all other cases, we show that the lower bound is attained.

Keywords: Genome rearrangements · Genome matrices

1 Introduction

The rank distance between matrices has been very successfully used in coding theory since at least 1985, when Gabidulin published his discoveries in matrix codes [5]. Recently, applications of the rank distance to genome evolution, specifically in the area of genome rearrangements, started to emerge [9]. In this context, the genome median problem, which takes a number of genomes A_1, A_2, ..., A_k and aims to find a genome M such that $d(A_1, M) + d(A_2, M) + \ldots + d(A_k, M)$ is minimized, is relevant. This problem can be stated for any genomic distance, not

© Springer Nature Switzerland AG 2020
J. C. Setubal and W. M. Silva (Eds.): BSB 2020, LNBI 12558, pp. 141–149, 2020.
https://doi.org/10.1007/978-3-030-65775-8_13

just the rank distance. In many cases, the genome median problem is NP-hard, but a number of approximate methods have been developed.

With regard to genome medians, much work has been published, especially in the case of exactly three inputs. This is one of the seminal steps in building phylogenetic trees. Finding a genome median is NP-hard for several genome distances, with the exception of SCJ and breakpoint for multichromosomal genomes. [4,8].

Center genomes, also called closest genomes or minimax genomes, are also aimed at somehow representing all the inputs, as a sort of average genome. The center genome problem takes genome inputs A_1 , A_2, ..., A_k and looks for a genome M minimizing $\max(d(A_1, M), d(A_2, M), \ldots, d(A_k, M))$. There is an important difference between using central genomes and median genomes as subroutines for ancestral reconstruction methods: when just two inputs are used for the median, the solution will probably be not very relevant, because many solutions exist, including both input genomes and anything in an optimal path from one to the other; on the other hand, the center genome, even with just two inputs, is already restricted enough to be relevant with respect to ancestral genomes.

For any distance defined as the minimum number of operations, when all operations have the same weight, clearly the theoretical lower bound for two genomes is readily achievable: it suffices to start at one of the genomes and walk towards the other, stopping when the right number of steps have been performed. However, if an arbitrary number of inputs is allowed, the problem becomes NP-hard, even for very simple distances such as the SCJ [2].

In contrast, distance measures where operations have distinct weights may not be able to always attain the lower bound. Here we concentrate on two inputs and examine the rank distance, which can be defined as the rank of $A - B$ for genomes (matrices) A and B, but also as the minimum number of cuts, joins, and double swaps, with weights 1, 1, and 2, respectively, that bring one genome to the other. Since we have different weights, it is not obvious the lower bound can be achieved. In fact, we show that it cannot in the case where $d(A, B) = 2n$ with n odd. In all other cases, the lower bound is achieved.

The rest of this paper is organized as follows. Section 2 contains the definitions used throughout the text. Section 3 presents the results. Finally, Sect. 4 summarizes our work and points to possible continuation of this research.

2   Definitions

We will represent genomes as matrices. For a genome G involving n genes and therefore $2n$ gene extremities, we choose an ordering for the extremities (any ordering is fine), and then define the corresponding **genome matrix** as follows:

$$G_{ij} = \begin{cases} 1 \text{ if } i \neq j \text{ and extremities } i \text{ and } j \text{ are adjacent in } G, \text{ or} \\ \quad \text{ if } i = j \text{ and extremity } i \text{ is a telomere in } G \\ 0 \text{ if } i \neq j \text{ and extremities } i \text{ and } j \text{ are } \textbf{not} \text{ adjacent in } G, \text{ or} \\ \quad \text{ if } i = j \text{ and extremity } i \text{ is } \textbf{not} \text{ a telomere in } G \end{cases}$$

For genomes with just one gene, we have just two extremities. There are only two genomes: one with an adjacency linking these two extremities, and the other with just telomeres. Here are some examples of genomes over two genes:

$$C = \begin{bmatrix} 0 & 1 & 0 & 0 \\ 1 & 0 & 0 & 0 \\ 0 & 0 & 1 & 0 \\ 0 & 0 & 0 & 1 \end{bmatrix}, D = \begin{bmatrix} 1 & 0 & 0 & 0 \\ 0 & 1 & 0 & 0 \\ 0 & 0 & 0 & 1 \\ 0 & 0 & 1 & 0 \end{bmatrix}$$

Genome matrices are therefore square matrices of size $(2n) \times (2n)$ and have the following properties:

- They are **binary** matrices, i.e., have 0's and 1's only.
- They are **symmetric** matrices, that is, they satisfy $A^\top = A$.
- They are **orthogonal** matrices, that is, they satisfy $A^\top = A^{-1}$.
- They are **involutions**, that is, they satisfy $A^2 = I$.

It is easy to verify that any two of the last three properties implies the third one. For binary matrices, being an orthogonal matrix is equivalent to having just one 1 in each row and in each column. Such binary matrices are called **permutation matrices**. We can then say that genome matrices are permutation matrices that are involutions.

Extremities x such that $Ax = x$ are called **telomeres** of A. A genome with no telomeres is called **circular**. Two genomes with exactly the same set of telomeres are called **co-tailed**.

3   Results

We recall a lower bound for the score relative to two genomes, and show exactly the cases where it is possible to achieve such a score. We also show that, in any case, it is always possible to find a genome within 1 unit of the lower bound.

We start by recalling the notion of **intermediate genomes**, defined as genomes that appear in an optimal scenario between two genomes A and B. The definition depends on A and B, so sometimes we will call them AB-intermediates for improved clarity. Although initially defined for DCJ [3], the definition works for any distance.

In addition to being optimal scenario members, intermediate genomes can be characterized as those for which the triangle inequality becomes an equality. They are also the medians of two genomes.

Given two genomes A, and B, a **center** genome for them is a genome M that minimizes the **score** $sc(M; A, B)$, defined as:

$$sc(M; A, B) = \max(d(A, M), d(B, M)).$$

The triangle inequality gives almost immediately a lower bound on the score:

Lemma 1. *For any three genomes A, B, and M we have:*

$$\mathrm{sc}(M; A, B) \geq \frac{d(A, B)}{2}.$$

Proof. Notice that:

$$d(A, B) \leq d(A, M) + d(B, M) \leq 2\max(d(A, M), d(B, M)) = 2\mathrm{sc}(M; A, B).$$

From this, the statement easily follows.

In fact, since the score is always an integer, we can strengthen this result and claim that:

$$\mathrm{sc}(M; A, B) \geq \left\lceil \frac{d(A, B)}{2} \right\rceil. \tag{1}$$

It would be tempting to state the following conjecture:

Conjecture 1. *For any two genomes A and B over the same genes, there is at least one genome M over the same genes that satisfies:*

$$d(A, M) = \lceil d(A, B)/2 \rceil$$

and

$$d(B, M) = \lfloor d(A, B)/2 \rfloor.$$

This genome would of course be a center genome, since it would attain the lower bound established in Eq. 1. However, this is false, as can be seen from the following example representing genomes that differ by a double swap:

$$A = \begin{bmatrix} 0 & 1 & 0 & 0 \\ 1 & 0 & 0 & 0 \\ 0 & 0 & 0 & 1 \\ 0 & 0 & 1 & 0 \end{bmatrix}, B = \begin{bmatrix} 0 & 0 & 1 & 0 \\ 0 & 0 & 0 & 1 \\ 1 & 0 & 0 & 0 \\ 0 & 1 & 0 & 0 \end{bmatrix}.$$

To compute their distance, let's subtract B from A:

$$A - B = \begin{bmatrix} 0 & 1 & -1 & 0 \\ 1 & 0 & 0 & -1 \\ -1 & 0 & 0 & 1 \\ 0 & -1 & 1 & 0 \end{bmatrix}.$$

This matrix has rank 2. Both A and B are circular genomes, since they do not have telomeres. Now for a circular genome such as A, the only genomes at distance 1 from it are the ones obtained by cutting an adjacency, since no extra adjacencies can be added to A. Genome A has only two adjacencies, so there are just two genomes at distance 1 from it, namely:

$$A_1 = \begin{bmatrix} 1 & 0 & 0 & 0 \\ 0 & 1 & 0 & 0 \\ 0 & 0 & 0 & 1 \\ 0 & 0 & 1 & 0 \end{bmatrix}, A_2 = \begin{bmatrix} 0 & 1 & 0 & 0 \\ 1 & 0 & 0 & 0 \\ 0 & 0 & 1 & 0 \\ 0 & 0 & 0 & 1 \end{bmatrix}.$$

However, it can be readily verified that none of these two genomes is at distance 1 from B. In fact, they are both at distance 3 from B. We conclude that the center conjecture is **not** true.

3.1   Co-tailed Genomes

When A and C are co-tailed, we do not always get a center genome satisfying the lower bound, but we can get within 1 unit of it. This case opens up the possibility of center genomes that are not intermediates, e.g., in the example of Sect. 3, the identity matrix is a center genome, but not an intermediate. Let's begin by studying properties of intermediate genomes between two co-tailed ones.

Lemma 2. *If A and C are co-tailed genomes and B is an intermediate genome between A and C, then B is co-tailed with A and C.*

Proof. It suffices to show that B is co-tailed with A. Suppose for a moment that B is not co-tailed with A. Then either A has a telomere that B doesn't, or B has a telomere that A doesn't. The first case is ruled out by Corollary 1 of a paper by Chindelevitch and Meidanis [1], because a telomere of A would also be a telomere of C, since they are co-tailed, and would have to be shared by all AC-intermediate genomes.

So let's assume that B has a telomere x not shared by A. In this case, at some point in an optimal operation series going from A to B, there must be a cut. However, any optimal such series can be extended to a sorting series going from A to C, since B is intermediate between A and C. However, no cuts can be present in an optimal scenario linking co-tailed genomes [6]. This shows that B cannot have telomeres not shared with A and C.

Only double swaps occur in optimal sorting scenarios of co-tailed genomes. This leads to a parity restriction.

Lemma 3. *If A and C are co-tailed genomes, and $\mathcal{L} = [B_0, B_1, \ldots, B_k]$ is an optimal scenario going from A to C, then $d(A, C) = 2k$.*

Proof. According to Lemma 2, all B_i's are co-tailed with A, so none of the operations $B_{i+1} - B_i$ can be cuts or joins. Therefore, we have $r(B_{i+1} - B_i) = 2$ for $0 \leq i \leq k - 1$. But then

$$d(A, C) = w(\mathcal{L}) = \sum_{i=0}^{k-1} r(B_{i+1} - B_i) = \sum_{i=0}^{k-1} 2 = 2k.$$

It is easy to find center genomes for co-tailed genomes whose distance is a multiple of 4. However, if their distance is not divisible by 4, we are forced to take the second best, which is 1 unit off the lower bound.

Lemma 4. *If A and C are co-tailed genomes and $d(A, C)/2$ is even, then there is a genome B satisfying the center lower bound.*

Proof. Let $[B_0, B_1, \ldots, B_k]$ be an optimal scenario going from A to C. We know that $d(A, C) = 2k$ from Lemma 3. Since $k = d(A, C)/2$ is even, we can write $k = 2m$ for some integer m. It is then straightforward to verify that B_m is the sought AC-intermediate genome satisfying the center lower bound.

Lemma 5. *If A and C are co-tailed genomes and $d(A, C)/2$ is odd, then there is no genome B satisfying the center lower bound.*

Proof. If such a genome B existed, then we would have:

$$d(A, B) = d(B, C) = d(A, C)/2.$$

This implies that B would be an intermediate genome between A and C. By Lemma 2, B would be co-tailed with A. But then, by Lemma 3, $d(A, B) = d(A, C)/2$ would have to be even, contradicting the hypothesis.

Lemma 6. *For any two genomes A and C, there is an intermediate genome B such that*

$$\lceil d(A, C)/2 \rceil \leq d(A, B) \leq \lceil d(A, C)/2 \rceil + 1.$$

Proof. If $A = C$ the result is clear taking $B = A$. If $A \neq C$, let $[B_0, B_1, \ldots, B_k]$ be an optimal scenario going from A to C and take i as the smallest index such that $d(A, B_i) \geq \lceil d(A, C)/2 \rceil$. We claim that $B = B_i$ is the sought genome. Notice that B is an intermediate genome between A and C because it is a member of an optimal scenario going from A to C. Moreover, the first inequality in the lemma statement is satisfied because of the choice of i.

For the second equality, notice that, by the minimality of i, we have:

$$d(A, B_{i-1}) < \lceil d(A, C)/2 \rceil.$$

Genome B_{i-1} exists since $A \neq C$ implies $\lceil d(A, C)/2 \rceil \geq 1$, so i cannot be zero. Given that in any scenario the steps have weight 1 or 2, we know that $d(B_{i-1}, B_i) \leq 2$. It follows that

$$d(A, B_i) \leq d(A, B_{i-1}) + d(B_{i-1}, B_i) \leq d(A, B_{i-1}) + 2 < \lceil d(A, C)/2 \rceil + 2$$

or

$$d(A, B_i) \leq \lceil d(A, C)/2 \rceil + 1,$$

since both sides are integers.

3.2   Genomes Not Co-tailed

If A and C are not co-tailed, then there are AC-intermediate genomes at any feasible distance between A and C. To ascertain that, we need a few preliminary lemmas on operation switch and other properties.

Lemma 7. *Let A be a genome, P a cut applicable to A, and Q a double swap applicable to $A + P$. Then Q is applicable to A.*

Proof. Let $Q = W(x, y, z, w)$. We know that Q is applicable to $A + P$, which means that $A + P$ has adjacencies xw and yz. Since P is a cut, which only removes adjacencies, xw and yz must have been present in A as well, leading to the conclusion that Q can be applied to A.

An analogous result is valid for joins, saying that joins can be brought back through double swaps, but we won't need it now.

Lemma 8. *Let A and C be two genomes not co-tailed. Then, for every integer i such that $0 \leq i \leq d(A, C)$ there is an intermediate genome B between A and C with $d(A, B) = i$.*

Proof. By induction on $d(A, C)$. The base case is $d(A, C) = 1$, because A and C are not co-tailed and hence cannot be equal. The statement is clearly true for $d(A, C)$ because in this case we only have two possibilities for i, namely, $i = 0$ or $i = 1$, and we can take $B = A$ for $i = 0$ and $B = C$ for $i = 1$.

Now assume $d(A, C) \geq 2$ and consider an integer i such that $0 \leq i \leq d(A, C)$. Since A and C are not co-tailed, there is either a telomere in A not shared by C or a telomere in C not shared by A. Without loss of generality, we may assume that there is a telomere in C not shared by A, otherwise we can just exchange A and C and i with $d(A, C) - i$.

Given that there is a telomere in C that is not an A-telomere, destroying the adjacency of x in A gives us a cut P applicable to A such that $A + P$ is an intermediate genome between A and C. If $A + P$ is not co-tailed with C, we can apply the induction hypothesis to $A + P$ and C and get intermediate genomes at an arbitrary distance j from $A + P$, provided that $0 \leq j \leq d(A + P, C) = d(A, C) - 1$, which will be at distance $j + 1$ from A. This covers all the distances we need except 0, for which we can take $B = A$.

Now if $A + P$ is co-tailed with C, then they are distinct, since $d(A, A + P) = 1$ and $d(A, C) \geq 2$. Co-tailed genomes can be sorted by double swaps, so there is a double swap Q applicable to A yielding an intermediate genome $A + P + Q$ between $A + P$ and C. However, according to Lemma 7, a cut can go forward past a double swap, which means that Q is applicable to A. The resulting genome, $A + Q$, is intermediate between A and C because $A + Q + P$ is just another way of getting to $A + P + Q$, which we know is intermediate between A and C. We can then apply the induction hypothesis to $A + Q$ and C, which are not co-tailed since $A + Q$ is co-tailed with A, obtaining intermediate genomes at distances i from A for $2 \leq i \leq d(A, C)$. For $i = 0$ we have A, and for $i = 1$ we have $A + P$. This completes the induction step and the proof of our lemma.

3.3   Main Result

Theorem 1. *Let A and C be arbitrary genome matrices over the same genes. Then:*

1. *If A and C are not co-tailed, then there is a genome matrix B such that:*

$$d(A, B) = \left\lceil \frac{d(A, C)}{2} \right\rceil$$

and

$$d(B, C) = \left\lfloor \frac{d(A, C)}{2} \right\rfloor .$$

2. *If A and C are co-tailed and d(A, C) is a multiple of 4, then there is a genome matrix B such that:*

$$d(A, B) = \frac{d(A, C)}{2}$$

and

$$d(B, C) = \frac{d(A, C)}{2}.$$

3. *If A and C are co-tailed and d(A, C) is not a multiple of 4, then there is no genome matrix B such that:*

$$d(A, B) = \frac{d(A, C)}{2}$$

and

$$d(B, C) = \frac{d(A, C)}{2}.$$

However, there is a genome matrix B such that:

$$d(A, B) = \frac{d(A, C)}{2} + 1$$

and

$$d(B, C) = \frac{d(A, C)}{2} - 1.$$

Proof. Part 1 is a consequence of Lemma 8, since $0 \leq \lceil d(A, C)/2 \rceil \leq d(A, C)$. Part 2 is a consequence of Lemma 4. Part 3 is a consequence of Lemmas 5 and 6.

4   Conclusions

In this paper we showed that center genomes do not always attain the theoretical lower bound in the case of two genomes, with respect to the rank distance. In spite of that, their are easy to calculate, and provide an attractive alternative to the median in ancestral genome reconstruction, even in the two-input version, which is already more restrictive than its median counterpart. Given that computing a median is NP-hard for the majority of relevant distances, its replacement by a center solution would bring a significant gain.

Nevertheless, it would be interesting to extend this analysis to three inputs, and determine what happens there. Probably the arbitrary input version is NP-hard, as similar problems with simpler distances have already been proved NP-hard [2,7]. In addition, considering genomes with unequal gene content would also be worthwhile.

Acknowledgements. We thank funding agency FAPESP (Brazil) for financial support (Grant numbers 2012/13865-7, 2012/14104-0, and 2018/00031-7). PB would also like to acknowledge the Canada 150 Research Chair program.

References

1. Chindelevitch, L., Zanetti, J.P.P., Meidanis, J.: On the rank-distance median of 3 permutations. BMC Bioinform. **19**(Suppl 6), 142 (2018). https://doi.org/10.1186/s12859-018-2131-4
2. Cunha, L.F.I., Feijão, P., dos Santos, V.F., Kowada, L.A., de Figueiredo, C.M.H.: On the computational complexity of closest genome problems. Discret. Appl. Math. **274**, 26–34 (2020)
3. Feijão, P.: Reconstruction of ancestral gene orders using intermediate genomes. BMC Bioinform. **16**(Suppl 14), S3 (2015). https://doi.org/10.1186/1471-2105-16-S14-S3
4. Feijão, P., Meidanis, J.: SCJ: a breakpoint-like distance that simplifies several rearrangement problems. Trans. Comput. Biol. Bioinform. **8**, 1318–1329 (2011)
5. Gabidulin, E.M.: Theory of codes with maximum rank distance. Probl. Peredachi Inf. **21**(1), 3–16 (1985)
6. Meidanis, J., Biller, P., Zanetti, J.P.P.: A matrix-based theory for genome rearrangements. Technical Report, IC-18-10. Institute of Computing, University of Campinas, August 2018
7. Popov, V.: Multiple genome rearrangement by swaps and by element duplications. Theor. Comput. Sci. **385**(1–3), 115–126 (2007)
8. Tannier, E., Zheng, C., Sankoff, D.: Multichromosomal median and halving problems under different genomic distances. BMC Bioinform. **10**(1), 120 (2009). https://doi.org/10.1186/1471-2105-10-120
9. Zanetti, J.P.P., Biller, P., Meidanis, J.: Median approximations for genomes modeled as matrices. Bull. Math. Biol. **78**(4), 786–814 (2016). A preliminary version appeared on the Proceedings of the Workshop on Algorithms for Bioinformatics (WABI) 2013

ImTeNet: Image-Text Classification Network for Abnormality Detection and Automatic Reporting on Musculoskeletal Radiographs

Leodécio Braz[ID], Vinicius Teixeira[ID], Helio Pedrini[(✉)][ID], and Zanoni Dias[ID]

Institute of Computing, University of Campinas, Campinas, Brazil
{leodeciobraz,viniciusteixeira}@liv.ic.unicamp.br,
{helio,zanoni}@ic.unicamp.br

Abstract. Deep learning techniques have been increasingly applied to provide more accurate results in the classification of medical images and in the classification and generation of report texts. The main objective of this paper is to investigate the influence of fusing several features of heterogeneous modalities to improve musculoskeletal abnormality detection in comparison with the individual results of image and text classification. In this work, we propose a novel image-text classification framework, named ImTeNet, to learn relevant features from image and text information for binary classification of musculoskeletal radiography. Initially, we use a caption generator model to artificially create textual data for a dataset lacking text information. Then, we apply the ImTeNet, a multi-modal information model that consists of two distinct networks, DenseNet-169 and BERT, to perform image and text classification tasks respectively, and a fusion module that receives a concatenation of feature vectors extracted from both. To evaluate our proposed approach, we used the Musculoskeletal Radiographs (MURA) dataset and compare the results obtained with image and text classification scheme individually.

Keywords: Deep learning · Musculoskeletal abnormalities · X-ray

1 Introduction

Musculoskeletal disorders represent a major health problem that affects a large part of the population. The X-ray images are one of the most commonly accessible radiological examinations used to detect and locate abnormalities in radiographic studies. The process of interpreting image examinations is a typically complex task. Specialist doctors are usually responsible for conducting the interpretation of these types of information [8]. The first step is to read and analyze the images in order to have a knowledge base to write a report on what is in the image and provide a diagnosis of normal or abnormal, for example.

Determining whether a radiographic study is normal or abnormal is a challenging problem. If a study is interpreted as normal, the possibility of disease is

© Springer Nature Switzerland AG 2020
J. C. Setubal and W. M. Silva (Eds.): BSB 2020, LNBI 12558, pp. 150–161, 2020.
https://doi.org/10.1007/978-3-030-65775-8_14

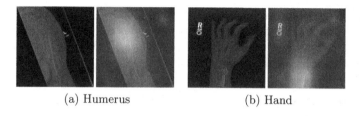

(a) Humerus   (b) Hand

Fig. 1. Original images from MURA dataset and corresponding heatmaps.

excluded, which may eliminate the need for the patient to undergo other tests or procedures [13]. In Fig. 1, we present two examples of images from MURA dataset and fracture location using activation maps.

To assist in the diagnostic process, many automated computational methods, such as Computer-Aided Diagnosis (CAD) have been explored [10]. With advances in techniques such as natural language processing (NLP) and computer vision (CV), various deep learning methods have been developed to automatically interpret medical images and reports in order to assist clinicians who are subjected to these tasks on a daily basis.

More recently, transformers have shown success in labeling radiological images [5,16]. However, using these methods, a large amount of resources are required to annotate the data manually to obtain a higher classification score. Several methods of deep learning have been developed for the task of classification, localization and interpretation of radiology images. Recent advances in deep convolutional neural network (DCNN) architectures have improved the performance of CAD systems, which support health experts [14]. The advancement in hardware and software development has made it possible that the amount of medical data collected per patient has increased considerably [12].

In this work, we propose an image-text classification network, called ImTeNet, for the automatic detection and notification of fractures on musculoskeletal radiographs. Our method uses information extracted from images and artificially generated captions to obtain a classification label. Initially, we use a caption generator model to create textual data for a dataset without text information. Then, we apply our multi-modal information model, which consists of two distinct networks, DenseNet-169 and BERT, to perform image and text classification tasks respectively, and a fusion module that receives a concatenation of feature vectors extracted from both.

Our main contributions are summarized as follows: (i) the proposed method is applied with a clinical objective for diagnostic abnormality recognition, (ii) distinct classifiers based on deep learning are trained with radiographs and texts for abnormality classification and (iii) experimental results on the MURA dataset show that the combination of image and text features can increase the classification results.

The text is organized as follows. Section 2 describes some relevant approaches related to the topic under investigation. Section 3 presents the proposed image-text classification network for musculoskeletal abnormality detection. Section 4 describes the dataset used in the experiments and some implementation details. Section 5 describes and evaluates the experimental results. Section 6 concludes the paper with some final considerations.

2   Related Work

Deep learning methods are the state of the art for classification tasks in the medical field. Moreover, Deep Convolutional Neural Networks (DCNNs) are widely used mainly for image domains [3,13,21]. Many Natural Language Processing (NLP) methods have been developed to extract structured labels from free-text medical reports [1,12]. Our work is closely related to approaches that explore the use of DCNNs on medical images, with the aim of extracting relevant information, as well as approaches that explore textual medical information and the combination of both.

Wang et al. [21] explored the use of many pre-trained DCNN models to perform a multi-label classification of thoracic disease on chest X-ray images. Chen et al. [3] proposed a dual asymmetric feature learning network called DualCheXNet to explore the complementarity and cooperation between the two networks to learn discriminative features. Rajpurkar et al. [13] proposed the Musculoskeletal Radiographs (MURA) dataset and explored the use of a DenseNet-169 [7] to perform a binary classification task.

Smit et al. [16] proposed a method, called CheXbert, for radiology report labeling, combining existing report labelers with hand-annotations to obtain accurate results. They used a BERT [4] model and obtained state-of-the-art results for report labeling on the chest X-ray datasets. Pelka et al. [12] proposed an approach to combine automatically generated keywords with radiographs, presenting a method to allow multi-modal image representations by fusing textual information with radiographic images to perform body part and abnormality recognition on MURA dataset. In contrast to these works, our approach explores the cooperation between image and text networks to learn the classification of abnormalities in a complementary and accurate way.

3   Proposed Method

In this section, we present details of the proposed method. We provide some notations used throughout this paper, as well as describe the different tasks performed and how we combine each one to produce our proposed architecture.

3.1   Caption Generation

Since text representations are not available on the MURA [13] dataset, we used a caption generator model, trained on the Radiology Objects in COntext (ROCO) [11] dataset, to automatically generate medical reports for the images on MURA dataset.

The ROCO dataset contains 81,825 radiology images including several medical imaging modalities, such as X-ray, Computer Tomography (CT), Ultrasound, Mammography (UM), Magnetic Resonance Imaging (MRI) and various other modalities. All images on ROCO have a corresponding caption information.

For this caption generation task, inspired by Pelka et al. [12], we propose an encoder-decoder framework [8,20]. As an encoder, we use a Convolutional Neural Network (CNN) model, called ResNet-152 [6], pretrained on ImageNet [15]. The encoder extracts the feature vector from an input image and this feature vector is linearly transformed to be the input to the decoder network. The decoder is a Long Short-Term Memory (LSTM) network, which receives as input the feature vector from the encoder and produces the image caption as output.

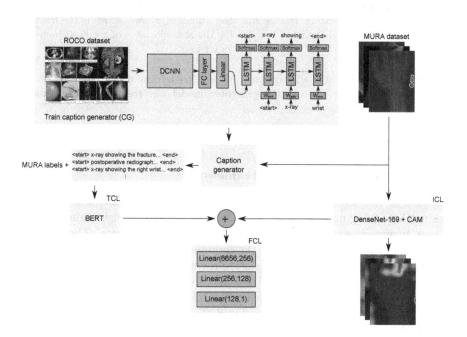

Fig. 2. Overview of the proposed method workflow. A DenseNet-169 network and a BERT model are used to learn from image and text representation, respectively, given an image and its corresponding caption. The output features of each network are concatenated in the fusion classifier and the outputs of each classifier are averaged to provide the final output prediction.

Figure 2 shows the caption generation model (CG) scheme. We trained the model using a corpus of paired image and captions from the ROCO dataset[1]. After the model training step, we construct a dataset, called MURA_caption, executing the caption generator for each image on MURA. At the end of this task, we have a combination of an original image and label from MURA and an

---

[1] No additional datasets were used for training.

automatically generated caption associated with it. Figure 3 shows examples of these captions for four radiographs from the MURA training set.

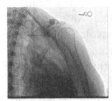  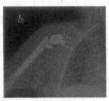 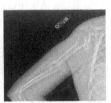

Set: Shoulder
Label: Normal
Caption: the right shoulder is normal

Set: Wrist
Label: Normal
Caption: X-ray showing the of the right wrist

Set: Shoulder
Label: Abnormal
Caption: X-ray of the shoulder showing the acromioclavicular joint

Set: Humerus
Label: Abnormal
Caption: X-ray showing the fracture of the right humerus

Fig. 3. Examples of generated captions. All of this information is present on the MURA_caption dataset. The caption generator model was trained using all images from the ROCO dataset.

3.2   Image-Text Classification Framework

Our focus in this work is to develop and evaluate an approach based on deep learning that, using multi-modal information, can make the detection of abnormalities in musculoskeletal radiograph samples more accurate. The intuition behind our proposed method is that the combination of visual and text information will benefit the classification task, improving its results. Figure 2 illustrates the main steps of our architecture. After the caption generation task, three steps are applied: (i) Image Classification Level (ICL), (ii) Text Classification Level (TCL), and (iii) Fusion Classification Level (FCL).

1. ICL Step: At this step, we trained a DCNN model, called DenseNet-169 [7], pre-trained on ImageNet [15], to perform an image classification. We also used this model as a feature extractor. In this model, we added an attention module called Class Activation Mapping (CAM) [22], which is employed to indicate the discriminative region detected by the DCNN model to identify the correct class. After this attention module, we applied an Average-Max (AVG-MAX) pooling layer to reduce computational complexity and extract low (average) and high (max) level features from the neighborhood. The AVG-MAX pooling is the concatenation of the average pooling and max pooling results.

2. TCL Step: In this step, we proposed a Natural Language Processing (NLP) approach to extract structured labels from free-text image captions. We fine-tuned the BERT [4] model to perform a text classification. The BERT model architecture is based on a multilayer bidirectional transformer [18], and pre-trained in an unsupervised way in two tasks: masked language model and next sentence prediction [2]. Our proposed method follows the same architecture as BERT. Each image caption text is tokenized, and the maximum number of tokens in each input sequence is limited to 64. The final-layer hidden state is then fed as input to each of the BERT linear heads. We changed the BERT output dimension to 1 to cover our binary classification problem. We also used this model as a feature extractor and then applied an Average-Max (AVG-MAX) pooling operation in hidden state layers.

3. FCL Step: At this step, the output features from ICL and TCL models are concatenated into a single feature vector used as input to the FCL. In the fusion classifier, the M input features are fed into three dense layers. As a loss function, we used the Binary Cross Entropy Loss (BCEL).

4 Experimental Setup

In this section, we present details about the dataset used in our experiments, as well as some implementation details.

4.1 Musculoskeletal Radiographs (MURA) for Abnormality Classification

Musculoskeletal Radiographs (MURA) is a large dataset of bone X-rays [13]. MURA dataset consists of 14,863 musculoskeletal studies of the upper extremity, from 12,173 patients, containing a total of 40,561 multi-view radiographic images. All radiographs presented on MURA belong to one of the seven standard upper extremities (elbow, finger, forearm, hand, humerus, shoulder and wrist) and were labeled as normal or abnormal by board-certified radiologists from the Stanford Hospital [13]. Figure 4 shows radiographs representing each of the seven anatomy classes.

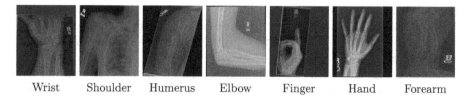

Wrist   Shoulder   Humerus   Elbow   Finger   Hand   Forearm

Fig. 4. Examples of radiographs present on MURA dataset. The images 'Forearm', 'Humerus' and 'Shoulder' belong to the abnormality positive class, whereas the images 'Elbow', 'Wrist', 'Finger' and 'Hand' belong to the abnormality negative class. All images were randomly chosen from the MURA training set.

The original dataset was split into training set(11,184 patients, 13,457 studies, 36,808 images), validation set (783 patients, 1,199 studies, 3,197 images), and test set (206 patients, 207 studies, 556 images).

For the purpose of comparing our results and evaluating our proposed method, as the test set is not public accessible, the validation set was adopted as a test set. In addition, we split stratified the original training set into 80% and 20% to generate new training and validation sets, respectively.

4.2 Implementation Details

We implemented the proposed image-text classification framework with the PyTorch deep learning toolbox on a GTX Titan V 12 GB GPU.

For the caption generation, we used the Adam [9] optimizer with a batch size of 64, learning rate of $lr = 1e-3$, with a vocabulary size of 16,380. We employed the same hyper-parameters as those based on the work described by Vinyals et al. [20]. For the image-text classification, in the image level step, we initially resized each original image to 384×384 pixels and randomly cropped it to 320×320 pixels. In the training stage, we applied a random horizontal flip. We used the Adam optimizer with a mini-batch size of 8 and learning rate of $lr = 1e-4$, reducing by a factor of $1e-1$ when the validation loss reaches a plateau.

In the text level step, the BERT model was trained using a Mean-Square loss, as we changed the default number of labels to 1 and Adam optimization with a learning rate of $lr = 2e-5$, as used by Devlin et al. [4] for fine-tuning tasks. Finally, in the fusion level step, we received the feature vectors from image and text levels and concatenated them, resulting in a single feature vector of size 6,656. For the training fusion model, we used the same hyper-parameter as the image level.

5 Results and Discussion

Our proposed model takes one or more views for a study case as input. In each view, our method predicts the probability of abnormality. We compute the abnormality classification for the study by taking the majority abnormality classification occurrence output by network for each image.[2] The model makes the binary prediction of abnormal if the probability of abnormality for the image is greater than 0.5.

To assess the effectiveness of the entire proposed method, we compare the results of the image, text and fusion classifier level individually and the proposed ImTeNet architecture, which performs a majority vote of labels of the three aforementioned classifiers. In Table 1, we report the performance of only image and text classification in the DenseNet-169 and BERT models, respectively. In addition, we report the results obtained with the proposed fusion classifier and with our final prediction approach.

For 'Elbow' and 'Wrist' studies, the performance of ImTeNet is lower than the fusion classifier individually and both performance rates are higher than

---

[2] If the normal and abnormal classification occurrences are equal, we perform an arithmetic mean of the probabilities.

Table 1. We compare the performance of the final predictions of DenseNet-169, BERT, Fusion and our ImTeNet, applying a voting method among the three proposed classifiers, with Balanced Accuracy and Cohen's kappa statistic. We highlight the best (**bold**) and the second best (<u>underline</u>) metric rates in each type of the studies and in aggregate.

	DenseNet-169		BERT		Fusion		ImTeNet	
	Balanced accuracy	Kappa	Balanced accuracy	Kappa	Balanced accuracy	Kappa	Balanced accuracy	Kappa
Elbow	0.8537	0.7214	0.6322	0.2562	**0.8731**	**0.7512**	<u>0.8634</u>	<u>0.7364</u>
Finger	0.8222	0.6521	0.6747	0.3446	**0.8300**	**0.6651**	<u>0.8294</u>	<u>0.6647</u>
Forearm	0.8292	0.6652	0.5354	0.0705	<u>0.8471</u>	<u>0.6974</u>	**0.8526**	**0.7114**
Hand	0.7680	0.5722	0.4578	-0.0882	**0.7835**	**0.5935**	<u>0.7707</u>	<u>0.5745</u>
Humerus	0.9038	0.8074	0.5868	0.1730	<u>0.9039</u>	<u>0.8075</u>	**0.9113**	**0.8222**
Shoulder	0.7724	0.5456	0.5473	0.0934	<u>0.7737</u>	<u>0.5467</u>	**0.7840**	**0.5673**
Wrist	0.8584	0.7390	0.5997	0.1973	**0.8806**	**0.7772**	<u>0.8738</u>	<u>0.7672</u>
Average	0.8296	0.6718	0.5762	0.1495	**0.8417**	<u>0.6912</u>	<u>0.8407</u>	**0.6920**

the results of DenseNet-169 and BERT individual. For 'Finger' study, the performance of ImTeNet is comparable to the fusion performance, which presents the best results. The performance of the ImTeNet presents the best results for 'Shoulder', 'Humerus' and 'Forearm' studies. In the latter, in particular, our ImTeNet provides the greatest gain compared to the DenseNet-169 model. For 'Hand' study, the BERT model presents a negative Kappa, indicating that there is no agreement on its results and the performance of ImTeNet is comparable to the performance of DenseNet-169 individually, with no relevant gain, the fusion classifier individually presents the best results. The results of the BERT model, as expected due the caption format, show the worst results in all studies.

Compared to the baseline results of DenseNet-169, the proposed method presented an overall performance gain, even with artificially generated text data where most contain texts with poor quality or lacking relevant information. In general, our model was able to extract and combine discriminative features from each proposed branch (ICL and TCL). Our method obtained an average balanced accuracy and Cohen's kappa statistic score of 0.8407 and 0.6920, respectively, with a significant difference of 0.0138 and 0.0202 for the DenseNet-169 individually. We also evaluate our proposed ImTeNet and compare with another existing method for musculoskeletal abnormality classification [12] on MURA dataset, under the same ruled test set, defined in Sect. 4.1.

To allow a fair comparison, we report the evaluation results of each method with the accuracy scores, as shown in Table 2. The 'Visual' column is related to the image classification task, the 'Text' column is associated with the text classification task, whereas the last column is related to the proposed method. Compared to this baseline, our proposed ImTeNet yields a high performance for musculoskeletal abnormality classification on the MURA dataset. Our method has an accuracy score of 0.8511 with a difference of 0.0356 for the best result method by Pelka et al. [12].

Table 2. Comparison against other approaches on the MURA dataset. We compute the mean accuracy score for each method. The best result is highlighted (**bold**).

Method	Visual	Text	Proposed
Pelka et al. [12]	0.7985	0.5424	0.8155
ImTeNet	0.8418	0.5807	**0.8511**

5.1   Analysis

We analyzed specific examples from the test set and their prediction on each level, as illustrated in Fig. 5. In addition, we applied the BERTViz [19] tool to visualize the BERT model attention produced in each input caption.

In the first two examples, all levels were able to correctly label the input image and text. The BERT model correctly detected relevant information and was able to discard spurious information present in both captions. The DenseNet-169 model could also identify relevant regions in the image. On the third example, which contains poor textual information with absolutely no semantic information, the BERT model incorrectly labeled the abnormality as negative, while the DenseNet-169 labeled the abnormality as positive with a high confidence. It is also possible to observe a well-defined highlighted area of the radiograph that was most important in making the prediction. In the last example, while the DenseNet-169 model was unable to detect the presence of abnormalities in the image, the BERT model was able to distinguish relevant information from the caption and correctly label the abnormality as positive. All features and predictions performed by DenseNet-169 and BERT model were considered in the fusion level classifier, which could increase the classification results.

Notwithstanding, our study has some limitations. First, our approach relies on the existence of a good caption generator model. Second, and very related to the first, our proposed generator model is designed to deal with musculoskeletal radiograph texts, however, MURA does not have this type of data. We tried to provide this information by artificially creating textual data, however, when observing examples of generated captions, many samples do not have relevant information, such as the caption of the third image in Fig. 5, in which we expected a text containing information such as "x-ray", "implant" and "humerus", however, <start> the of the the of the the of the <end> was obtained. Third, the average length of our caption texts is lower than other datasets usually performed with the BERT model [17]. We conjecture that longer text information can increase the BERT results.

Original image	Heatmap	Generated caption and classification	Reasoning
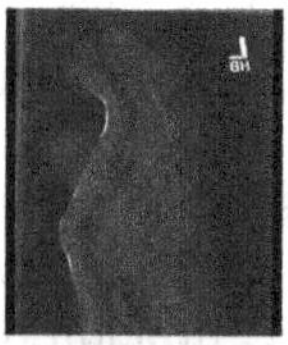	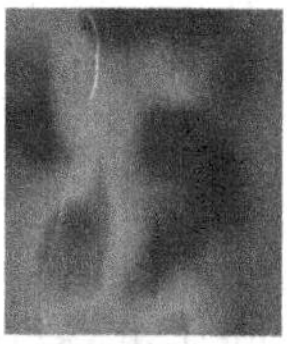	\<start\> postoperative x-ray of the patient showing the fracture of the right femur \<end\> abnormal TCL: abnormal ✓ ICL: abnormal ✓ FCL: abnormal ✓	All levels can identify the abnormality in the input data. Even with the caption missing some information, such as saying right femur, the visualization of attention shows that words such as "x-ray" and "fracture" were relevant for the BERT model.
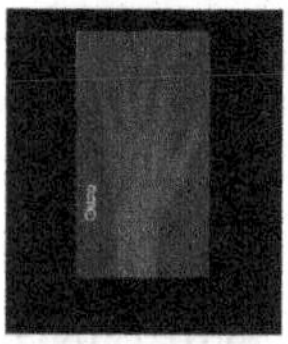	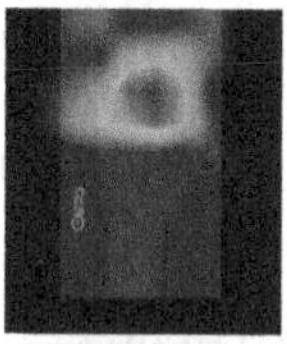	\<start\> the same case as in figure 1 the right hand is normal \<end\> normal TCL: normal ✓ ICL: normal ✓ FCL: normal ✓	All levels correctly labeled this example. The entire caption has no relevant information, but it contains the correct location of the body part and the word "normal", which was relevant for the BERT model.
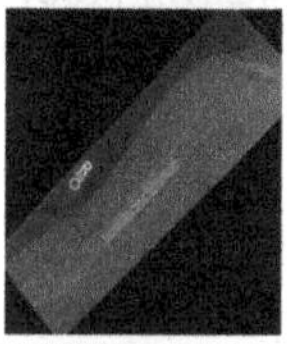	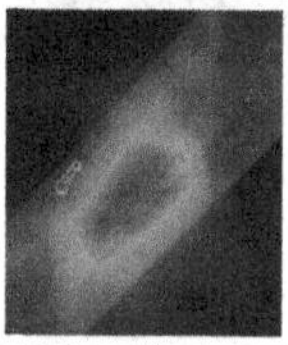	\<start\> the of the the of the the of the \<end\> abnormal TCL: normal ✗ ICL: abnormal ✓ FCL: abnormal ✓	The caption has absolutely no semantic information. The BERT model misspelled this example. DenseNet-169 produced a well-defined heatmap and correctly predicted this example with high confidence.
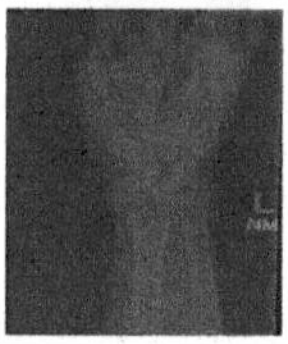	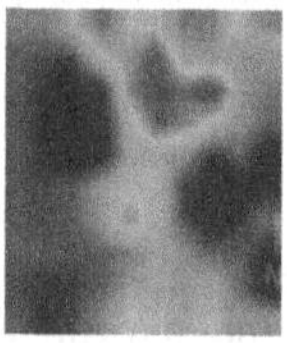	\<start\> lateral view of the ankle showing a lytic lesion in the distal tibia\<end\> abnormal TCL: abnormal ✓ ICL: normal ✗ FCL: normal ✓	DenseNet-169 was unable to identify an abnormality in this sample, not producing a well-defined heatmap. However, even the caption with some incorrect information, the word "lesion" was relevant for the Bert model to correctly label this example.

Fig. 5. Original image, DenseNet-169 heatmap and artificially generated caption from test set. A check mark indicates the correct label prediction for each level. The BERTViz tool was used to visualize the BERT attention maps for each caption.

6 Conclusions and Future Work

In this study, we proposed an automatic caption generator for musculoskeletal radiograph images and a method for automatically combining these artificially generated captions with radiographs for accurate musculoskeletal abnormality classification. A DCNN model was trained in musculoskeletal radiograph images. In parallel, the BERT model was trained in automatically generated captions associated with each musculoskeletal radiograph image. Finally, we proposed a data fusion approach, that is carried out by combining features from different heterogeneous modalities, to concatenate features extracted from DCNN and BERT models. This process allowed for an enriched multimodal data representation used as input to a fusion module. This multi-modal data representation presented the highest prediction results.

To create a caption generation model, image-caption pairs from the Radiology Objects in COntext (ROCO) dataset were adopted to train our encoder-decoder model framework. We use this caption generation model to generate text representations for the Musculosketal Radiograph (MURA) dataset. Through the proposed multimodal data representation method (ImTeNet), we outperformed the results achieved only with image and text features individually with DenseNet-169 and BERT, respectively.

The proposed work can be further improved through the exploration of better caption generator models trained with specific musculosketal text representations to avoid text quality problems, as mentioned in Sect. 5.1. In addition, other fusion methods to combine features of different heterogeneous modalities could be explored. Since there are several tasks and sources in medicine, the proposed work has great potential to be applied in other medical areas by combining metadata with other imaging modalities.

Acknowledgments. The authors would like to thank FAPESP (grants #2015/11937-9, #2017/12646-3, #2017/16246-0, #2017/12646-3 and #2019/20875-8), CNPq (grants #304380/2018-0 and #309330/2018-1) and CAPES for their financial support.

References

1. Annarumma, M., Withey, S.J., Bakewell, R.J., Pesce, E., Goh, V., Montana, G.: Automated triaging of adult chest radiographs with deep artificial neural networks. Radiology **291**(1), 196–202 (2019)
2. Beltagy, I., Lo, K., Cohan, A.: SciBERT: A Pretrained Language Model for Scientific Text. arXiv preprint arXiv:1903.10676 (2019)
3. Chen, B., Li, J., Guo, X., Lu, G.: DualCheXNet: dual asymmetric feature learning for thoracic disease classification in chest X-rays. Biomed. Signal Process. Control **53**, 101554 (2019)
4. Devlin, J., Chang, M.W., Lee, K., Toutanova, K.: BERT: pre-training of deep bidirectional transformers for language understanding. arXiv preprint arXiv:1810.04805 (2018)
5. Drozdov, I., Forbes, D., Szubert, B., Hall, M., Carlin, C., Lowe, D.J.: Supervised and unsupervised language modelling in chest X-ray radiological reports. PLoS ONE **15**(3), e0229963 (2020)
6. He, K., Zhang, X., Ren, S., Sun, J.: Deep residual learning for image recognition. In: IEEE Conference on Computer Vision and Pattern Recognition (CVPR) (2016)
7. Huang, G., Liu, Z., van der Maaten, L., Weinberger, K.Q.: Densely connected convolutional networks. In: IEEE Conference on Computer Vision and Pattern Recognition (CVPR) (2017)
8. Jing, B., Xie, P., Xing, E.P.: On the automatic generation of medical imaging reports. In: 56th Annual Meeting of the Association for Computational Linguistics - Proceedings of the Conference (Long Papers), vol. 1, pp. 2577–2586 (2018)
9. Kingma, D.P., Ba, J.: Adam: a method for stochastic optimization, pp. 1–15. arXiv preprint arXiv:1412.6980 (2014)
10. Kooi, T., et al.: Large scale deep learning for computer aided detection of mammographic lesions. Med. Image Anal. **35**, 303–312 (2017)
11. Pelka, O., Koitka, S., Rückert, J., Nensa, F., Friedrich, C.M.: Radiology objects in context (ROCO): a multimodal image dataset. In: Stoyanov, D., et al. (eds.) LABELS/CVII/STENT -2018. LNCS, vol. 11043, pp. 180–189. Springer, Cham (2018). https://doi.org/10.1007/978-3-030-01364-6_20

12. Pelka, O., Nensa, F., Friedrich, C.M.: Branding - fusion of meta data and musculoskeletal radiographs for multi-modal diagnostic recognition. In: International Conference on Computer Vision Workshop (ICCV), pp. 467–475 (2019)
13. Rajpurkar, P., et al.: MURA: large dataset for abnormality detection in musculoskeletal radiographs. arXiv preprint arXiv:1712.06957 (2017)
14. Ranjan, E., Paul, S., Kapoor, S., Kar, A., Sethuraman, R., Sheet, D.: Jointly learning convolutional representations to compress radiological images and classify thoracic diseases in the compressed domain. In: 11th Indian Conference on Computer Vision, Graphics and Image Processing, pp. 1–8 (2018)
15. Russakovsky, O., et al.: ImageNet large scale visual recognition challenge. Int. J. Comput. Vis. **115**(3), 211–252 (2015). https://doi.org/10.1007/s11263-015-0816-y
16. Smit, A., Jain, S., Rajpurkar, P., Pareek, A., Ng, A.Y., Lungren, M.P.: CheXbert: combining automatic labelers and expert annotations for accurate radiology report labeling using BERT. arXiv preprint arXiv:2004.09167 (2020)
17. Sun, C., Qiu, X., Xu, Y., Huang, X.: How to fine-tune BERT for text classification? In: Sun, M., Huang, X., Ji, H., Liu, Z., Liu, Y. (eds.) CCL 2019. LNCS (LNAI), vol. 11856, pp. 194–206. Springer, Cham (2019). https://doi.org/10.1007/978-3-030-32381-3_16
18. Vaswani, A., et al.: Attention is all you need. In: Guyon, I., et al. (eds.) Advances in Neural Information Processing Systems 30, pp. 5998–6008 (2017)
19. Vig, J.: A Multiscale visualization of attention in the transformer model, pp. 1–6. arXiv preprint arXiv:1906.05714 (2019)
20. Vinyals, O., Toshev, A., Bengio, S., Erhan, D.: Show and tell: a neural image caption generator. IEEE Comput. Soc. Conf. Comput. Vis. Pattern Recognit. **7**(12), 3156–3164 (2015)
21. Wang, X., Peng, Y., Lu, L., Lu, Z., Bagheri, M., Summers, R.M.: ChestX-ray8: hospital-scale chest X-ray database and benchmarks on weakly-supervised classification and localization of common thorax diseases. In: IEEE Conference on Computer Vision and Pattern Recognition (CVPR) (2017)
22. Zhou, B., Khosla, A., Lapedriza, A., Oliva, A., Torralba, A.: Learning deep features for discriminative localization. In: IEEE Conference on Computer Vision and Pattern Recognition (CVPR) (2016)

A Scientometric Overview of Bioinformatics Tools in the *Pseudomonas Putida* Genome Study

Alex Batista Trentin[(⊠)] [iD], Beatriz Thomas Metzner[iD], Glecio Oliveira Barros[iD], and Deborah Catharine de Assis Leite[iD]

Federal University of Technology – Paraná, Dois Vizinhos, Brazil
a.trentinx@gmail.com

Abstract. *Pseudomonas putida* is a microorganism widely used in environmental science due to its high degradation power of recalcitrant compounds. This study aimed to perform a scientometric analysis of the global panorama of publications on the *P. putida* genome, in addition to the bioinformatics tools mainly used. The growth of publications on the *P. putida* genome is continuous until 2020, being France, Spain and USA the main countries with publications on the subject. Illumina was the main sequencing platform used and, in this set of articles, 120 genomes were sequenced, 106 complete and 14 drafts. The main assembly software was SPAdes, comprising 23.3% of the articles, and NCBI PGAP was the main genome annotation tool, in 25% of the documents. Thus, this study allowed the visualization of the main bioinformatics tools used for the analysis of the *P. putida* genome, besides presenting the advance of the research on this subject, and also supporting next studies with *P. putida*.

Keywords: Bioinformatic tools · Scientometric analysis · *Pseudomonas putida*

1 Introduction

The planet is fighting against all types of environmental pollution, with the soil being one of the places most affected by environmental degradation. There is, therefore, a need to control soil pollution in order to maintain fertility as well as productivity [1, 2]. In this context, the development of technologies to remedy these degraded environments becomes indispensable [3].

Pseudomonas genera, belonging to the *Gammaproteobacteria* class, have been extensively studied over time due to their great potential for degradation and biotransformation of xenobiotic and recalcitrant compounds, and are likely to be used in various biotechnological processes of environmental recovery [4]. *Pseudomonas putida* is an ubiquitous bacterium, mainly in soil, classified as chemoorganotrophic, capable of metabolizing large carbon chains, as well as several recalcitrant pollutants [5].

From the sequencing of the genome, as well as from bioinformatics analyses, it is possible to know the proteins and metabolic pathways of *Pseudomonas putida* applied to environmental remediation and degradation of xenobiotic compounds [6]. Thus,

© Springer Nature Switzerland AG 2020
J. C. Setubal and W. M. Silva (Eds.): BSB 2020, LNBI 12558, pp. 162–167, 2020.
https://doi.org/10.1007/978-3-030-65775-8_15

using molecular level analysis it is possible to study new techniques and environmental applications for this organism [7].

This work aimed to perform a systematic analysis, by means of the scientometric methodology, of the global publications on the genome of *Pseudomonas putida*, analyzing mainly the bioinformatics tools used for the assembly and annotation of the sequenced genomes.

2 Methods

The searches for performing the scientometric analysis were performed in the Web of Science database using the terms "Genome" AND "*Pseudomonas putida*" AND "soil". Initially 159 documents were found on the researched terms, however, a manually filtering was performed in order to find only those documents that reported the sequencing process of the microorganism. After the filtering, 29 papers were left, being 100% journal articles, containing the sequencing of the genome.

After filtering, the data was extracted to *Microsoft Excel* and *Citespace* software in order to produce graphs of data and connections on: main countries with publications on the subject, knowledge areas applied, keywords most used, number of publications and citation per year. Also, all the documents were read, in order to extract the quantity of sequenced genomes, the sequencing platforms, besides the genome annotation and assembly softwares, these data were tabulated and had graphs generated in Excel.

3 Results and Discussion

The first publication studying the genomic sequencing of *Pseudomonas putida* bacterium associated with soil, within the database, was in 2002, followed by a gap of 11 years, only starting the publications again in 2014, however, this time with a continuous growth of publications until 2020. The rate of citations on the subject grew continuously from 2002, totaling 1129 citations until August 2020, as shown in the Fig. 1. The growth in genomic studies of this soil-related organism is due to the extremely versatile metabolism of *P. putida*, it's great capacity of adaptation in several environments, the resistance to physical-chemical stresses, and the genes associated with degradation of recalcitrant compounds. This way, studies using this organism in biological remediation processes have maintained constant growth [8, 9]. Also, the top 5 of the main countries that published on the subject are: Spain, Japan, USA, Germany and France. The number of publications by country each year is also shown in Fig. 1.

3.1 Knowledge Areas

From the analysis performed at CiteSpace Software it is possible to see that Microbiology, Environmental Sciences and Biotechnology are the main areas of knowledge associated with the theme, fact based on the search direction of the research, aiming at the application of *P. putida* in soils, due to the high diversity of metabolisms of this microorganism associated with bioremediation [10]. The Fig. 2 presents the network of

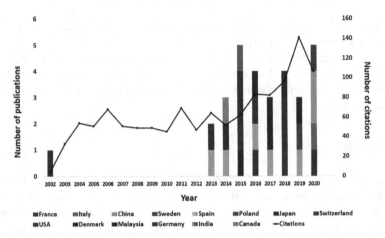

Fig. 1. Number of publications by country and year, and the number of citations per year

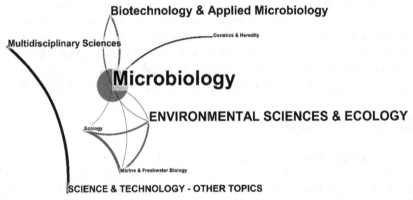

Fig. 2. Network of connections over knowledge areas. Yellow circles represent the centrality. (Color figure online)

connections related to the areas of study, and the higher the font of the letter, the more frequently this area is used in these researches.

Also, the yellow circle represents the centrality, a factor that indicates the amount of connections that the theme performs [11], thus, it is visualized that Microbiology is the only area that presents a significant centrality, due to the fact that this set of data approaches the use of a microorganism, therefore, even the most specific researches are still interconnected with microbiology, making microbiology coherent as a central area of knowledge [12].

3.2 Main Journals

In this data set 12 journals published the articles, most of them related to microbiology, environmental science or genomics. The main journal was Microbiology Resource

Announcements, with 9 articles published, 31% of the 29 documents. Table 1 presents all journals, along with the number of documents and the impact factor.

Table 1. Journals and number of published articles

Journal	Papers	Impact factor
Microbiology Resource Announcements	9	0.89
Environmental Microbiology	3	4.93
Environmental Microbiology Reports	3	2.97
Amb Express	2	2.49
Applied And Environmental Microbiology	2	4.016
Current Microbiology	2	1.73
Environmental Science And Pollution Research	2	3.30
Plos One	2	2.74
Biotechnology And Bioengineering	1	4.00
Microbial Ecology	1	3.86
Scientific Reports	1	3.99
Standards In Genomic Sciences	1	1.44

3.3 Sequencing Data

In total, 117 genomes of *Pseudomonas putida* were sequenced in the 26 documents, 106 of them complete and 14 drafts. Four sequencing platforms were used for these works, and some of these documents used more than one platform, the main one being Illumina, in 17 papers; followed by Roche 54 GS FLX, in 5 papers; Pac Bio, in 4 of the 26 documents; and Ion Torrent, in 1 document. Figure 3A and B presents the percentages of whole/draft genome and the main platforms.

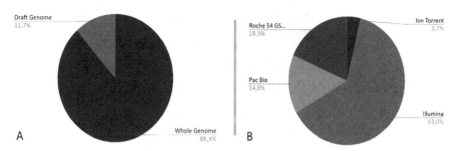

Fig. 3. (A) Percentage of 120 whole/draft genome. (B) Percentage of the main sequencing platforms used.

3.4 Assembly

In total, 9 genome assembly softwares were used, the most used being SPAdes, one of the main bacterial genome assembly softwares [13]. Other software such as CLC Genomics Workbench, Newbler and HGAP were used, as shown in Fig. 4. However, many papers (6 out of 29) did not inform the assembly software.

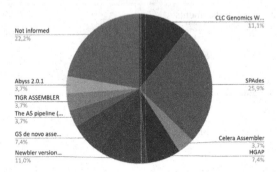

Fig. 4. Genome assembly softwares used

3.5 Annotation

In the same way as in the assembly, most of the articles did not inform the genome annotation software (7 of the 26 documents). There were 5 softwares used in the 26 papers, the main one being the NCBI Prokaryotic Genome Annotation Pipeline, a tool in constant modification and evolution for the annotation process [14]. Figure 5 shows the software used along with the percentage within the 26 papers.

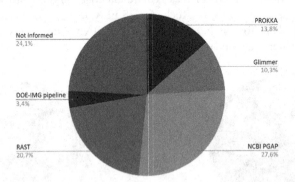

Fig. 5. Genome annotation softwares used in the data set

4   Conclusions

Pseudomonas putida is a widely studied organism with an extreme potential in the area of bioremediation of affected environments. In this context, the genomic study of this microorganism is of extreme relevance, since from the analyses at molecular level it is possible to know the main metabolic pathways, as well as the genes and proteins used by *P. putida* to perform the degradation processes of recalcitrant compounds.

In this way, the bioinformatics tools are able to analyze these molecular data and allow the development of new technologies using this microorganism, besides allowing the dissemination of genomic data. Thus, this systematic study allowed the visualization of the main bioinformatics tools used for the analysis of the *P. putida* genome, besides presenting the advance of the research on the subject, helping possible future studies with *P. putida* with application in soil environmental science and other environments.

References

1. Ashraf, A.M., Maah, J.M., Yusoff, I.: Soil contamination, risk assessment and remediation, vol. 1, no. 1, pp. 3–5. Intech (2014)
2. Ranga, P., Sharma, D., Saharan, S.B.: Bioremediation of azo dye and textile effluents using Pseudomonas putida MTCC 244. Asian J. Microbiol. Biotechnol. Environ. Sci. **22**(2), 88–94 (2019)
3. Horemans, B., Breugelmans, P., Saeys, W., Springael, D.: A soil-bacterium compatibility model as a decision-making tool for soil bioremediation. Environ. Sci. Technol. **51**(3), 1605–1615 (2017)
4. Loh, K.C., Cao, B.: Paradigm in biodegradation using Pseudomonas putida—A review of proteomics studies. Enzyme Microbial Technol. **43**(1), 1–12 (2008)
5. Tsirinirindravo, H.L., et al.: Bioremediation of soils polluted by petroleum hydrocarbons by Pseudomonas putida. Int. J. Innov. Eng. Sci. Res. **2**(5), 9–18 (2018)
6. Iyer, R., Iken, B., Damania, A., Krieger, J.: Whole genome analysis of six organophosphate-degrading rhizobacteria reveals putative agrochemical degradation enzymes with broad substrate specificity. Environ. Sci. Pollut. Res. **25**(14), 13660–13675 (2018)
7. Nelson, K.E., et al.: Complete genome sequence and comparative analysis of the metabolically versatile Pseudomonas putida KT2440. Environ. Microbiol. **4**(12), 799–808 (2002)
8. Crovadore, J., et al.: Whole-genome sequence of Pseudomonas putida strain 1312, a potential biostimulant developed for agriculture. Microbiol. Resour. Announc. **10**(7), 1–2 (2018)
9. Weimer, A., et al.: Industrial biotechnology of Pseudomonas putida: advances and prospects. Appl. Microbiol. Biotechnol. **104**(1), 7745–7766 (2020)
10. Abyar, H., et al.: The role of Pseudomonas putida in bioremediation of naphthalene and copper. World J. Fish Marine Sci. **3**(5), 444–449 (2011)
11. Ping, Q., He, J., Chen, C.: How many ways to use CiteSpace? A study of user interactive events over 14 months. J. Assoc. Inf. Sci. Technol. **68**(5), 1234–1256 (2017)
12. Chen, C.: Manual do CiteSpace, 1st edn. Drexel University, Philadelphia (2014)
13. Prjibelski, A., et al.: Using SPAdes de novo assembler. Curr. Prot. Bioinform. **70**(1), 1–29 (2020)
14. Tatusova, T., et al.: NCBI prokaryotic genome annotation pipeline. Nucl. Acids Res. Adv. **1**(1), 1–11 (2016)

Polysome-seq as a Measure of Translational Profile from Deoxyhypusine Synthase Mutant in *Saccharomyces cerevisiae*

Fernanda Manaia Demarqui[✉][iD], Ana Carolina Silva Paiva[iD],
Mariana Marchi Santoni[iD], Tatiana Faria Watanabe,
Sandro Roberto Valentini[iD], and Cleslei Fernando Zanelli[iD]

Department of Biological Sciences, School of Pharmaceutical Sciences,
São Paulo State University (UNESP), Araraquara – SP 14800-903, Brazil
fernanda.demarqui@unesp.br

Abstract. The profile of proteins observed in a cell is characterized by the control of gene expression, which has several regulation points acting individually or in concert, such as epigenetic, transcriptional, translational, post-transcriptional or post-translational modification. Copulating the total mRNA data and mRNAs actively translated can facilitate the identification of the key regulatory points of gene expression. Here, we analyze the transcriptional and translational profiles of the deoxyhypusine synthase mutant *dys1-1* in yeast. This enzyme is involved in the post-translational modification of translation factor eIF5A, which has an important role in the elongation translational process. This work presents gene expression data from the total mRNA levels and the polysomally-loaded mRNAs for the *Saccharomyces cerevisiae DYS1* and *dys1-1* strains, based on RNA-seq and Polysome-seq. Our results showed that for this mutant, most of the changes in the transcripts forwarded for translation are due to transcriptional control; and, to solve translation problems, cell responds with positive regulation of ribosome biogenesis. Besides, polysome-seq as a tool to study translation profiles is useful to understand gene expression changes.

Keywords: eIF5A · Gene regulation · Ribosome biogenesis

1 Introduction

Protein synthesis consists of decoding the messenger RNA. This process is catalyzed by ribosomes and mediated by translation factors. The regulation of the repertoire of proteins expressed in a cell is determined by the selective control of gene expression by several cellular mechanisms, such as epigenetic, transcriptional, translational, post-transcriptional or post-translational modification [4,18,21].

© Springer Nature Switzerland AG 2020
J. C. Setubal and W. M. Silva (Eds.): BSB 2020, LNBI 12558, pp. 168–179, 2020.
https://doi.org/10.1007/978-3-030-65775-8_16

The eukaryotic translation elongation factor 5A (eIF5A - ortholog elongation factor P (EF-P) of bacteria) is a highly conserved protein in eukaryotes and archaea [5,7,19]. In addition, eIF5A is essential for cell viability in all tested eukaryotes [3,20].

eIF5A undergoes a post-translational modification which leads to hypusine biosynthesis, called hypusination. This process is irreversible and involves two enzymatic steps. In the first one, a deoxyhypusine synthase catalyzes the modification of a specific lysine residue (K51 in *Saccharomyces cerevisiae*) to a hypusine in a spermidine-dependent manner. In the second one, it occurs a hydroxylation by deoxyhypusine hydroxylase with molecular oxygen as the source. Both enzymes are also evolutionarily conserved [1,15]. Hypusinated eIF5A is described to aid in the efficiency of peptide binding of motifs that tend to induce ribosomes stalling and also assists with translational termination [22]. In this study, by measuring the total mRNAs of cells (transcriptome) and the polysomally-loaded mRNAs (translatome) for the yeast deoxyhypusine synthase mutant *dys1-1* and its wild-type counterpart [9], we obtained a picture of overall relationship between the two changes for the majority of genes. Polysome-seq can explain the regulation of post-transcriptional gene expression, as a reliable measure for a translational profiling study, showing the mRNA recruited for translation. We show that the majority of statistically significant differences at RNA-seq level correspond to similar differences at Polysome-seq level, suggesting that, in most transcripts for this mutant, changes in translation are due to a transcriptional control and ribosome biogenesis is the main response to translational problems.

2   Materials and Methods

2.1   Strain and Growth Conditions

Saccharomyces cerevisiae strains SVL613 (MATa leu2 trp1 ura3 his3 dys1::HIS3 [DYS1/TRP1/CEN - pSV520]) and SVL614 (MATa leu2 trp1 ura3 his3 dys1::HIS3 [dys1 W75R T118A A147T /TRP1/CEN - pSV730]), *DYS1* and *dys1-1*, respectively, were used to RNA highthroughput experiments. Cells were grown under previously described conditions [9].

2.2   Polysome Profilling

For the polysome profiling assay, cell extracts from *DYS1* and *dys1-1* strains were prepared as described in [9]. Briefly, the cell cultures were grown to mid-log phase ($OD600\,nm = 0.6$) and cross-linked with 1% formaldehyde for 1 h in ice bath. 15 A260 nm units of cell lysates were layered onto 10–50% (w/w) sucrose gradients and centrifuged for 3 h (39.000 rpm at 4 °C in Beckman SW41-Ti rotor). The absorbance at 254 nm of gradient fractionation was continuously measured. Fractions corresponding to mRNA populations bound by 3 ribosomes were pooled and stored at −80 °C for future RNA isolation.

2.3 RNA Isolation

For total RNA isolation, *DYS1* and *dys1-1* strains were grown in exponential phase an OD600   0.6. Cultures were centrifuged and cell pellets were stored at −80 °C. Cell lysis was conducted with zymolyase and total RNA was extracted using the RNeasy mini kit (cat. number 74104, Qiagen). The polysome-associated RNA from pooled fractions was extracted using TRIzol® Reagent, following the manufacturer's protocol (cat. n 15596026, ThermoFisher Scientific). Both total RNA and polysome-associated RNA were quantified using a NanoDrop 2000 Spectrophotometer (ThermoFisher) and the integrity was verified by electrophoresis gel on 2100 Bioanalyzer equipment (Agilent, Santa Clara, CA), using a High Sensitivity Total RNA Analysis Chip.

2.4 Library Preparation and Sequencing

Library preparation and sequencing (RNA-seq) for total and polysome-associated RNA were conducted by Life Sciences Core Facility (LaCTAD) from State University of Campinas (UNICAMP). Three biological replicates for transcriptome analysis (RNA-seq of total RNA) or translatome analysis (RNA-seq of polysome-associated RNA) from *DYS1* and *dys1-1* strains were carried out according to the manufacturer's guidelines for TrueSeq kit (catalog number RS-1222001, Illumina) by selection of mRNA by poly-A tail. These 12 libraries were sequenced for 51 cycles paired-end on a Illumina HiSeq 2500 platform.

2.5 RNA-seq Data Analysis

The public server (usegalaxy.org/) was used to process the highthroughput data. FASTQ files had their quality checked by the FastQC tool (Galaxy Version 0.72). TrimGalore! (Galaxy Tool Version: 0.4.3.1 + galaxy1) was used to remove reads with Phred quality score <25 and adapter strings. Files were mapped against a *S. cerevisiae* non-coding RNA (ncRNA) sequence file (downloads.yeastgenome.org/sequence/S288C_reference/rna/archive/rna_coding_R64-1-1_20110203.fasta.gz), by Bowtie software (Galaxy Tool Version: 1.1.2) with the parameters −v 2 −y −a −m 1 −best −strata −S −p 4. The mapping and quantification of reads was performed by Stringtie software (Galaxy Tool Version: 1.3.4) with standard parameters. Only genes in which the median read count of the three replicates was larger than 10 in all conditions (*dys1-1* and *DYS1* strain, for RNA-seq and for Ribo-Seq from polysome-profiling) were kept. The filtered table of counts contained data for 5.334 genes. Count of reads was converted into RPM (reads per million).

2.6 Ribo-Seq and Protein Abundance Comparative Analysis

We used the table of counts converted in $\log_2$RPM to compare the relative abundance of total or polysome-bound mRNAs in wild-type strain between two published ribosome profiling data: RPM normalized data from Ribo-Seq [23] and Ribo-Seq from polysome-profiling [10]; and protein abundance estimation [6].

2.7   Differential Expression Analysis

Non-normalized RNA-seq count tables were used as input in anota2seq (ver. 1.2.0; datatype = "RNA-seq", normalize = TRUE, transformation = "TMM-log2") and normalized using Trimmed Mean of M-values (TMM). Changes in translational efficiencies were assessed using the *anota2seqAnalyze* function. We applied e*anota2seqSelSigGenes* function to identify differentially expressed genes, separately for RNA-seq and polysome-profiling RNA-seq data and analysis of partial variance for identification of gene expression modes from both profiles. Significance was determined using an adjusted p-value limit of 0.05.

2.8   Enrichment of Gene Ontology and Enrichment Analysis of Transcription Factors

For the regulatory gene groups, we performed gene ontology (GO) analysis with terms of biological process to determine whether specific biological functions were enriched using Yeastmine database [8]. Fisher's exact test was used to test for statistically significant differences, and the Holm-Bonferroni correction test procedure to adjust for the effects of multiple tests [2]. GO terms were considered significant when FDR <0.05. Gene lists obtained via the statistical differential from transcriptome profile were submitted to the PSCAN (v1.5, http://159.149. 160.88/pscan/) online tool.

3   Results and Discussion

3.1   RNA-seq and Polysome-Seq Experiments in *DYS1* and *dys1-1* Strains

We conducted transcriptional and translational profiling (Fig. 1A) for *S. cerevisiae dys1-1* strain and its wild-type counterpart. The number of RNA-seq reads mapping to a gene was used to quantify the relative abundance of the transcript, whereas the Polysome-seq provided a quantification of the translatome (Table 1).

Table 1. Number of mapped reads for each sample

Profile	$DYS1_1$	$DYS1_2$	$DYS1_3$	$dys1-1_1$	$dys1-1_2$	$dys1-1_3$
Transcriptional	55801619	34587137	116149587	30329854	45292306	51380070
Translational	1214644	1221375	1070191	6150284	1936203	6720246

After filtering out non-expressed genes (see Methods), the table of read counts per gene contained data for 5,334 *S. cerevisiae* annotated ORFs. Both transcriptional and translational profiles results were highly reproducible among biological replicates for each strain (Fig. 1B and 1C) (Table 2 and 3).

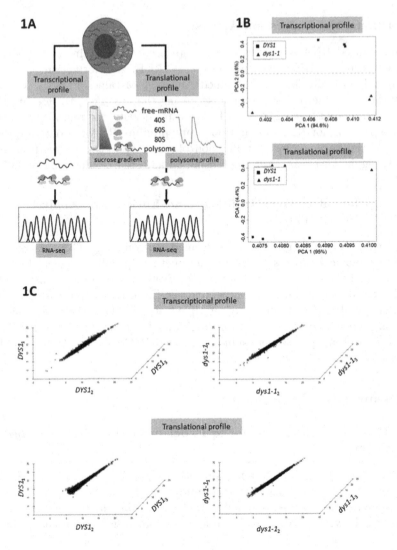

Fig. 1. (A) Experimental approaches for studying the transcribed and recruited mRNAs for translation. Transcriptional profile: the total RNA is extracted, the mRNAs are separated and subjected to large-scale sequencing. Translational profile: extracts are separated by ultracentrifugation through sucrose gradient which is then fractionated while its absorbance is continuously monitored at 254 nm (A254), allowing the separation of free RNA, the 40S and 60S ribosomal subunits, the 80S monosomes and the polysomes. The RNA is isolated from individualized gradient fractions and pooled for further large-scale analysis. (B) Principal Component Analysis indicating the distribution of replicates in the plan. Three biological replicates independent of the *DYS1* and *dys1-1* strains are represented in the distribution graphs along two main components, from the normalized RPM values of the genes sequenced by RNA-seq of each profile. (C) Linear correlation between replicates of $\log_2$RPM values of genes sequenced by RNA-seq. The linear correlation of the $\log_2$RPM values of experimental replicates for the transcriptional profile varied between 0.94 and 0.98 whereas for the translational profile this value varied between 0.98 and 0.99.

Table 2. Pearsons correlation values for $\log_2$RPM values from transcriptional profile for each replicate

	$DYS1_1$	$DYS1_2$	$DYS1_3$	$dys1-1_1$	$dys1-1_2$	$dys1-1_3$
$DYS1_2$	0.973	1				
$DYS1_3$	0.983	0.992	1			
$dys1-1_1$	0.940	0.929	0.935	1		
$dys1-1_2$	0.935	0.916	0.922	0.995	1	
$dys1-1_3$	0.925	0.906	0.914	0.986	0.990	1

Table 3. Pearsons correlation values for $\log_2$RPM values from translational profile for each replicate

	$DYS1_1$	$DYS1_2$	$DYS1_3$	$dys1-1_1$	$dys1-1_2$	$dys1-1_3$
$DYS1_2$	0.973	1				
$DYS1_3$	0.968	0.973	1			
$dys1-1_1$	0.845	0.853	0.847	1		
$dys1-1_2$	0.845	0.855	0.854	0.986	1	
$dys1-1_3$	0.854	0.867	0.865	0.989	0.988	1

3.2   Polysome-seq as a Measure for Translational Profile

One technique aimed for studying the composition of mRNAs recruited for translation by large-scale analysis is the polysome profiling, which segregates mRNAs associated with polysomes from ribosome-free mRNAs, associated with RNA-seq (Fig. 1A). In addition to Polysome-seq, Ribo-seq methodology, or ribosome profiling, is based on the sequencing of ribosome-protected fragment (RPF) mRNAs [12]. We observed high Pearson correlations with the $\log_2$RPM wild-type data from this study to ribosome profiling wild-type data available in the literature [10,23] and (Fig. 2A and 2B).

Next, we compared the wild-type strain quantification of gene expression by RNA-seq and Polysome-seq to published proteomic data [6]. The correlation and coefficient of determination from translatome (Polysome-seq) to the proteome normalized abundances (Fig. 2C) was higher than the transcriptome measurements (Fig. 2D), indicating that this former quantification of gene expression provides a more accurate picture of protein abundance, since translation is regulated by (1) translation rate, (2) translation rate modulation, (3) modulation of a protein's half-life, (4) protein synthesis delay, (5) protein transport [17,18]. So Polysome-seq allows a better understanding of regulatory mechanisms that involves post-transcriptional gene expression programs [11,13], as regulation via tuning transcript levels alone [16], resulting in a profile of selected mRNAs recruited for translation.

3.3 Yeast Hypusination Mutant *dys1-1* Responds Transcriptionally for Gene Regulation

We first calculated the gene expression level fold change (FC) between the two strains using RNA-seq and Polysome-seq data separately and we observed similar numbers of differentially expressed genes (DEGs) for both profiles - 2432 and 2826 DEGs for transcriptional and translational level, respectively - (Fig. 3A and 3B), however, Polysome-seq data had a higher variance than RNA-Seq data for the significant $\log_2$FC distribution (Fig. 3C), a consistent result for a mutant involved with a translational factor.

To establish the relationship between mRNA and polysome-associated mRNA changes when comparing *DYS1* and *dys1-1*, we categorized DEGs into gene expression modes by computing analysis of partial variance with transcriptome and translatome (Fig. 3D): (1) Homodirectional DEGs, significantly change in both profiles in a concordant way, indicating a transcriptional regulation; (2) Polysome-only DEGs, up or down polysome-associated mRNA with no significant changes in mRNA levels, a result of translation regulatory mode; (3) Transcriptome only DEGs, differences in mRNA levels not followed by a significant change in polysome-associated mRNA, a result of buffering regulatory mode; (4) Antidirectional DEGs, significantly change in both profiles but antidirectional ways. Most DEGs (67%) showed a coupled significant change, i. e., genes with significant homodirectional change in both the transcriptome and the translatome (Fig. 3E). This result is in accordance with the fact that under stress conditions, differential expressed proteins correlated strongly with the corresponding mRNA level, indicating that transcriptional control seems to be the major driver behind changes in protein levels [14].

Transcriptionally regulated genes were significantly enriched for Gene Ontology (GO) biological process terms as "maturation of SSU-rRNA" (GO:0030490), "transposition" (GO:0032196), "RNA modification" (GO:0009451) (Table 4) and Transcription Factors (TF) as Tod6, Dot6 and Stb3 (Table 5). Additionally, BUD27, the gene that encodes a protein which impacts the homeostasis of the ribosome biogenesis process by regulating the activity of the three RNA polymerases [17], is classified as an homodirectional gene and upregulated in both profiles. Taking together, these results revealed a cell response to ribosome biogenesis, a high-energy consumption process that requires stringent regulation to ensure proper ribosome production to deal with cell growth and protein synthesis in different environmental and metabolic situations [17].

The results of this study illustrate the use of Polysome-seq as a measurement of mRNAs recruited for translation. We identified for a deoxyhypusine synthase mutant *dys1-1*, a protein involved in translation, a pattern of gene expression control that is transcription dependent and upregulation of ribosome synthesis is one of the cell responses to translation impairment.

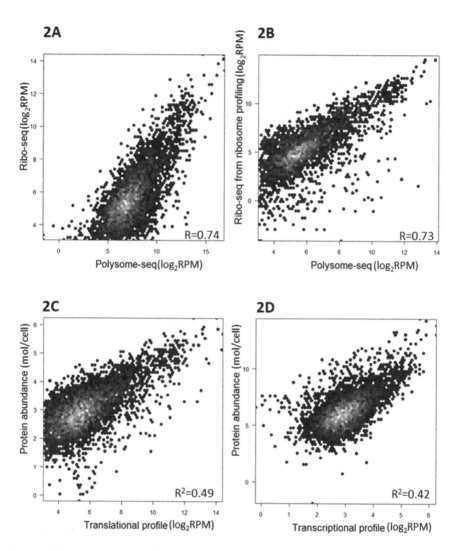

Fig. 2. Polysome-seq correlates satisfactorily to Ribo-seq data and is a good predictor of protein abundance. (A) Correlation between the translational profile ($\log_2$RPM) of this study and the translational profile of obtained by Ribo-seq ($\log_2$RPM) [23]. (B) Correlation between the translational profile of this study ($\log_2$RPM) and the translational profile of obtained by a combination of polysomal profile followed by Ribo-seq ($\log_2$RPM) [6]. C) Distribution between protein abundance (molecules per cell) and the translational profile ($\log_2$RPM) of this study. D) Distribution between protein abundance (molecules per cell) and the transcriptional profile ($\log_2$RPM) of this study. Protein abundance data are indicated in molecules per cell according to [6].

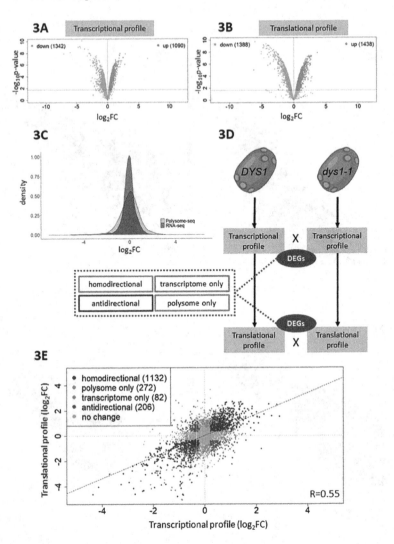

Fig. 3. Volcano plot of the distribution of the transcripts differentially expressed in the transcriptional profile (A) and translational profile (B). The values of $-\log_1 0$ p-value were plotted according to the differencial expression between *DYS1* and *dys1-1* ($\log_2$ fold change). Downregulated genes are highlighted in blue (left), upregulated genes, in orange (right); dashed horizontal line indicates an adjusted p-value of 0.05. (C) Distribution of gene expression fold change (FC) values. FC was calculated as the ratio between the number of reads in *dys1-1* and *DYS1* strains. We took the average number of reads per gene among the replicates. (D) Scheme of differential expression analysis between the transcriptional and translational profile of the *dys1-1* mutant. Genes classified as differentially expressed were called transcriptome only (blue), polysome only (orange), antidirectional (purple) - significantly opposite variations between transcriptional and translational profiles - and homodirectional (green) - variations significantly converging between both profiles. (E) Distribution of the $\log_2$ fold change of the transcriptional and translational profile. Genes showing statistical differences between *dys1-1* and *DYS1* were simultaneously compared in the two profiles. Categories are defined in 3D. (Color figure online)

Table 4. Gene Ontology (GO) analysis of transcriptionally regulated mRNAs from *dys1-1* mutant as determined by anota2seq

Analysis by Anota2seq - "Homodirectional"				
Term ID	Description	$\log_1 0$ p-value	Dispensability	N genes
GO:0030490	Maturation of SSU-rRNA	−3.000.000	0.00	53
GO:0032196	Transposition	−53.792	0.03	42
GO:0032197	Transposition, RNA-mediated	−34.237	0.12	40
GO:0006278	RNA-dependent DNA biosynthetic process	−76.021	0.24	39
GO:0090305	Nucleic acid phosphodiester bond hydrolysis	−61.694	0.28	98
GO:0001510	RNA methylation	−45.888	0.29	24
GO:0006396	RNA processing	−64.134	0.37	175
GO:0009451	RNA modification	−27.911	0.40	43
GO:0034660	ncRNA metabolic process	−28.050	0.48	166
GO:0000966	RNA 5'-end processing	−62.684	0.48	28

Table 5. Transcriptional factor (TF) enrichment analysis of differentially expressed genes in the transcriptional profile from *dys1-1* mutant as determined by anota2seq

MatrixID	Matrix name	p-value
MA 0350.1	TOD6	1,92E−23
MA 0351.1	DOT6	1,72E−21
MA 0390.1	STB3	1,14E−05
MA 0378.1	SFP1	4,49E−03
MA 0398.1	SUM1	4,64E−02
MA 0345.1	NHP6A	1,12E−01
MA 0346.1	NHP6B	2,73E−01
MA 0386.1	SPT15	0.000184904
MA 0418.1	YAP6	0.000859765
MA 0435.1	YPR015C	0.00121852

Acknowledgement. This study was financially supported by grant #10/50044-6, São Paulo Research Foundation (FAPESP). This study was financed in part by the Coordenação de Aperfeiçoamento de Pessoal de Nível Superior - Brasil (CAPES) - Finance Code 001.

References

1. de Almeida, O.P., et al.: Hypusine modification of the ribosome-binding protein eIF5A, a target for new anti-inflammatory drugs: understanding the action of the inhibitor GC7 on a murine macrophage cell line. Curr. pharm. Des. **20**(2), 284–92 (2014)
2. Benjamini, Y., Yekutieli, D.: The control of the false discovery rate in multiple testing under dependency. Ann. Stat. (2001). https://doi.org/10.1214/aos/1013699998
3. Buskirk, A.R., Green, R.: Ribosome pausing, arrest and rescue in bacteria and eukaryotes. Philos. Trans. R. Soc. Lond. Ser. B Biol. Sci. **372**, 20160183 (2017). https://doi.org/10.1098/rstb.2016.0183
4. Chassé, H., Boulben, S., Costache, V., Cormier, P., Morales, J.: Analysis of translation using polysome profiling. Nucleic Acids Res. (2017). https://doi.org/10.1093/nar/gkw907
5. Chen, K.Y., Liu, A.Y.: Biochemistry and function of hypusine formation on eukaryotic initiation factor 5A. NeuroSignals (1997). https://doi.org/10.1159/000109115
6. Csárdi, G., Franks, A., Choi, D.S., Airoldi, E.M., Drummond, D.A.: Accounting for experimental noise reveals that mRNA levels, amplified by post-transcriptional processes, largely determine steady-state protein levels in yeast. PLoS Genet. **11**(5), e1005206 (2015). https://doi.org/10.1371/journal.pgen.1005206
7. Dever, T.E., Ivanov, I.P.: Roles of polyamines in translation. J. Biol. Chem. **293**(48), 18719–18729 (2018). https://doi.org/10.1074/jbc.TM118.003338. http://www.jbc.org/
8. Engel, S.R., et al.: The reference genome sequence of saccharomyces cerevisiae: then and now. G3: Genes Genomes Genetics **4**(3), 389–398 (2014). https://doi.org/10.1534/g3.113.008995
9. Galvão, F.C., Rossi, D., Silveira, W.D.S., Valentini, S.R., Zanelli, C.F.: The deoxyhypusine synthase mutant dys1-1 reveals the association of eIF5A and Asc1 with cell wall integrity. Plos One **8**(4), e60140 (2013). https://doi.org/10.1371/journal.pone.0060140
10. Heyer, E.E., Moore, M.J.: Redefining the translational status of 80S monosomes. Cell **164**(4), 757–769 (2016). https://doi.org/10.1016/j.cell.2016.01.003
11. Ingolia, N.T.: Ribosome profiling: new views of translation, from single codons to genome scale. Nat. Rev. Genet. **15**, 205–213 (2014). https://doi.org/10.1038/nrg3645
12. Ingolia, N.T., Ghaemmaghami, S., Newman, J.R., Weissman, J.S.: Genome-wide analysis in vivo of translation with nucleotide resolution using ribosome profiling. Science **324**(5924), 218–223 (2009). https://doi.org/10.1126/science.1168978
13. Jin, H.Y., Xiao, C.: An integrated polysome profiling and ribosome profiling method to investigate in vivo translatome. In: Methods in Molecular Biology, vol. 1712, pp. 1–18. Humana Press Inc. (2018). https://doi.org/10.1007/978-1-4939-7514-31
14. Lahtvee, P.J., et al.: Absolute quantification of protein and mRNA abundances demonstrate variability in gene-specific translation efficiency in yeast. Cell Syst. **4**(5), 495–504 (2017). https://doi.org/10.1016/J.CELS.2017.03.003. https://www.sciencedirect.com/science/article/pii/S2405471217300881#mmc4
15. Landau, G., Bercovich, Z., Park, M.H., Kahana, C.: The role of polyamines in supporting growth of mammalian cells is mediated through their requirement for translation initiation and elongation. J. Biol. Chem. **285**(17), 12474–12481 (2010). https://doi.org/10.1074/jbc.M110.106419

16. Liu, Y., Beyer, A., Aebersold, R.: On the dependency of cellular protein levels on mRNA abundance. Cell **165**, 535–550 (2016). https://doi.org/10.1016/j.cell.2016.03.014

17. Martínez-Fernández, V., et al.: Prefoldin-like Bud27 influences the transcription of ribosomal components and ribosome biogenesis in Saccharomyces cerevisiae. RNA (2020). https://doi.org/10.1261/rna.075507.120

18. Piccirillo, C.A., Bjur, E., Topisirovic, I., Sonenberg, N., Larsson, O.: Translational control of immune responses: from transcripts to translatomes. Nat. Immunol. **15**(6), 503–511 (2014). https://doi.org/10.1038/ni.2891. http://www.nature.com/doifinder/10.1038/ni.2891

19. Rossi, D., Kuroshu, R., Zanelli, C.F., Valentini, S.R.: eIF5A and EF-P: two unique translation factors are now traveling the same road. Wiley Interdiscip. Rev.: RNA **5**(2), 209–222 (2014). https://doi.org/10.1002/wrna.1211

20. Schnier, J., Schwelberger, H.G., Smit-McBride, Z., Kang, H.A., Hershey, J.W.: Translation initiation factor 5A and its hypusine modification are essential for cell viability in the yeast Saccharomyces cerevisiae. Mol. Cell. Biol. (1991). https://doi.org/10.1128/MCB.11.6.3105

21. Schuller, A.P., Green, R.: Roadblocks and resolutions in eukaryotic translation. Nat. Rev. Mol. Cell Biol. **19**(8), 526–541 (2018). https://doi.org/10.1038/s41580-018-0011-4. http://www.nature.com/articles/s41580-018-0011-4

22. Schuller, A.P., Wu, C.C.C., Dever, T.E., Buskirk, A.R., Green, R.: eif5a functions globally in translation elongation and termination. Mol. Cell **66**(2), 194–205 (2017). https://doi.org/10.1016/j.molcel.2017.03.003

23. Sen, N.D., Zhou, F., Ingolia, N.T., Hinnebusch, A.G.: Genome-wide analysis of translational efficiency reveals distinct but overlapping functions of yeast DEAD-box RNA helicases Ded1 and eIF4A. Genome Res. **25**(8), 1196–1205 (2015). https://doi.org/10.1101/gr.191601.115

Anti-CD3 Stimulated T Cell Transcriptome Reveals Novel ncRNAs and Correlates with a Suppressive Profile

Manuela M. do Almo[1,2], Isabel G. Sousa[1], Waldeyr Mendes Cordeiro da Silva[1,3], Thomas Gatter[4], Peter F. Stadler[4,5,6,7,8], Steve Hoffmann[4,9,10], Andrea Q. Maranhão[1,11], and Marcelo Brigido[1,11(✉)]

[1] Department of Cell Biology, Institute of Biological Science,
University of Brasília, Brasília, Brazil
mmbrigido@gmail.com
[2] Molecular Pathology Graduation Program, Medicine Faculty,
University of Brasília, Brasília, Brazil
[3] NEPBio - Federal Institute of Goiás, Formosa, Brazil
[4] Bioinformatics Group, Department of Computer Science;
Interdisciplinary Center for Bioinformatics,
University of Leipzig, Leipzig, Germany
[5] Institute for Theoretical Chemistry,
University of Vienna, Vienna, Austria
[6] Facultad de Ciencias,
Universidad Nacional de Colombia, Bogotá, Colombia
[7] Max Planck Institute for Mathematics in the Sciences, Leipzig, Germany
[8] Santa Fe Institute, Santa Fe, USA
[9] Computational Biology Group, Leibniz Institute on Ageing -
Fritz Lipmann Institute (FLI), Jena, Germany
[10] Friedrich-Schiller-University Jena, Jena, Germany
[11] Instituto de Investigação em Imunologia,
Instituto Nacional de Ciência e Tecnologia (iii-INCT), São Paulo, Brazil

Abstract. T lymphocytes are key players in immunity. Anti-CD3 antibodies activate T cells and promote a suppressive phenotype *in vivo*. Although the pharmacological use of these antibodies is widely studied, the underlying mechanisms are still poorly understood. Here we describe the response of the non-coding RNA transcriptome of T cells after *in vitro* stimulation of peripheral blood mononuclear cells (PBMC) by anti-CD3 and demonstrate that several novel long non-coding RNA are associated with antibody treatment. Regulated long intergenic non-coding RNAs were associated with lymphocyte activation and signaling pathways. In particular the lncRNA transcripts WFDC21P and GAPLINC are regulated in stimulated T cell.

Keywords: Anti-CD3 · lncRNA · Gene regulation · Alternative splicing · Transcriptome

© Springer Nature Switzerland AG 2020
J. C. Setubal and W. M. Silva (Eds.): BSB 2020, LNBI 12558, pp. 180–191, 2020.
https://doi.org/10.1007/978-3-030-65775-8_17

1   Introduction

Targeting T lymphocytes was among the first monoclonal antibodies (mAb) associated immunotherapies. Anti-CD3 therapy was used to control T cell activity, suppress the immune response, and substitute the polyclonal anti-thymocyte antibody preparation, previously used for graft rejection [17]. Muronomab, a monoclonal antibody specific for the human CD3 antigen, was the first mAb used in clinical studies, but its use was abolished due to its overall toxicity [21]. Nowadays, novel and less toxic CD3 specific antibodies have reemerged as promising therapeutics for controlling autoimmune diseases.

To better understand the human T cell reprogramming after anti-CD3 treatment, we previously investigated the protein-coding (PTC) genes [25] regulated *ex vivo* in a PBMC milieu. Based on a new transcript prediction algorithm, we reannotated the non-coding transcriptome of human T cells treated with anti-CD3 antibodies to unveil differentially expressed lncRNA (DEL) that may be involved in CD3 targeted antibody therapy. We observed several novel non-coding transcripts along with previously annotated ones, and we discuss their possible participation in T cell fate and the suppressive phenotype.

2   Materials and Methods

2.1   Sample Preparation and Sequencing

We extend here analysis of the RNA-seq data GEO database (GSE112899) originally described in [25]. In brief, Ficoll-Paque purified volunteer human PBMC were cultured with or without anti-CD3 antibodies. We used three antibody preparations: an anti-CD3ϵ Monoclonal Antibody - OKT3, and two Recombinant Antibodies derived from OKT3, and produced in a human IgG1 scFvFc format: a humanized (FvFcR) and a chimeric antibody (FvFcM). After 72 h in culture, CD3$^+$ cells were enriched by negative selection. The RNA-seq data were produced with an Illumina HiSeq in 2×150 bp paired-end mode [25]. All human blood experiments were performed in accordance with the Ethics Committee of the University of Brasilia guidelines, which approved the study protocol (CAAE: 32874614.4.0000.0030).

2.2   Genome Mapping and Transcript Prediction

All sequenced reads produced by Illumina were analyzed for quality control using FASTQC [1]. The reads were filtered using BBDuk [3] at $k = 31$ to a reference of ribosomal kmers provided by the developers. Adapters were trimmed by cutadapt [15], and reads were then aligned to the HG19/GRCh37 Human Genome using HISAT2 [13] at standard settings. Ryūtō [7] was run on the alignments to predict individual transcripts for each set. Transcript predictions were joined using the TACO meta assembler [16]. Results were compared to the Gencode V19 annotation to identify novel transcripts. Salmon [18] was used to realign the filtered reads for each sample to the full set of GENCODE. A count matrix was created from transcripts together with the additional predicted novel transcripts. DESeq2 [14] was used to identify differentially expressed genes (**DEGs**) on all replicates, applying a significance threshold for the adjusted p-values of 0.05.

2.3 Data Mining and Ontology Classification

The annotated transcriptome and DEGs were explored using in-house awk scripts and BEDTools [19] commands and visualized with Integrative Genomics Viewer (IGV) [22] and Ensembl browser (ENSEMBL) [12]. Venn diagrams were designed using UGENT tools[1]. The protein-coding (PTC) transcripts (Ensembl transcript type) were excluded from analysis, and only non-coding transcripts from non-protein-coding and protein-coding genes were considered. Only transcripts with base mean above zero were considered. The overlaps between different transcriptomes were analyzed using bedtools intersect [19]. Non-coding potentials were calculated using Portrait [2] using a cutoff of 60%.

2.4 cDNA Synthesis Transcript Testing

The cDNA was synthesized with the SuperScript IV Reverse Transcriptase kit (Invitrogen, Carlsbad, CA, USA) from the total RNA extracted using miRNeasy® Mini Kit (Qiagen, Valencia, CA, USA) [25] from T lymphocytes from three donors, treated or not with anti-CD3 antibodies, following the manufacturer's guidelines. Primers were designed for the predicted transcripts lnc-DC and GAPLINC, that were amplified by PCR with Taq Platinum DNA Polymerase (Invitrogen, Carlsbad, CA, USA) under the following conditions: 35 cycles where the DNA was denatured at 94 °C for 30 s, annealed at 52 °C for 30 s and elongated at 72 °C for 1 min. The amplification cycles were performed in the SimpliAmp thermocycler (Applied Biosystems). The PCR products were ligated into the pGEM-T Easy Vector (Promega, Madison, Wisconsin) and transformed into XL1Blue competent cells. The cloning was confirmed by plasmid enzymatic digestion with Nco I and Not I. Plasmids were sequenced by the Sanger method.

2.5 Gene Expression Analysis by qPCR Assays

The quantitative PCR assays were performed in triplicate with total RNA isolated from T cells utilized for cDNA synthesis using RT2 First Strand Kit (Qiagen, Valencia, CA, USA). As previously described [25], expression genes were quantified using RT2 qPCR SYBR Green/ROX Master-Mix (Qiagen, Valencia, CA, USA) following the manufacturer's instructions. The reference gene Beta-2 microglobulin (B2M) was used as the endogenous control. The assays were performed using the ABI Step One Plus Real-Time PCR System (Applied Biosystems). The 2-$\Delta\Delta$Ct method was used to calculate the levels of the lncRNA

---

[1] bioinformatics.psb.ugent.be/webtools/Venn.

transcripts (fold change) and for analysis of the obtained data, the RT2 Profiler PCR Array Data Analysis (SABiosciences Frederick, MD, USA) software was used. Real-Time qPCR p-values were calculated based on Student's t-test using RT2 Profiler PCR Array Data Analysis software.

3   Results

The transcriptome reconstructed from the union of all mapped reads comprises 174,649 transcripts of which about a third are protein-coding, nearly half (44.1%) are correspond to known non-coding Ensembl/VEGA transcripts, and 20% previously unannotated transcripts (Fig. 1). The distribution of Ensembl/VEGA transcript types among the known ncRNA is summarized in Fig. 1-A. The most abundant types are transcripts with retained introns (IR) and Processed Transcripts (PT) that are not classified as belonging to one of the lncRNAs classes.

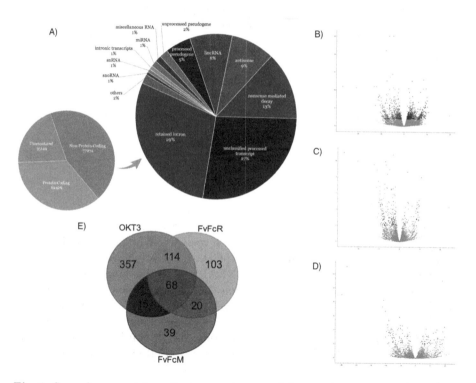

Fig. 1. Several non-protein coding transcripts are differentially expressed in anti-CD3 stimulated cells. (A) Gene type distribution of the observed ncRNA in all experimental samples. The distribution of non-protein-coding transcripts is detailed in the right; Volcano plot of DEL transcripts regulated by (B) OKT3; (C) FvFcR; and (D) FvFcM. Each group is compared to the non-treated sample; (E) Venn diagram comparing statistically significative DEL in each experimental group and their intersections.

Long intergenic ncRNA (LincRNA), Antisense RNA (AS), and pseudogenes were also significantly observed. The three antibody treatments resulted in a total of 726 differentially expressed lncRNA (DELs). OKT3 induced a larger number of ncRNA compared to FvFcR and FvFcM, as observed in the volcano plots (Fig. 1-D). The Venn diagram (Fig. 1-E) summarizes the DEL set seen in each treatment and their intersections. The number of statistically significant, differentially expressed transcripts after each antibody treatment category is summarized in Table 1 for the five major categories of lncRNA. Anti-CD3 antibodies regulate only a small fraction of total observed lncRNA, ranging from 0,9% of retained intron transcripts (IR) after OKT3 stimulation to 0,02% of pseudogenes after FvFcM stimulation.

Table 1. Differentially expressed lncRNA after anti-CD3 treatment

Transcript type*	OKT3		FvFcR		FvFcM		Total transcripts#
	Up	Down	Up	Down	Up	Down	
Intron Retained	90	132	76	36	30	18	22,360
Unclassified Processed Transcript	85	46	69	10	27	13	21,074
Antisense	11	13	8	1	2	6	6,737
LincRNA	17	19	15	4	3	1	6,552
Pseudogene&	4	4	4	3	0	3	5,696

*Transcript type classification after ENSEMBL.
&Pseudogene includes: pseudogene, processed pseudogene, transcribed unprocessed pseudogene, unprocessed pseudogene and polymorphic pseudogene type.
#Total number of transcripts in the T cell transcriptome.

3.1   LincRNAs

LincRNAs are known to be involved in critical processes in the differentiation of T cells. The whole T cell transcriptome reveals 6,552 LincRNA, and a small proportion of them are regulated by anti-CD3 stimulation. Some of them are known to be expressed T cells, such as NEAT1, a nuclear long ncRNA associated with Th2 [10] and Th17 differentiation [23]. NEAT1 is down-regulated after OKT3 and FvFcR treatment, but only OKT3 induces a statistically significant three-fold reduction in transcript level. Linc00861 was found to be expressed in several CD4 and CD8 cells. OKT3 stimulation also diminished linc00861 with high confidence ($p < 10^{-13}$), while repression by FvFcR shows a barely significant adjusted p-value ($< 4 \times 10^{-3}$).

Three lincRNA were observed to be activated in all antibodies treatment, AC017002.1, LINC00339, and LINC01132. Only the first two were previously found in T cells [5]. AC017002.1 was mostly but not exclusively detected in memory Treg. This lincRNA is in close genomic proximity to BCL2L11, a proapoptotic gene involved in the T cell negative selection in thymus associated with T cell activation by high-affinity antigens [8].

3.2   Novel Predicted Transcripts

Among the reconstructed transcripts, about a fifth (35,149) remained unannotated transcripts (TU) (Fig. 1-A). Most of these TU overlapped with protein-coding and non-protein coding genes (Fig. 2-A), and thus most likely constitute previously undescribed isoforms of known genes. However, 575 TUs do not overlap any annotated gene and are novel genes candidate. The TU set was further tested for their coding potential that suggested that at least 413 (72%) of them are novel lncRNA transcripts (Fig. 2-B). These transcripts contained between 1 and 15 exons, with a modal exon number of 2 per gene (Fig. 2-C). Among these putative novel transcripts revealed by RNA-seq, 77 are regulated by anti-CD3 stimulation, 72 by OKT3, 19 by FvFcR, and 13 by FvFcM (Fig. 2-D). All three antibodies regulated eight of them.

Interestingly, some genomic regions accumulate novel TU with CD4 and CD8 specific transcripts suggesting a regulatory role for such loci. An example is the TU35249, which appears close to KLF3-AS1 and KLF3 loci (data not shown). All three transcripts are repressed during anti-CD3 treatment. TU35249 also seems to be coregulated with the antisense RNA and overlaps a CD4 and a CD8 specific transcript previously reported in [11, 20].

3.3   Testing Novel LncRNA Isoforms

Two of the computationally predicted TU were tested experimentally for differential expression after T cell stimulation with anti-CD3. These transcripts were cloned and sequenced, and their expression levels were quantified by qPCR. The data on T cell RNA-seq suggested transcriptional activity in the locus WFDC21P on Chromosome 17. This locus is described as an ancient pseudogene with coding capacity in mammals, including primates, but not in humans [6]. In the genus *Homo*, this locus was reported to code for a lincRNA, Lnc-DC, found to regulate STAT3 activity in dendritic cells (DC) [27]. We observed several transcripts in this locus, and two novel isoforms TU20859 and TU20860 were found repressed in OKT3 stimulated T cell ($p_{\text{adj}} < 0.05$). These isoforms differ from the Lnc-DC due to the use of novel exons at the 5' end (Fig. 3). Transcriptional activity was validated in T cell RNA by qPCR analyses that corroborate the RNA-seq data to demonstrate the presence of WFDC21P transcripts in resting T cell, besides their repression after anti-CD3 treatment. Moreover, sequence analyses of cDNA amplicon are compatible with the TU20859 and TU20861 transcripts, but no cDNA clone was found consistent with the presence of TU20860, the transcript Lnc-DC (Genbank NR_030732.1). Though, the WFDC21P transcript that is repressed after anti-CD3 stimulation may be a novel transcript distinct from the previously described Lnc-DC [27].

GAPLINC is a lincRNA that has been described as a marker for gastric adenocarcinoma [9] and was not reported to be expressed in lymph nodes [5]. The data on T-cell RNA-seq presented here suggested the presence of GAPLINC transcripts in the untreated cells and significant reduction after anti-CD3 treatment. All four reference transcripts of GAPLINC were observed, but no significant differential expression was observed, except for the GAPLINC-204[2], barely significantly repressed by OKT3 with a $p_{adj} = 0.0135$. Along with them, two novel isoforms, TU21901 and TU21904, were detected. TU20901 was the most abundant transcript predicted in this locus and is a DEL, repressed as a result of OKT3, FvFcR, and FvFcM treatment (p_{adj} value of 0.0001, 0.0487, and 0.0502, respectively). The qPCR quantification suggested that all antibodies induced certain repression, especially for a particular primer pair, which detects all variants except for GAPLINC-201 (Fig. 4).

The analysis of cDNA clones obtained from PCR for exons 1 to 4 and 3 to 4 yielded sequences that confirmed the presence of predicted transcripts. Three independent clones could unequivocally validate TU21901 with the same exon-

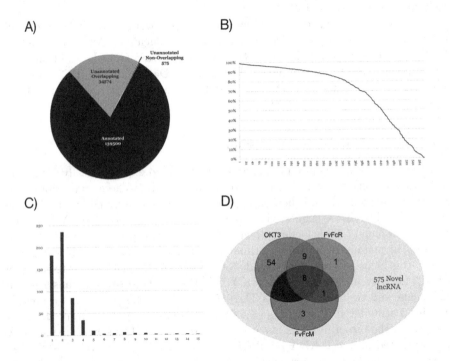

Fig. 2. A set of unannotated predicted transcripts may correspond to novel lncRNA. (A) a large fraction of the predicted transcriptome could be machine annotated (blue), from the unannotated transcripts most overlaps known gene loci (salmon), except for a small fraction (red) of them. (B) Non-coding probability of the unannotated nonoverlapping transcripts. (C) Histogram of exon content of the unannotated nonoverlapping transcripts. The number of transcripts is quoted following the number of predicted exons. (D) Venn diagram showing that part of the unannotated nonoverlapping transcripts is regulated by the anti-CD3 treatment. In blue are transcripts regulated by OKT3, red, FvFcR and green, FvFcM. Yellow eclipse reflects the total number of unannotated nonoverlapping transcripts. (Color figure online)

---

[2] ENST00000581442.1.

exon junction. Two clones showed the unique junction of GAPLINC-204, where exon 2 uses an alternative donor splice site compared to TU21901. Three other predicted clones showed an exon-exon junction that is shared either by TU21904 or by the previously reported transcript GAPLINC-202. Therefore, the data showed that along with GAPLINC-202 and GAPLINC-204, at least the novel regulated transcript TU21901 could be found in non-stimulated T cells.

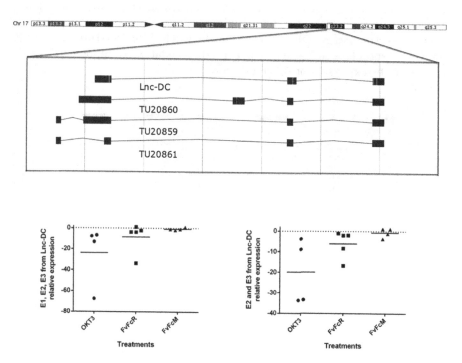

Fig. 3. The WFDC21P transcript is depicted to reveals the Lnc-DC along with TU20859, TU20860, and TU20861 intron-exon structure. (A) The transcripts in the opposite strand of chromosome 17 were rotated to facilitate visualization. Primers used to check for transcripts are marked in red and green. Quantitative expression by qPCR assay of lncRNAs was performed with total RNA extracted from T cells stimulated with anti-CD3 antibodies. The results are expressed as the fold change relative to unstimulated T cells ($n = 5$; $p < 0.05$). (B) Expression of transcripts detected with the primer pair for the first and third exon of Lnc-DC (red). (C) Expression of transcripts detected with the primer pair for the second and third exon of Lnc-DC (green). (Color figure online)

4   Discussion

The stimulation of T lymphocytes with anti-CD3 antibodies induces a change in transcriptional landscape. In this new study, we reanalyzed the data on the anti-CD3 induced T cells [25] to unveil the lncRNA transcriptome. We produced a new read mapping of the reads and focused on the reconstruction of splicing isoforms. The number of annotated non-protein-coding transcript observed in the cell transcriptome was close to that in the hg19 human genome assembly [12], suggesting a good coverage of the total universe of lncRNAs. The focus of the research presented here, however, was not the complete lncRNA set, but the differentially expressed transcriptome; trying to figure out the changes in genetic programming that is achieved after antibody stimulation. Due to limitation of our model system, we are not able to pinpoint specific T cell subpopulations. Nevertheless, considering DEL observed in the whole T cell mixture in a PBMC context, we speculate that a particular T cell population is imposed (or polarized), either by expansion or differentiation.

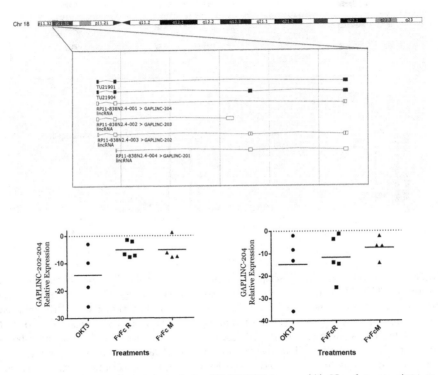

Fig. 4. Transcriptional activity of the GAPLINC locus. (A) Novel transcripts are depicted along with annotated transcripts. In the top, the genomic view of transcripts with exon-intron structured. Primers used to check for transcripts are marked in green and magenta. Quantitative expression by qPCR assay of lncRNAs was performed with total RNA extracted from T cells stimulated with anti-CD3 antibodies. The results are expressed as the fold change relative to unstimulated T cells (n = 5; $p < 0.05$). (B) qPCR with a primer to the junction of the first and second exon and another for the third exon of GAPLINC-204 (red), detecting all transcripts except GAPLINC 201. (C) Expression of transcripts detected with a primer pair for the third and fourth exon of GAPLINC-202/TU21904 (green). (Color figure online)

LincRNA expression seems to be more cell type-specific than protein-coding genes [4]. After that, we inspect the ncRNA transcriptome for lincRNA DEL, which could support discrete changes in cell differentiation, explaining the upraise of a suppressive phenotype. Among the annotated lincRNA, only three (AC017002.1, LINC00339, LINC01132) were consistently activated with the three anti-CD3 molecules. AC017002.1 was associated with memory CD4 cells, and its upregulation may reflect the expansion of memory cells, a reported effect of commercial anti-CD3 [26].

The antibody molecule format has a significant impact on the expression profile. OKT3 is far the most effective in regulating lncRNA. It chimeric format FvFcM and the humanized antibody fragment FvFcR nevertheless display a similar, although generally less intense response. The former, a mouse mAb, has a much stronger mitogenic response than FvFc format [24,25]. Despite the differences in the regulated gene set, several of DEL are consistently regulated after all antibody molecules. Yet, other regulated lncRNA is antibody specific, such as FvFcM specific DEL. The chimeric antibody is the only antibody that significantly regulates TU13951. Maybe variation in the antibody's paratope and the Fc component affect the strength and quality of TCR signaling, and further engineering may improve regulatory bias reducing the inflammatory response.

5   Conclusion

We investigate the lncRNA transcriptome of T lymphocyte cells to uncover the changes incurred by anti-CD3 immunotherapy, which could reflect in a suppressive phenotype. Several lncRNAs and known lincRNAs were observed, and its role in the reversal of inflammation may be associated with the induction of a regulatory phenotype.

The successful release of novel anti-CD3 therapeutics will readdress the investigation of the novel suppressive and tolerogenic effect of these immune pharmaceuticals in humans. The selective immunoregulation of the anti-CD3 treatment observed in clinics may become the basis of novel therapy for autoimmune disease. In this sense, the development of clinical-grade markers could help this development. The data on lncRNA revealed in this work may not only contribute to novel markers to follow immunoregulation on the whole T cell context but also contribute to potential non-coding RNA regulation. Association of anti-CD3 data may yield new disease markers and treatment targets among the differentially expressed lncRNA. Beyond simplifying therapeutics monitoring, biological relevant ncRNA could become a pharmacological target for future therapies. As a result, a new generation of more powerful pharmaceuticals to immunosuppress and control the immune response.

Acknowledgments. We are thankful to CAPES and CNPq for scholarship funding and to FAPDF for financial support of this project. We are also grateful to Prof. Concepta M. McManus for English correction.

References

1. Andrews, S., et al.: FastQC: a quality control tool for high throughput sequence data (2010). https://www.bioinformatics.babraham.ac.uk/projects/fastqc/
2. Arrial, R.T., Togawa, R.C., Brigido, M.M.: Screening non-coding RNAs in transcriptomes from neglected species using PORTRAIT: case study of the pathogenic fungus *Paracoccidioides brasiliensis*. BMC Bioinform. **10**(1), 239 (2009). https://doi.org/10.1186/1471-2105-10-239
3. BBTools: BBDuk. http://jgi.doe.gov/data-and-tools/bb-tools/
4. Cabili, M.N., et al.: Integrative annotation of human large intergenic noncoding RNAs reveals global properties and specific subclasses. Genes Dev. **25**(18), 1915–1927 (2011)
5. GTEx Consortium et al.: Genetic effects on gene expression across human tissues. Nature **550**(7675), 204–213 (2017)
6. Dijkstra, J.M., Ballingall, K.T.: Non-human lnc-DC orthologs encode Wdnm1-like protein. F1000Research **3**, 160 (2014)
7. Gatter, T., Stadler, P.F.: Ryūtō: network-flow based transcriptome reconstruction. BMC Bioinform. **20**(1), 190 (2019). https://doi.org/10.1186/s12859-019-2786-5
8. Hojo, M.A., et al.: Identification of a genomic enhancer that enforces proper apoptosis induction in thymic negative selection. Nat. Commun. **10**(1), 1–15 (2019)
9. Hu, Y., et al.: Long noncoding RNA GAPLINC regulates CD44-dependent cell invasiveness and associates with poor prognosis of gastric cancer. Cancer Res. **74**(23), 6890–6902 (2014)
10. Huang, S., et al.: NEAT1 regulates Th2 cell development by targeting STAT6 for degradation. Cell Cycle **18**(3), 312–319 (2019)
11. Hudson, W.H., et al.: Expression of novel long noncoding RNAs defines virus-specific effector and memory CD8+ T cells. Nat. Commun. **10**(1), 1–11 (2019)
12. Hunt, S.E., et al.: Ensembl variation resources. Database **2018** (2018)
13. Kim, D., Langmead, B., Salzberg, S.L.: HISAT: a fast spliced aligner with low memory requirements. Nat. Methods **12**(4), 357–360 (2015). https://doi.org/10.1038/nmeth.3317
14. Love, M.I., Huber, W., Anders, S.: Moderated estimation of fold change and dispersion for RNA-seq data with DESeq2. Genome Biol. **15**(12), 550 (2014). https://doi.org/10.1186/s13059-014-0550-8
15. Martin, M.: Cutadapt removes adapter sequences from high-throughput sequencing reads. EMBnet. J. **17**(1), 10–12 (2011)
16. Niknafs, Y.S., et al.: Taco produces robust multisample transcriptome assemblies from RNA-seq. Nat. Methods **14**(1), 68–70 (2017)
17. Norman, D.J., et al.: The use of OKT3 in cadaveric renal transplantation for rejection that is unresponsive to conventional anti-rejection therapy. Am. J. Kidney Dis. **11**(2), 90–93 (1988). https://doi.org/10.1016/S0272-6386(88)80186-0
18. Patro, R., et al.: Salmon provides fast and bias-aware quantification of transcript expression. Nat. Methods **14**(4), 417–419 (2017)
19. Quinlan, A.R.: BEDTools: the Swiss-army tool for genome feature analysis. Curr. Protocols Bioinform. **47**(1), 11–12 (2014)
20. Ranzani, V., et al.: The long intergenic noncoding RNA landscape of human lymphocytes highlights the regulation of T cell differentiation by linc-MAF-4. Nat. Immunol. **16**(3), 318 (2015)
21. Reichert, J.M.: Marketed therapeutic antibodies compendium. MAbs **4**(3), 413–415 (2012). https://doi.org/10.4161/mabs.19931

22. Robinson, J.T., et al.: Integrative genomics viewer. Nat. Biotechnol. **29**(1), 24–26 (2011)
23. Shui, X., et al.: Knockdown of lncRNA NEAT1 inhibits Th17/CD4+ T cell differentiation through reducing the STAT3 protein level. J. Cell. Physiol. **234**(12), 22477–22484 (2019)
24. Silva, H.M., et al.: Novel humanized anti-CD3 antibodies induce a predominantly immunoregulatory profile in human peripheral blood mononuclear cells. Immunol. Lett. **125**(2), 129–136 (2009)
25. Sousa, I.G., et al.: Gene expression profile of human T cells following a single stimulation of peripheral blood mononuclear cells with anti-CD3 antibodies. BMC Genomics **20**(1), 593 (2019). https://doi.org/10.1186/s12864-019-5967-8
26. Tooley, J.E., et al.: Changes in T-cell subsets identify responders to FcR-nonbinding anti-CD3 mAb (teplizumab) in patients with type 1 diabetes. Eur. J. Immunol. **46**(1), 230–241 (2016)
27. Wang, P., et al.: The STAT3-binding long noncoding RNA lnc-DC controls human dendritic cell differentiation. Science **344**(6181), 310–313 (2014)

A Simplified Complex Network-Based Approach to mRNA and ncRNA Transcript Classification

Murilo Montanini Breve[ID] and Fabrício Martins Lopes[✉][ID]

Departamento Acadêmico de Computação (DACOM),
Universidade Tecnológica Federal do Paraná (UTFPR) Câmpus Cornélio Procópio,
Av. Alberto Carazzai,
1640, Cornélio Procópio, PR CEP 86300-000, Brazil
`murilobreve@alunos.utfpr.edu.br`, `fabricio@utfpr.edu.br`

Abstract. Bioinformatics is an interdisciplinary area that presents several important computational challenges. These challenges are usually related to the large volume of biological data generated and that needs to be analyzed for information discovery. An important challenge is the need to distinguish mRNAs and ncRNAs in an efficient and assertive way. The correct identification of these transcripts is due to the existence of thousands of non-coding transcripts, whose function and meaning are not known, as well as the challenge to understand the expression and regulation of genetic information. On the other hand, the complex network theory has been successfully applied in many real-world problems in different contexts. Therefore, this work presents a simplified and efficient complex network-based approach for the classification of mRNA and ncRNA sequences. Experiments were performed to evaluate the proposed approach considering a dataset with six different species and with important methods in the literature such as CPC, CPC2 and PLEK. The results indicated the assertiveness of the proposed approach achieving average accuracy rates higher than 98% in the classification of mRNA and ncRNA considering all compared species. Besides, the proposed approach presents fewer variations on its results when compared to competitor methods, indicating its robustness and suitability for the classification of transcripts.

Keywords: RNA classification · Complex networks · Feature extraction · Bioinformatics · Pattern recognition

1 Introduction

After more than 150 years of the discovery of nucleic acids, the interest in the study of these molecules has been growing over the years [1]. Since the sequencing of bacteriophage ϕX174 in 1977 [2], a high amount of organisms have been sequenced and stored in databases. The advances in the development of sequencing technologies has led to the generation of a huge volume of biological data.

J. C. Setubal and W. M. Silva (Eds.): BSB 2020, LNBI 12558, pp. 192–203, 2020.
https://doi.org/10.1007/978-3-030-65775-8_18

Thus, it became essential that computational tools were developed to analyze these data. This need led to the emergence of a new interdisciplinary research field: Bioinformatics, which had its beginning in the 1970's, defined as "the study of informatic processes in biotic systems"[3]. Since then, bioinformatics has become an important interdisciplinary research field.

Nowadays, the high-throughput sequencing methods, such as RNA-seq, allows the generation of a massive volume of transcritomic data [4]. Besides the quantification of transcriptomic sequences [5], it is important to assign an annotation to these sequences (reads), e.g. their functionality [6].

Two classes of transcripts have received a lot of attention due to their functionality in organisms. The first is the mRNAs that carry information for protein synthesis. The second are the non-coding RNAs (ncRNAs), for which the functionality of only a few is known. In fact, the functionality of the vast majority of ncRNA remains unknown [7].

Therefore, it is important to identify the classes of RNA in order to better understand the mechanisms of action and their functionality and thus contribute to effective analysis and annotation of these sequences. In particular, ncRNAs are very heterogeneous in terms of length, conformation and cellular function. The ncRNA can be organized into long non-coding RNA (lncRNA) with sequences >200 nucleotides and small non-coding RNA (sncRNA) [8,9]. In addition, recent findings indicate that lncRNAs has an important participation in many biological processes, such as transcriptional regulation [10], complex diseases [11] including cancer [12,13], and also analyzed regarding preserved regions in their structure due to the evolution of living beings [14].

In this context, some methods have been proposed in the literature with the objective of classifying transcripts, such as: Coding Potential Calculator (CPC) [15] and its updated version CPC2 [16], PLEK [17] and BiologicAl Sequences NETwork (BASiNET) [18]. However, the CPC and CPC2 methods use biological information about the composition of sequences such as open reading frame (ORF) features and also sequence alignment scores from UniProt protein database. PLEK is a predictor of long non-coding RNAs and messenger RNAs based on an improved k-mer scheme that considers the nucleotide frequency (k-mer) as feature taking into account a sliding window to count the k-mer ranging from 1 to 5, not considering sequence alignment or features related to the structure of the molecule, such as the position or adjacency between the nucleotides. BASiNET presents a methodology for feature extraction without considering any biological information a priori about the sequence. This method performs a mapping from biological sequences to networks (vertices and edges) and topological measurements are extracted from these networks through the complex networks theory[19,20].

This work proposes a method for the classification of coding (mRNA) and non-coding (ncRNA) sequences, based on the BASiNET method, thus considering the mapping of biological sequences in networks and extracting topological measures, however improving the learning process and the feature selection leading to more efficient and accurate classification. The proposed method was evaluated and compared with the methodologies CPC, CPC2 and PLEK, considering CPC2 dataset[16] with six different species in which methodology obtained adequate results and with greater accuracy than the competing methods considering all the species evaluated.

2   Graphs and Complex Networks

The study of graphs can be seen as a subarea of mathematics, having as function the analysis of the relations between the objects of a certain context [21]. For this purpose structures called graphs are used, in short, a graph can be seen as a set of nodes connected by edges, which represent connections between the nodes. For example, if a sequence composed of the 'ACG' nucleotides is considered, and the neighborhood of the nucleotides is used as a criterion for its connection, this sequence can be mapped in a graph such as the one shown in Fig. 1.

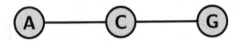

Fig. 1. Graph generated from the sequence 'ACG', considering the nucleotides as nodes and its neighborhood as edges.

If the frequency of neighborhood of a given nucleotide is considered, the graph may contain weights on its edges, maintaining the unique occurrence of the nodes and traversing the sequence for the identification of neighborhoods. Besides, instead of considering a single nucleotide as a graph node, it can be considered k-mers, with different values of k. Figure 2 shows an example of a graph generated from the sequence 'GCACCGGCCG' considering $k = 2$.

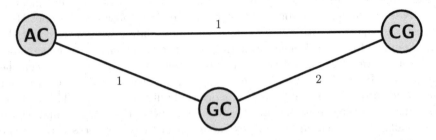

Fig. 2. Graph generated from the sequence 'GCACCGGCCG', considering k-mers, with $k = 2$ as nodes and its neighborhood as edges.

Complex networks can be seen as graphs with non-trivial topologies [22]. These complex networks are present in several areas of knowledge, having been successfully applied in the representation of many real-world relationships in different areas [18,23–32].

An important aspect of the complex networks is their dynamics, i.e. how the topological structure of networks changes by some factor such as time, threshold, perturbation, etc. leading to a dynamics for networks [19,20].

The theory of complex networks presents a variety of topological measurements that can be used to characterize and to represent the topologies of these networks [19].

3   Materials and Methods

3.1   Materials

This work adopts a dataset in order to validate the proposed method as well as to compare its results with the competitor methods. The adopted dataset was obtained from CPC2 [16], which contains six species of organisms: *Arabidopsis thaliana*, human, zebrafish, fruitfly, mouse and worm. The CPC2 dataset presents transcripts (mRNA), small non-coding transcripts (sncRNAs) and long non-coding transcripts (lncRNAs). In this work the sncRNA and lncRNA sequences were grouped in a ncRNA subset. Redundant sequences were removed from the dataset. The adopted dataset, the number of samples per class and species are presented in Table 1.

Table 1. Description of the number of samples per class of the dataset adopted in this work.

Species	Class of RNAs	Number of samples
Arabidopsis	mRNA	15931
	ncRNA	3853
Human	mRNA	6142
	ncRNA	12015
Zebrafish	mRNA	2344
	ncRNA	1528
Fruitfly	mRNA	3680
	ncRNA	3556
Mouse	mRNA	10638
	ncRNA	12251
Worm	mRNA	3551
	ncRNA	9313

3.2   Methods

The first step is to map the sequences in complex networks. Therefore, the sequences were organized in FASTA files, and the messenger RNA and non-coding RNA sequences were separated, that is, they were previously classified in a supervised learning model.

For the mapping of sequences in complex networks, two parameters were adopted: k and *step*. The *step* parameter defines the distance traveled in the sequence to define the neighborhood of each iteration. The k parameter defines the amount of nucleotides for forming the k-mer. Figure 3 shows an example in which the network was generated from the sequence 'ACCGATG' with the parameters $step = 1$ and $k = 3$, as defined by the BASiNET [18] method.

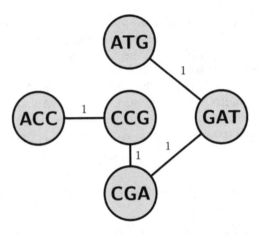

Fig. 3. Graph generated from the sequence 'ACCGATG', considering k-mers, with $k = 3$ as nodes and its neighborhood with $step = 1$ as edges.

The second step is to consider a feature selection approach in order to reduce noise and improve network representativeness leading to reduced complexity and contributing to the classification step. The feature selection consists of identifying the exclusive edges for each type of sequences: mRNA and ncRNA, so that before performing the feature extraction (topological measurements), filter the edges avoiding the appearance of edges which are repeated in both classes. Figure 4 presents an overview of the proposed feature selection approach. It is important to note that the network's adjacency matrix is transformed into a binary matrix by filtering the edges according to a percentage of exclusivity (defined as parameter), leading to the identification of the unique edges between the RNA classes and providing the filtering in a simplified way by a subtraction of matrices.

To identify the exclusivity parameter that produces an adequate selection, the adopted dataset was analyzed and the exclusivity rates were identified in an iterative way, producing the plot shown in Fig. 5. It is possible to observe that

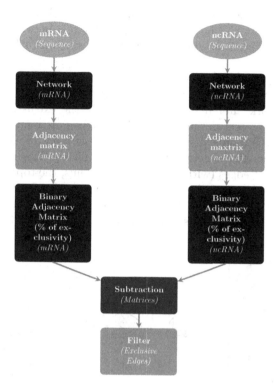

Fig. 4. Overview of the proposed feature selection approach.

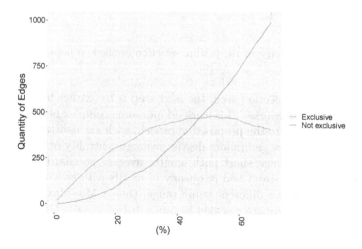

Fig. 5. Quantity of exclusive and non-exclusive edges identified in the dataset in both classes of RNA.

the higher point on the curve of the exclusive edges is between 40% and 60%, and that the saturation point of the exclusive curve is between 40% and 50%. Thus, the exclusivity parameter adopted in this work was 45% for the feature selection step.

Figure 6 presents an overview of the feature selection step. It is possible to notice that (a) presents the exclusive edges, (b) presents the original network and (c) the network filtered by considering only the exclusive edges, i.e. the edges that are not present in the filter (exclusive edge) are removed. This feature selection reduces the complexity of the BASiNET method [18]. More specifically, BASiNET considers all the network edges and applies a threshold iteratively to remove less frequent edges (with lower weights). Thus, the topological measurements are extracted at each iteration at different levels of network resolution. As a consequence, the proposed feature selection approach eliminates the need to apply the threshold and leads to an improved and simplified approach.

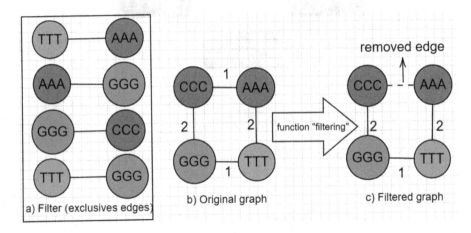

Fig. 6. Overview of the feature selection applied on networks.

After the feature selection step, the next step is to extract topological measurements from the networks. This work adopts some complex network measurements commonly used in the proposed approach, such as: assortativity, medium degree, maximum degree, minimum degree, average centrality of intermediation, cluster coefficient, average short path length, average standard deviation, frequency of motifs with size 3 and frequency of motifs with size 4 [20]. However, each measurement has a different value range, thus a Min-Max is applied and the measurements values are rescaled in range [0,1].

The last step consists of classifying the sequences based on their topological features extracted from their respective networks. For this step, the supervised learning was adopted taking into account the Random Forest [33] classifier. The R project [34] was adopted and the rfUtilities [35] package was also considered for 10-fold cross-validation.

4   Results and Discussion

To evaluate the proposed approach, the adopted dataset (Sect. 3.1) was considered in the same way for all the methods. The proposed approach was performed considering the following parameters values: edge exclusivity = 45%, $step = 1$ and $k = 3$, which were presented and their values justified in Sect. 3.2.

Table 2 presents the accuracy rates of classifications regarding the mRNAs and ncRNAs using the 10-fold cross validation. It is possible to verify the adequacy of the proposed approach for the correct identification between ncRNA and mRNA sequences with a high accuracy rate achieving accuracy rates higher than 98,7% for all species.

Table 2. Accuracy rates in the classification of mRNA and ncRNA sequences using the proposed method for different species.

Species	Accuracy	
	mRNA	ncRNA
Human	6142 (100%)	12005 (99.9%)
Fruitfly	3665 (99,5%)	3551 (99,7%)
Mouse	10638 (100%)	12251 (100%)
Zebrafish	2313 (98,7%)	1528 (100%)
Arabidopsis	15930 (100%)	3834 (100%)
Worm	3523 (99,7%)	9312 (100%)

Figure 7 presents the average accuracy of the adopted methods for the classification of the mRNA and ncRNA sequences, considering each species of the dataset. It can be noted that the CPC and PLEK present greater variations in their results. While the proposed approach and the CPC2 show more stable behaviors, the proposed approach presents superior results when compared to all competitor methods, indicating the suitability in the classification of the mRNA and ncRNA sequences.

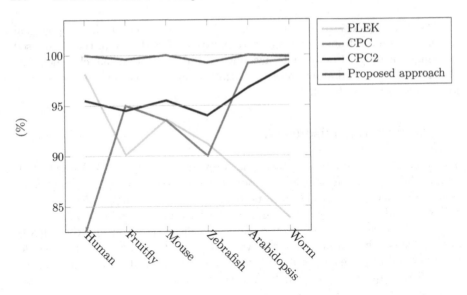

Fig. 7. Comparison of the accuracy of the methods considering the species separately.

Table 3. Comparison of classifiers by class and average accuracy.

	PLEK	CPC	CPC2	Proposed
mRNA	79.9%	99,50%	95,20%	99,65%
ncRNA	92.31%	87,30%	97,14%	99,92%
Average accuracy	**90.75%**	**93,20%**	**96,1%**	**99,78%**

Table 3 presents the average results considering all the species available in the dataset. It can be noted that the classification of mRNA and ncRNA sequences by the proposed approach presents results superior to competing methods, again indicating their adequacy in the classification of transcripts.

The results indicated a high accuracy in the identification of RNA sequences by the proposed approach. The average results obtained both considering the species individually and when grouped indicate the robustness of the proposed approach, with small variations and superior accuracies when compared with competitor methods. Therefore, the feature selection approach by filtering the exclusive edges proved to be adequate for the correct identification of the features, reducing the complexity of the classification and with a high accuracy rate for transcripts identification.

5   Conclusion

Complex network theory has been successfully applied in modeling various real-world problems, in particular in bioinformatics. This work applies the theory of complex networks by adopting the mapping of transcripts in networks and performing the extraction of topological measurements from these networks. These measurements are used as features vector for the classification of mRNA and ncRNA sequences. More specifically, this work is based on the BASiNET method [18], and proposes a feature selection approach by filtering exclusive edges of the networks, leading to a reduction in the complexity and obtaining robust and adequate results.

Experiments were performed to evaluate the proposed approach considering the CPC2 dataset [16] with mRNA and ncRNA sequences with six different species. The experiments were performed comparing the classification results with important methods in the literature such as CPC [15], CPC2 [16] and PLEK [17]. The results were obtained considering the 10-fold cross-validation. In general, the proposed approach achieved average accuracy rates above 98% in the classification of mRNA and ncRNA considering the all compared species, indicating superior results when compared with competitor methods. Besides, the proposed approach presents less variations on its results when compared to competitor methods, indicating its robustness and suitability for the classification of transcripts.

As a future work, a further study considering the proposed approach can lead to a better understanding of the structure of non-coding RNA. In addition, the proposed approach can be applied to biological sequences considering other contexts and of different types of biological sequences.

Acknowledgments. This work was financed in part by the Coordenação de Aperfeiçoamento de Pessoal de Nível Superior - Brasil (CAPES), Conselho Nacional de Desenvolvimento Científico e Tecnológico - CNPq (Grant number 406099 /2016-2) and the Fundação Araucária e do Governo do Estado do Paraná/SETI (Grant number 035/2019).

References

1. Dahm, R.: Friedrich Miescher and the discovery of DNA. Dev. Biol. **278**(2), 274–288 (2005)
2. Sanger, F., et al.: Nucleotide sequence of bacteriophage ϕ X174 DNA. Nature **265**(5596), 687–695 (1977)
3. Hogeweg, P.: The roots of bioinformatics in theoretical biology. PLOS Comput. Biol. **7**(3), 1–5 (2011)
4. Wang, Z., Gerstein, M., Snyder, M.: RNA-Seq: a revolutionary tool for transcriptomics. Nat. Rev. Genet. **10**(1), 57–63 (2009)
5. Costa-Silva, J., Domingues, D., Lopes, F.M.: RNA-Seq differential expression analysis: an extended review and a software tool. PLOS One **12**(12), 1–18 (2017)
6. Garber, M., Grabherr, M.G., Guttman, M., Trapnell, C.: Computational methods for transcriptome annotation and quantification using RNA-Seq. Nat. Methods **8**(6), 469–477 (2011)
7. Panwar, B., Arora, A., Raghava, G.P.S.: Prediction and classification of ncRNAs using structural information. BMC Genomics **15**(1), 1–13 (2014)

8. Wang, K.C., Chang, H.W.: Molecular mechanisms of long noncoding RNAs. Mol. Cell **43**(6), 904–914 (2011)
9. Peng, Y., Li, J., Zhu, L.: Chapter 8 - cancer and non-coding RNAs. In: Ferguson, B.S. (ed.)Nutritional Epigenomics, volume 14 of Translational Epigenetics, pp. 119–132. Academic Press (2019)
10. Long, Y., Wang, X., Youmans, D.T., Cech, T.R.: How do lncRNAs regulate transcription? Sci. Adv. **3**(9), p. eaao2110 (2017)
11. Chen, X., Yan, G.-Y.: Novel human lncRNA-disease association inference based on lncRNA expression profiles. Bioinformatics **29**(20), 2617–2624 (2013)
12. Huarte, M.: The emerging role of lncRNAs in cancer. Nat. Med. **21**(11), 1253–1261 (2015)
13. Peng, W.-X., Koirala, P., Mo, Y.-Y.: LncRNA-mediated regulation of cell signaling in cancer. Oncogene **36**(41), 5661–5667 (2017)
14. Kung, J.T., Colognori, D., Lee, J.T.: Long noncoding RNAs: past, present, and future. Genetics **193**(3), 651–669 (2013)
15. Kong, L., et al.: CPC: assess the protein-coding potential of transcripts using sequence features and support vector machine. Nucleic Acids Res. **35**(suppl_2), W345–W349 (2007)
16. Kang, Y.J., et al.: Cpc2 a fast and accurate coding potential calculator based on sequence intrinsic features. Nucleic Acids Res. **45**(W1), W12–W16 (2017)
17. Li, A., Zhang, J., Zhou, Z.: PLEK: a tool for predicting long non-coding RNAs and messenger RNAs based on an improved k-mer scheme. BMC Bioinformatics **15**(1), 311 (2014)
18. Ito, E.A., Katahira, I., Vicente, F.F.D.R., Pereira, L.F.P., Lopes, F.M.: BASiNET - biological sequences network: a case study on coding and non-coding RNAs identification. Nucleic Acids Res. **46**(16), e96–e96 (2018)
19. Boccaletti, S., Latora, V., Moreno, Y., Chavez, M., Hwang, D.-U.: Complex networks: structure and dynamics. Phys. Rep. **424**(4), 175–308 (2006)
20. Costa, L.D.F., Rodrigues, F.A., Travieso, G., Villas Boas, P.R.: Characterization of complex networks: a survey of measurements. Adv. Phys. **56**(1), 167–242 (2007)
21. Diestel, R.: Graph Theory, 3rd edn. Springer-Verlag, Heidelberg (2005)
22. Barabási, A.L.: Linked: How Everything Is Connected to Everything Else and What It Means. Plume, New York (2003)
23. Watts, D.J., Strogatz, S.H.: Collective dynamics of small-world networks. Nature **393**, 440–442 (1998)
24. Barabási, A.-L., Albert, R.: Emergence of scaling in random networks. Science **286**(5439), 509–512 (1999)
25. Newman, M.E.J.: The structure and function of complex networks. SIAM Rev. **45**(2), 167–256 (2003)
26. Lopes, F.M., Cesar, R.M., da F. Costa, L.: AGN simulation and validation model. In: Bazzan, A.L.C., Craven, M., Martins, N.F. (eds.) BSB 2008. LNCS, vol. 5167, pp. 169–173. Springer, Heidelberg (2008). https://doi.org/10.1007/978-3-540-85557-6_17
27. Yi Ming Zou: Modeling and analyzing complex biological networks incooperating experimental information on both network topology and stable states. Bioinformatics **26**(16), 2037–2041 (2010)
28. Lopes, F.M., Cesar Jr, R.M., Costa, L.D.F.: Gene expression complex networks: synthesis, identification, and analysis. J. Comput. Biol. **18**(10), 1353–1367 (2011)
29. Lopes, F.M., Martins Jr., D.C., Barrera, J., Cesar Jr., R.M.: A feature selection technique for inference of graphs from their known topological properties: revealing scale-free gene regulatory networks. Inf. Sci. **272**, 1–15 (2014)

30. da Rocha Vicente, F.F., Lopes, F.M.: SFFS-SW: a feature selection algorithm exploring the small-world properties of GNs. In: Comin, M., Käll, L., Marchiori, E., Ngom, A., Rajapakse, J. (eds.) PRIB 2014. LNCS, vol. 8626, pp. 60–71. Springer, Cham (2014). https://doi.org/10.1007/978-3-319-09192-1_6

31. de Lima, G.V.L., Castilho, T.R., Bugatti, P.H., Saito, P.T.M., Lopes, F.M.: A complex network-based approach to the analysis and classification of images. CIARP 2015. LNCS, vol. 9423, pp. 322–330. Springer, Cham (2015). https://doi.org/10.1007/978-3-319-25751-8_39

32. de Lima, G.V., Saito, P.T., Lopes, F.M., Bugatti, P.H.: Classification of texture based on bag-of-visual-words through complex networks. Expert Syst. Appl. **133**, 215–224 (2019)

33. Liaw, A., Wiener, M.: Classification and regression by randomforest. R News **2**(3), 18–22 (2002)

34. R Core Team. R: A Language and Environment for Statistical Computing. R Foundation for Statistical Computing, Vienna, Austria (2014)

35. Evans, J.S., Murphy, M.A.: rfUtilities. R package version 2.1-3 (2018)

A Systems Biology Driven Approach to Map the EP300 Interactors Using Comprehensive Protein Interaction Network

Shivananda Kandagalla[1]([✉]), Maria Grishina[1], Vladimir Potemkin[1]([✉]),
Sharath Belenahalli Shekarappa[2], and Pavan Gollapalli[3]

[1] Laboratory of Computational Modeling of Drugs, Higher Medical and Biological School,
South Ural State University, Chelyabinsk 454080, Russia
kandagallas@gmail.com, potemkinva@susu.ru
[2] Department of PG Studies and Research in Biotechnology and Bioinformatics, Kuvempu
University, JnanaSahyadri, Shankaraghatta, Shivamogga 577451, Karnataka, India
[3] Central Research Lab, K.S. Medical Academy, Nitte University (Deemed To Be University),
Deralakatte, Mangalore 575018, Karnataka, India

Abstract. EP300 is one of the putative tumor-suppressor genes and is mutated/deleted, under expressed/overexpressed in several types of cancer. The role of EP300 and its interactions during cancer is crucial to explore its reprogramming events that lead to malignant phenotype and acquisition of drug resistance. In this context, all the experimentally valid EP300 interactors were collected from the primary protein-protein interaction (PPI) databases and followed by tracing their subcellular location using the UniProtKB database. Further, all the EP300 interactors were categorized based on their subcellular location and functionally annotated with the DAVID gene ontology tool. Subsequently, the interactome of EP300 with its interactors was constructed and identified TP53, CREBBP, JUN, HDAC1, CTNNB1, MYC, PCNA, HDAC2, FOS, and KAT2B as the top first neighbors of EP300. Together, the present analysis gives a comprehensive overview on EP300 interactors located in different subcellular locations.

Keywords: EP300 · Interactome · Cytoscape

1 Introduction

EP300(p300) is a ubiquitously expressed transcriptional coactivator and a member of the EP300/CBP family of type 3 major lysine (K) acetyltransferases (KAT3), present in all mammals and found in many multicellular organisms, such as flies, worms, and plants. In humans, 31 exons in chromosome 22 (locus 22q13) codes for the p300 gene, and gene size spans approximately 90 kb. Overexpression and inappropriate activation of EP300 are associated with malignancy, tumor size, poor differentiation, tumor progression, and poor prognosis [1–3]. Increased expression of EP300 has been observed in advanced human malignancies, such as liver, prostate cancers, primary human breast cancers, etc., [4].Recent reports highlight EP300 as a central regulator of angiogenesis, hypoxia,

© Springer Nature Switzerland AG 2020
J. C. Setubal and W. M. Silva (Eds.): BSB 2020, LNBI 12558, pp. 204–214, 2020.
https://doi.org/10.1007/978-3-030-65775-8_19

and EMT pathway in esophageal squamous carcinoma [5]. The increased expression of cancer stem cell markers, tumorsphere formation was observed in EP300-depleted cells and diminished in EP300-overexpressing cells [6]. Apart from cancer, EP300 is a key player in Rubinstein − Taybi syndrome (RTS or RSTS) disease [7].

EP300 shares high sequence homology with CBP (CREBBR or KAT3A), and less with other acetyltransferases [8]. Both proteins have almost 86% amino acid residue identity in the catalytic domain and significant sequence homology was found in several types of protein-protein interacting motifs, and other non-catalytic domains [9]. In EP300, the acetyltransferase domain spans from residues 1284 to 1673, and IBiD (Interferon Binding Domain) located at the C-terminal side. The IBiD contains an NCBD (Nuclear Coactivator Binding Domain) and glutamine-rich domain, followed by a proline-containing PxP motif. There are three cysteine/histidine-rich domains (C/H) like C/H1, C/H2 (which is part of the catalytic domain), and C/H3. The C/H1 and C/H3 domains contain transcriptional adaptor zinc fingers (TAZ1 and TAZ2), and additionally, C/H3 domain contains a ZZ zinc finger. The C/H2 domain contains a plant homeodomain (PHD) and the domains such as interferon binding homology domain (IHD), KIX domain, and bromodomain is located between the C/H1 and C/H2 domains [10, 11].

EP300 functions as acetyltransferase by facilitating transcription through acetylation of histones, transcription factors, sequence-specific DNA binding factors, and basal transcriptional machinery. During intracellular or extracellular signaling the cell must turn different subsets of genes to regulate different cellular functions accomplished by acetylation of histone proteins during transcription. Most of the cellular signaling pathway such as cAMP signaling pathway, HIF-1 signaling pathway, FoxO signaling pathway, cell cycle, Wnt signaling pathway, Notch signaling pathway, TGF-beta signaling pathway, adherens junction signaling, Jak-STAT signaling pathway, DNA damage pathways, and other pathways use EP300 as downstream effector protein [12, 13]. EP300 responds to those signaling pathways differently, which mainly depends on the cell environment and its phosphorylation state. Various proteins such as PKC, cyclin E/CDK-2, CaMKIV, IKK, and AKT phosphorylate EP300 at different sites which ultimately impact on its acetyltransferase activity. Along with, self-modification (auto-acetylation) of EP300 also influences on the acetyltransferase activity. EP300 contain methylation sites near the KIX domain and lysine SUMOylation site near the bromodomain. EP300 also has acetylation site (17 lysine residues) in the regulatory loop of acetyltransferase domain and their acetylation is essential for its acetyltransferase activity, and for binding with other proteins. In addition, EP300 through protein interacting domains binds to the different proteins and thereby it regulates wide variety of signaling pathways [14, 15]. All these reports clearly show that EP300 regulates signaling pathways by interacting with multiple proteins and targeting these interactions during disease conditions could be a good solution. In this concern, all the experimentally valid datasets of EP300 interactors were collected from primary protein interaction databases, followed by tracing their subcellular locations and functional annotations. Finally, the interactome of EP300 with its interactors was developed and first-degree interactors were identified.

2   Materials and Method

2.1   Collection of EP300 Interactors

The experimentally detected proteins having the interaction with EP300 were extracted from the public databases such as, IntAct [16], BioGRID [17], APID [18], PINA [19], Mentha [20], HitPredict [21], WiKi-Pi [22], PIPs [23], PPI-finder [24] and PrePPI [25]. Non-human interactors of EP300 were excluded from the study. Using the UniProt Knowledge base (UniProtKB) Id mapping, the gene symbols and protein symbols were identified [26].

2.2   Protein Class and Subcellular Location Analysis

The subcellular location of EP300 interactors was explored using the UniProtKB database based on the record "Subcellular location". UniProtKB database act as a central hub in identifying functional information of proteins with accurate annotations, and also it includes widely accepted biological ontologies, classifications and cross-references, and clear indications of the quality of annotation in the form of evidence attribution of experimental and computational data (https://www.uniprot.org/help/uniprotkb). The PANTHER classification system was used to identify the protein classes of EP300 interactors. The PANTHER (Protein Analysis THrough Evolutionary Relationships) database contains comprehensive information on the evolution and function of protein-coding genes from 104 completely sequenced genomes. PANTHER classification tools allow users to classify new protein sequences and to analyze gene lists obtained from large scale genomics experiments [27, 28].

2.3   Functional Annotation and Pathway Enrichment Analysis

The EP300 interactors located in the different subcellular location were functionally annotated with gene ontology (GO) terms in the PANTHER database and the pathway enrichment analysis was performed in the DAVID database (The Database for Annotation, Visualization, and Integrated Discovery) against PANTHER and KEGG pathways with a p-value < 0.05 [29].

2.4   Construction of EP300 Interactome

The primary protein interaction data of EP300 interactors were extracted from STRING database v10.5 [31] with a high confidence score of 0.9. The interactions in the STRING database are derived from different sources: text mining, experiments, co-expression, neighborhood, gene fusion, and co-occurrence. The high confidence interaction of score above 0.9 indicates, all the interactions are validated in all the above-mentioned sources. The low confidence (score 0.7) interactions were considered for N4BP2, MSTO1, MYB, HOXD10, and KLF16 interactors. The interactome of EP300 with its interactors was constructed using Cytoscape 3.4.0 [30] based on the subcellular location and the first-degree interactors of EP300 were identified from the core interactome.

3 Results and Discussion

A total of 854 predicted or experimentally validated EP300 interactors were obtained from public databases as illustrated in the methods section. The predicted EP300 interactors in many of the above primary databases are mainly from indirect clues such as data mining, Bayesian prediction, structural information, and more information can be found in [32]. It is difficult to maintain the accuracy of predicted results from the indirect clue and Zhang QC et al. [25] reported that the interactions from indirect methods are often more indicative of functional associations between two proteins than of direct physical interactions. Hence, the interactors only with experimental evidence were selected to maintain accuracy and other computational predictions without experimental validations were excluded for the analysis. A total of 540 EP300 interactors were included for further analysis and a complete list is provided in the supplementary file [37].

3.1 Analysis of Subcellular Location and Protein Class

The subcellular location analysis shows that EP300 interactors were located in different cell locations, and further based on location we categorized EP300 interactors into three broad classes: (i) cytoplasm; (ii) nucleus; (iii) both in cytoplasm and nucleus. Among 540 EP300 interactors, 202 interactors present in both cytoplasm and nucleus, 72 interactors present in the cytoplasm, 263 interactors in the nucleus, and the remaining interactors subcellular location is not available in the UniprotKB database (excluded for further analysis). Cytoplasm location includes apical cell membrane, cell membrane, cytoskeleton, focal adhesion, mitochondrion outer mem-brane, etc., and nucleus location is found to include chromosome, centromere, nucleus matrix, PML body, etc. Further analysis of the protein class of these EP300 interactors shows that they are mainly associated with the protein class nucleic acid binding and transcription factors. Other enriched top protein classes are transferase, hydrolase, enzyme modulator, receptor, etc., are shown in subsequent figures (Fig. 1D; 2D; 3D). All these results give a comprehensive overview of EP300 interactors protein classes and their subcellular locations.

3.2 Functional Enrichment Analysis of EP300 Interactors

The functional enrichment analysis of EP300 interactors present in the different subcellular locations was done separately. The analysis shows that EP300 interactors present in the cytoplasm enriched in various biological and molecular functions. As shown in Fig. 1A, the interactors present in the cytoplasm are mainly participating in response to the extracellular stimulus, positive regulation of apoptosis, positive regulation of programmed cell death, positive regulation of cell death, and regulation of apoptosis. Further, these EP300 interactors are mainly enriched in the molecular function of ribonucleotide binding, nucleoside binding, and ATP binding process (Fig. 1B). The EP300 interactors present in the nucleus mainly participate in the regulation of transcription, regulation of transcription (DNA-dependent), regulation of transcription from RNA polymerase II promoter, and regulation of RNA metabolic process(Fig. 2A). DNA binding, transcription regulator activity, and transcription factor activity are the top enriched GO molecular function are shown in Fig. 2B.

Finally, the analysis of EP300 interactors present in both cytoplasm and nucleus revealed that most of the interactors are participate in the regulation of transcription, positive regulation of macromolecule metabolic process, and regulation of RNA metabolic process (Fig. 3B). The transcriptional regulator activity, transcription factor binding, transcriptional activator activity, and transcription factor activity are the top enriched molecular function (Fig. 3B). Together, this analysis provides the functional significance of EP300 interactors present in the different subcellular locations. Interactors present in the cytoplasm were mainly involved in the biological process such as extracellular stimulus, positive regulation of apoptosis, positive regulation of programmed cell death, positive regulation of cell death, etc. Whereas interactors present in both cytoplasm and nucleus were engaged in almost similar biological processes.

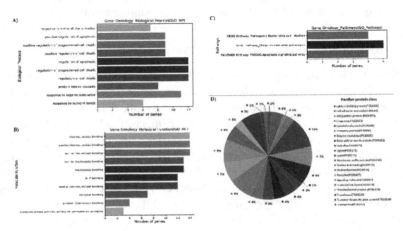

Fig. 1. Functional enrichment of EP300 interactors present in the Cytoplasm: Top annotated EP300 interactors involved in A) Biological Process, B) Molecular function, C) Pathways (KEGG and PANTHER), D) PANTHER protein class.

3.3 Pathway Enrichment Analysis of EP300 Interactors

The EP300 interactors present in the cytoplasm were enriched during pathogenic Escherichia coli infection and ubiquitin-mediated proteolysis based on KEGG pathway analysis. The PANTHER pathway results suggest that EP300 interactors are mainly associated with the apoptosis signaling pathway (Fig. 1C). The EP300 interactors present in nucleus were enriched in the cell cycle, DNA replication in KEGG pathways and the p53, p53 pathway feedback loop 2, oxidative response, and Wnt signaling are the top PANTHER enriched pathways (Fig. 2C).

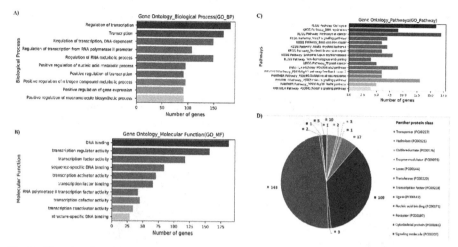

Fig. 2. Functional enrichment of EP300 interactors present in the Nucleus: Top annotated EP300 interactors involved in A) Biological Process, B) Molecular function, C) Pathways (KEGG and PANTHER), D) PANTHER protein class.

Finally, EP300 interactors present in both cytoplasm and nucleus are enriched during pathways in cancer, chronic myeloid leukemia, prostate cancer, acute myeloid leukemia, pancreatic cancer, cell cycle, and ErbB signaling pathway are among the top enriched KEGG pathways. The PDGF signaling pathway, JAK/STAT signaling pathway, B cell activation, T cell activation, p53 pathway, EGF receptor signaling pathway, p53 pathway feedback loops 2 and TGF-beta signaling pathway are among the top enriched PAN-THER pathways (Fig. 3C). Collectively these analysis provides the details on EP300 interactors associated pathways. From the results, it can be seen that apart from normal pathways, EP300 interactors also enriched in associated disease related pathways such as cancer, infection, etc. Further analysis of EP300 interactors associated with disease related pathways gives broad insights on the role of EP300 and also, it provides a new avenue in developing new drugs.

3.4 EP300 Interactome and Identification of First-Degree Nodes

The interactome of EP300 with its interactors was constructed to check the influence of EP300 based on the analysis of the first-degree nodes. First, the network of EP300 interactors present in the nucleus was constructed and the network consists of 1165 nodes and 6046 edges (Fig. 4A). Further first-degree nodes of EP300 were identified and these nodes have direct contact with EP300 and any alteration in these nodes changes the signaling pattern. A total of 135 nodes form the direct connection with EP300 and the top nodes based on the degree are TP53, CREBBP, HIST2H2BE, HDAC1, JUN, HIST2H2AC, H2AFZ, and MYC. Next, interactome of EP300 interactors present in the cytoplasm were constructed and the network has 635 nodes with 2331 edges (Fig. 4A). Further identified first-degree nodes of the EP300 in the network and with 18 nodes

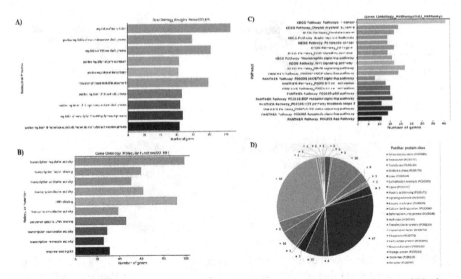

Fig. 3. Functional enrichment of EP300 interactors present in the both Cytoplasm and Nucleus: Top annotated EP300 interactors involved in A) Biological Process, B) Molecular function, C) Pathways (KEGG and PANTHER), D) PANTHER protein class.

EP300 has a direct connection. The UBA52, NR3C1, GRIP1, SREBF1, and JUN are the top nodes based on the degree. Further, the network of EP300 interactors present in the cytoplasm and nucleus were constructed and the network has 1214 nodes with 5443 edges (Fig. 4A). Total 89 nodes have a direct connection with EP300 and among TP53, JUN, AKT1, CREBBP, HDAC2, HDAC1, and MYC are top nodes based on degree (complete list is provided in the supplementary file [37]).

Altogether, the final interactome consists of 2388 nodes with 12577 edges. Among 2388 nodes Ep300 form the direct interaction with only 186 nodes (Fig. 4C) and among TP53, CREBBP, JUN, HDAC1, CTNNB1, MYC, PCNA, HDAC2, FOS and KAT2B are the top interactors. Previously several reports show the significance of EP300 interactions with TP53, CREBBP, JUN, CTNNB1, and MYC in several pathophysiological conditions [33–36], and still, their role is not clearly understood. Further, *in vitro* validation of these interactors is required to understand the role of EP300 in different cancer conditions and which ultimately helps in developing the novel inhibitor also, these interactors act as potential biomarkers.

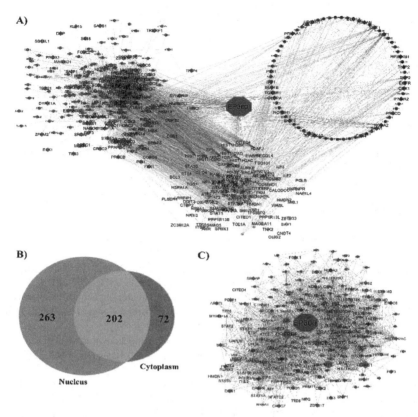

Fig. 4. Interactome of EP300 with it interactors, A) The PPI network of EP300 interactors present in nucleus are colored in red, green nodes corresponds to cytoplasm, and grey node corresponds to both nucleus and cytoplasm. B) Venn diagram showing number of EP300 interacting partners present in different subcellular location. C) First neighbors of EP300 in the interactome.

4   Conclusion

The evaluation of EP300 interactors present in different subcellular locations provides a broad sense to the role of EP300 in complex disease or cellular events. The functional and pathways enrichment analysis of EP300 interactors clearly shows their involvement in several pathological conditions and mainly in cancer. Among the EP300 interactors, TP53, CREBBP, JUN, HDAC1, CTNNB1, MYC, PCNA, HDAC2, FOS, and KAT2B are the top first-degree nodes, and these interactors are the key players with which EP300 interact and perform its functions. Further, *in vitro* validation of these interactors with EP300 is required in different cancer conditions. Altogether, the present analysis gives the complete overview on EP300 interactors presents different subcellular locations.

Acknowledgement. The work was supported by Act 211 Government of the Russian Federation, contract 02.A03.21.0011 and by the Ministry of Science and Higher Education of Russia (Grant FENU-2020–0019).

Conflict of Interest. All authors declared that they have no competing interest.

References

1. Karamouzis, M.V., Konstantinopoulos, P.A., Papavassiliou, A.G.: Roles of CREB-binding protein (CBP)/p300 in respiratory epithelium tumorigenesis. Cell Res. **17**, 324–332 (2007). https://doi.org/10.1038/cr.2007.10
2. Dutto, I., Scalera, C., Prosperi, E.: CREBBP and p300 lysine acetyl transferases in the DNA damage response. Cell. Mol. Life Sci. **75**(8), 1325–1338 (2017). https://doi.org/10.1007/s00 018-017-2717-4
3. Mees, S.T., Mardin, W.A., Wendel, C., et al.: EP300-A miRNA-regulated metas-tasis suppressor gene in ductal adenocarcinomas of the pancreas. Int. J. Cancer **126**, 114–124 (2010). https://doi.org/10.1002/ijc.24695
4. Yang, H., Pinello, C.E., Luo, J., et al.: Small-molecule inhibitors of acetyltrans-ferase p300 identified by high-throughput screening are potent anticancer agents. Mol. Cancer Ther. **12**, 610–620 (2013). https://doi.org/10.1158/1535-7163.MCT-12-0930
5. Bi, Y., Zhang, L., et al.: EP300 promotes tumor development and correlates with poor prognosis in esophageal squamous carcinoma. Oncotarget **9**(1), s376–s392 (2018)
6. Asaduzzaman, M., et al.: Tumour suppressor EP300, a modulator of paclitaxel resistance and stemness, is downregulated in metaplastic breast cancer. Breast Cancer Res. Treat. **163**(3), 461–474 (2017). https://doi.org/10.1007/s10549-017-4202-z
7. Babu, A., et al.: Chemical and genetic rescue of an ep300 knockdown model for Rubinstein Taybi Syndrome in zebrafish. Biochim. Biophys. Acta - Mol. Basis Dis. **1864**, 1203–1215 (2018). https://doi.org/10.1016/j.bbadis.2018.01.029.
8. Gayther, S.A., Batley, S.J., Linger, L., et al.: Mutations truncating the EP300 acetylase in human cancers. Nat. Genet. **24**, 300–303 (2000). https://doi.org/10.1038/73536
9. Chan, H.M., La Thangue, N.B.: p300/CBP proteins: HATs for transcriptional bridges and scaffolds. J. Cell. Sci. **114**, 2363–2373 (2001)
10. Ogryzko, V.V., Schiltz, R.L., Russanova, V., et al.: The transcriptional coactiva-tors p300 and CBP are histone acetyltransferases. Cell **87**, 953–959 (1996)
11. Yang, X.J., Seto, E.: Lysine acetylation: codified crosstalk with other post-translational modifications. Mol. Cell. **31**, 449–461 (2008). https://doi.org/10.1016/j.molcel.2008.07.002
12. Vo, N., Goodman, R.H.: CREB-binding protein and p300 in transcriptional regulation. J. Biol. Chem. **276**, 13505–13508 (2001). https://doi.org/10.1074/jbc.R000025200https://doi.org/10. 1074/jbc.R000025200
13. Bedford, D.C., Brindle, P.K.: Is histone acetylation the most important physio-logical function for CBP and p300? Aging (Albany NY) **4**, 247–55 (2012). https://doi.org/10.18632/aging. 100453
14. Attar, N., Kurdistani, S.K.: Exploitation of EP300 and CREBBP lysine acetyltransferases by cancer. Cold Spring Harb. Perspect. Med. **7**, a026534 (2017). https://doi.org/10.1101/cshper spect.a026534
15. Dancy, B.M., Cole, P.A.: Protein lysine acetylation by p300/CBP. Chem. Rev. **115**, 2419–2452 (2015). https://doi.org/10.1021/cr500452k
16. Hermjakob, H., Montecchi-Palazzi, L., Lewington, C., et al.: IntAct: an open source molecular interaction database. Nucleic Acids Res. **32**, D452–D455 (2004). https://doi.org/10.1093/nar/ gkh052
17. Stark, C., Breitkreutz, B.J., Reguly, T., et al.: BioGRID: a general repository for interaction datasets. Nucleic Acids Res. **34**, D535–D539 (2006). https://doi.org/10.1093/nar/gkj109

18. Alonso-López, D., Gutiérrez, M.A., Lopes, K.P., et al.: APID interactomes: providing proteome-based interactomes with controlled quality for multiple spe-cies and derived networks. Nucleic Acids Res. **44**, W529–W535 (2016). https://doi.org/10.1093/nar/gkw363
19. Cowley, M.J., Pinese, M., Kassahn, K.S., et al.: PINA v2.0: mining interactome modules. Nucleic Acids Res. **40**, D862–D865 (2012). https://doi.org/10.1093/nar/gkr967
20. Calderone, A., Castagnoli, L., Cesareni, G.: mentha: aresource for browsing integrated protein-interaction networks. Nat. Methods **10**, 690–691 (2013). https://doi.org/10.1038/nmeth.2561
21. Patil, A., Nakai, K., Nakamura, H.: HitPredict: a database of quality assessed protein-protein interactions in nine species. Nucleic Acids Res. **39**, D744–D749 (2011). https://doi.org/10.1093/nar/gkq897
22. Orii, N., Ganapathiraju, M.K.: Wiki-Pi: a web-server of annotated human protein-protein interactions to aid in discovery of protein function. PLoS ONE **7**, e49029 (2012). https://doi.org/10.1371/journal.pone.0049029
23. McDowall, M.D., Scott, M.S., Barton, G.J.: PIPs: human protein-protein interac-tion pre-diction database. Nucleic Acids Res. **37**, D651–D656 (2009). https://doi.org/10.1093/nar/gkn870
24. He, M., Wang, Y., Li, W.: PPI Finder: a mining tool for human protein-protein interactions. PLoS ONE **4**, e4554 (2009). https://doi.org/10.1371/journal.pone.0004554
25. Zhang, Q.C., Petrey, D., Garzón, J.I., et al.: PrePPI: a structure-informed data-base of protein-protein interactions. Nucleic Acids Res. **41**, D828–D833 (2013). https://doi.org/10.1093/nar/gks1231
26. Bateman, A., Martin, M.J., O'Donovan, C., et al.: UniProt: the universal protein knowl-edgebase. Nucleic Acids Res. **45**, D158–D169 (2017). https://doi.org/10.1093/nar/gkw1099
27. Thomas, P.D., Campbell, M.J., Kejariwal, A., et al.: PANTHER: a library of pro-tein families and subfamilies indexed by function. Genome Res **13**, 2129–2141 (2003). https://doi.org/10.1101/gr.772403
28. Mi, H., Dong, Q., Muruganujan, A., et al.: PANTHER version 7: improved phy-logenetic trees, orthologs and collaboration with the gene ontology consortium. Nucleic Acids Res. **38**, D204–D210 (2010). https://doi.org/10.1093/nar/gkp1019
29. Huang, D.W., Sherman, B.T., Tan, Q., et al.: The DAVID gene functional classification tool: a novel biological module-centric algorithm to functionally analyze large gene lists. Genome Biol. **8**, R183 (2007). https://doi.org/10.1186/gb-2007-8-9-r183
30. Shannon, P., Markiel, A., Ozier, O., et al.: Cytoscape: a software environment for integrated models of biomolecular interaction networks. Genome Res. **13**, 2498–2504 (2003). https://doi.org/10.1101/gr.1239303
31. Szklarczyk, D., Morris, J.H., Cook, H., et al.: The STRING database in 2017: quality-controlled protein-protein association networks, made broadly accessible. Nucleic Acids Res. **45**, D362–D368 (2017). https://doi.org/10.1093/nar/gkw937
32. Kong, F.Y., et al.: Bioinformatics analysis of the proteins interacting with LASP-1 and their association with HBV-related hepatocellular carcinoma. Sci. Rep. **7**, 1–5 (2017). https://doi.org/10.1038/srep44017
33. Kaypee, S., et al.: Mutant and wild-type tumor suppressor p53 induces p300 autoacetylation. iScience **4**, 260–272 (2018). https://doi.org/10.1016/j.isci.2018.06.002
34. Attar, N., Kurdistani, S.K.: Exploitation of EP300 and CREBBP lysine acetyltransferases by cancer. Cold Spring Harb. Perspect. Med. 7 (2017). https://doi.org/10.1101/cshperspect.a026534.

35. Yu, W., et al.: Cellular Physiology and Biochemistry Cellular Physiology and Biochemistry Yu et al.: B-catenin cooperates with CREB binding protein β-catenin cooperates with CREB binding protein to promote the growth of tumor cells cellular physiology and biochemistry cellular physiology and biochemistry. Cell Physiol. Biochem. **44**, 467–478 (2017). https://doi.org/10.1159/000485013.

36. Wang, Y.N., Chen, Y.J., Chang, W.C.: Activation of extracellular signal-regulated kinase signaling by epidermal growth factor mediates c-Jun activation and p300 recruitment in keratin 16 gene expression. Mol. Pharmacol. **69**, 85–98 (2006). https://doi.org/10.1124/mol.105.016220

37. Kandagalla, S., Grishina, M., Potemkin, V., Shekarappa, S.B., Gollapalli, P.: A systems biology driven approach to map the EP300 interactors using comprehensive protein interaction network (2020). https://doi.org/10.5281/ZENODO.4112838

Analyzing Switch Regions of Human Rab10 by Molecular Dynamics Simulations

Levy Bueno Alves[1](✉) ⬤, William O. Castillo-Ordoñez[2] ⬤, and Silvana Giuliatti[1] ⬤

[1] Department of Genetics, University of São Paulo, Ribeirão Preto, Brazil
levybuenoalves@usp.br
[2] Department of Biology, University of Cauca, Popayán, Colombia

Abstract. Rab10 is a small GTPase that regulates cellular processes by alternating between its GDP-bound inactive and the GTP-bound active states. Studies have shown that functional deficiencies in the Rab10 pathways are implicated in ciliophaties, gliobastomas and neurodegenerative diseases. Thus, the modulation of Rab10 activity may represent an interesting strategy in drug discovery. In order to identify potential Rab10 inhibitors for the treatment of Alzheimer's disease, we studied the mobility of the switch1-interswitch-switch2 surface to understand the active "ON" and inactive "OFF" states of this enzyme. Even today, no in silico study on Rab10 linked to GTP and GDP has been carried out. We used molecular dynamics simulations to investigate the atomic movements of the Rab10 switch regions associated with these nucleotides. We found noticeable differences in the local flexibility of switch 1 when Rab10 was linked to GDP. However, the heuristic method used was not able to successfully differentiate the flexibility of switch 2 region. We hypothesized that the flexibility of the switch 1 region can be used as an indicator of in silico studies that search potential competitive inhibitors based on nucleotides against Rab10. Furthermore, the present study can be useful for research that involves the description on-to-off process of other target proteins.

Keywords: Small GTPases · Structural flexibility · In silico

1 Introduction

Rab10 is a small monomeric enzyme that belongs to the Rab GTPases family [1]. It is responsible for regulating intracellular traffic in various pathways of different cellular sublocations, having roles in the endoplasmic reticulum, trans-Golgi network, endosomes, lysosomes and primary cilium [2]. Functional deficiencies in the Rab10 pathways are implicated in ciliopathies [3], glioblastomas [4] and neurodegenerative diseases [5]. Studies have shown that Rab10 has a relevant role in Alzheimer disease (AD), helping in the process of the amyloid precursor protein (APP) and in the production of Aβ through intracellular vesicle transport [6]. Such evidence paves the way for the application of new strategies for targeting drugs in the treatment of AD. Thus, modulation of Rab10 activity may represent an alternative to reduce the proportion of neurotoxic Aβ, making it a potential therapeutic target for the prevention and treatment of AD [5].

© Springer Nature Switzerland AG 2020
J. C. Setubal and W. M. Silva (Eds.): BSB 2020, LNBI 12558, pp. 215–220, 2020.
https://doi.org/10.1007/978-3-030-65775-8_20

Rab GTPases regulate cellular processes by alternating the nucleotides GTP and GDP. When linked to GTP, its switch 1, interswitch and switch 2 regions interacts with a series of effector proteins promote downstream signaling events. On the other hand, the hydrolysis of GTP results in conformational changes in the G domain of these enzymes, inactivating them [7]. The differences between the conformations of the G domain linked to GDP and GTP suggest that after the hydrolysis of GTP the switch 1 and switch 2 regions show a high degree of flexibility and disorder. In contrast, such regions are stabilized in the active state, which favors Rab10 to be recognized by effector proteins [8].

Therefore, the present study aimed to detail the structural flexibility of Rab10 and its switch regions, considering simulations of 200 ns of molecular dynamics (MD). In the past, 10 ns MD simulations have been done to investigate the internal movements of Rab5a of wild and mutant type [9]. However, even today, no in silico study on Rab10 linked to GTP and GDP has been carried out. Due to the unavailability of Rab10 crystallographic models associated with GTP and GDP, we used the molecular docking technique to form complexes with such nucleotides. Thus, it was possible to analyze atomic movements using classical mechanics and verify whether the heuristic method used was able to describe the active "ON" and inactive "OFF" state of this enzyme. The results discussed here may be useful for MD studies that aim to identify potential competitive inhibitors based on nucleotides against Rab10.

2 Methodology

2.1 Molecular Docking

Rab10 (PDB ID: 5SZJ) [10], GDP and GTP structures were downloaded from the Protein Data Bank (PDB) [11]. Modeller software v9.23 [12] was used to fill the missing atoms of Rab10. The addition of hydrogen in each structure, considering the protonation state of the atoms at physiological pH, was performed using the Open Babel 3.0.0 software [13]. The Autodock Vina 1.1.2 software [14] was used to docking the nucleotides at the active site of Rab10. The grid box with a size of 22 Å^3, was defined by the average of the Cartesian coordinates of the co-crystallized GNP nucleotide, being 34.692, 26.252 and –46.027 for the x, y and z axes, respectively. The GNP compound was redocked to validate the docking study. The poses of each nucleotide were chosen by means of the lowest binding energy and the highest number of intermolecular bonds. The interactions between the ligands and receptor were calculated using the Maestro 12.3 interface [15].

2.2 Molecular Dynamics

The GROMACS package version 2019.3 [16] was used in the MD simulations of complexes with GDP and GTP. The force field used was CHARMM36 [17]. The ligand parameters were obtained by the CGenFF server [18]. The complexes were centralized in cubic boxes, where the distance between the solute and the edge was 14 Å. The molecules were solvated with TIP3P water molecules and neutralized by adding the appropriate number of Na + Cl- ions considering the ionic concentration of 0.15 M. The energy minimization was performed using the steepest descent method with a maximum force of $1000 \text{ kJ.mol}^{-1}.\text{nm}^{-1}$. After minimization, the systems were equilibrated in two stages: a canonical NVT set (number of particles, volume and temperature) followed by an isothermal-isobaric NPT set (number of particles, pressure and temperature). The NVT equilibrium was performed with a constant temperature of 300 K for 500 ps. The NPT equilibrium was performed with a constant pressure of 1 bar and a constant temperature of 300 K for 500 ps. The production step was carried out at 300 K for 200 ns and the trajectories were saved every 10 ps. The tools of the root mean square deviation (RMSD), root mean square fluctuation (RMSF), radius of gyration (R_g) and solvent accessible surface area (SASA) were used for the trajectory analysis.

3   Results and Discussion

3.1   Molecular Docking Study

The lowest energy values for each test were grouped and their molecular interactions were analyzed. The most promising poses of each ligand are described in Table 1. The comparison between the co-crystallized GNP ligand pose and all docking poses indicated RMSD ≤ 0.60 Å. These values are lower than the tolerance level of 2.0 Å, indicating that the docking protocol has been validated. The GDP and GTP nucleotides were successfully docking at the active site of Rab10. These complexes presented intermolecular bonds in common: 2 saline bridges are formed with residue K22 and another with D125; 6 hydrogen bonds involve residues G21, K22, T23, C24, D125 and K154; and 2 π stacking interactions are found in residues F34. Moreover, the presence of γ-phosphate in GTP guarantees four more bonds of hydrogen with residues G19, T41, G67 and D125. These binding modes are consistent with the interactions found in the crystal of the Rab10 structure associated with GNP [10].

Table 1. Score by the Autodock Vina and the number of interactions calculated by Maestro.

Ligand	Score (Kcal.mol^{-1})	Salt Bridge	HBonds	Stacking
GDP	– 10,7	3	6	2
GTP	– 11,5	3	10	2

3.2   Molecular Dynamics Study

The RMSD is a crucial parameter to analyze the stability of biomolecular simulations along the trajectories of MD. Based in our findings, we observed that the stability of the trajectories of the two systems is only achieved after 100 ns of simulation. The results of the RMSD of these systems are shown in Table 2. considering all residues of the enzyme and those present in the switch regions. In these results, the switch 1 region showed significant differences when Rab10 was associated with the tested nucleotides. Here, the Rab10_GDP system showed greater fluctuations compared to Rab10_GTP. This is explained by the absence of γ-phosphate in GDP, which makes the switch 1 region more flexible due to the lack of stabilizing bonds between the enzyme and this nucleotide. In contrast, the results found for switch 2 did not reflect the nature of the structural flexibility of Rab10. High RMSD values were expected in switch 2 when Rab10 was associated with GDP, indicating the high fluctuations resulting from disordered movements. However, in this region the Rab10_GDP system showed a lowest RMSD value. In the case of interswitch, although the RMSD has been higher in the Rab10_GTP system, the variations in flexibility were not significant.

Table 2. Analysis of RMSD (nm) for the entire enzyme, switch 1, interswitch and switch 2.

Systems	Entire enzyme Res. M1–P175	Switch 1 Res. D31–I44	Interswitch Res. D45–T65	Switch 2 Res. A66–A82
Rab10_GDP	0.34 ± 0.07	0.49 ± 0.08	0,33 ± 0.06	0,47 ± 0.07
Rab10_GTP	0.29 ± 0.10	0.26 ± 0.09	0,27 ± 0.05	0,49 ± 0.10

Figure 1 shows the results of the RMSF, SASA and Rg analyzes of the enzyme. The RMSF allows analyzing the amino acid residues that contributed most to the fluctuations during the simulation. As it can be seen in Fig. 1A, the residues composes the switch 1 region showed greater fluctuations when Rab10 is linked to GDP and lesser fluctuations when linked to GTP. Although the Rab10_GDP system had predominantly higher peaks than Rab10_GTP in the interswitch region, the difference in RMSF values between the two systems was subtle. In relation to the switch 2 region, the RMSF values reflected greater fluctuations when Rab10 is associated with GTP, which is erroneous and does not represent Rab10's biological behavior.

The values of the radius of gyration of switch 1 (see Fig. 1B) confirm that this region has more disordered movements when Rab10 is associated with GDP. Here, the Rab10_GDP system showed an average of 0.94 ± 0.03 nm, while Rab10_GTP 0.91 ± 0.01 nm. Constant values of Rg indicate structures folded in a stable way, this indicates that the switch 1 region of Rab10 has greater flexibility when linked to GDP. The analysis of the interswitch radius of gyration (see Fig. 1C) showed that for the two systems studied, the flexibility is stable, where the calculated mean of the interswitch region was 1.07 ± 0.01 nm for all systems. In switch 2 (see Fig. 1D), the Rab10_GTP system showed greater disorder in fold movements (0.79 ± 0.03 nm), while Rab10_GDP had better stability, with Rg of 0.81 ± 0.01 nm.

The SASA analysis allowed to quantify the molecular surface and describe the contact between Rab10 and the solvent. The systematic increase in SASA indicates the destabilization of the biomolecule, which can expose its hydrophobic regions to the solvent [19]. Figure 1E shows that the SASA of the Rab10_GDP system has predominantly higher peaks than Rab10_GTP. When Rab10 is linked to GDP, the average of SASA was $105.92 ± 2.40$ nm^2; when connected to GTP, it was $101.91 ± 2.86$ nm^2. Thus, we

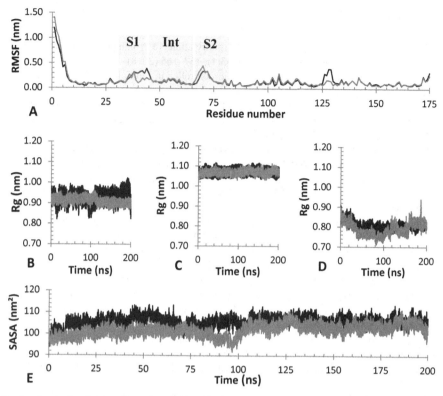

Fig. 1. Analysis of the trajectories obtained in the MD simulation: the grey line represents the Rab10_GTP system and the black line Rab10_GDP system. (A) RMSF of the entire enzyme: Switch 1 (S1) region is defined by positions 31–44, while interwitch (In) and Switch 2 (S2), 45–65, 66–82, respectively. (B) Rg of switch 1. (C) Rg of Interswitch. (D) Rg of switch 2. (E) SASA of the entire enzyme.

can infer that the disordered movements of switch 1 and the absence of γ-phosphate contribute to the increase in SASA of Rab10.

4   Conclusions

In short, the MD simulations used in this study were able to obtain notable differences in the switch 1 region of Rab10, enabling the identification of its active "ON" and inactive "OFF" states. However, the classical mechanics method was unable to accurately predict the disordered movements of the switch 2 region. We hypothesized that the flexibility of the switch 1 sensitive region can be used as an indicator of in silico studies that search potential competitive inhibitors based on nucleotides against Rab10. Our findings suggest that the in silico study of the flexibility of sensitive regions involved in the on-to-off mechanism of other protein targets may be useful in the discovery of potential drug candidates.

References

1. Yan, T., Wang, L., Gao. J., et al.: Rab10 Phosphorylation is a Prominent Pathological Feature in Alzheimer's Disease. J. Alzheimer Dis. **63**(1), 157–165 (2018)
2. Chua, C.E.L., Tang, B.L.: Rab10 – a traffic controller in multiple cellular pathways and locations. J. Cellular Phys. **233**(9), 6483–6494 (2018)
3. Ordónez, A.J.L., Fernández, B., Fdez, E., et al.: RAB8, RAB10 and RILPL1 contribute to both LRRK2 kinase–mediated centrosomal cohesion and ciliogenesis deficits. Human Mol. Genetycs **28**(21), 3552–3568 (2019)
4. Shen, G., Mao, Y., Su, Z., et al.: PSMB8-AS1 activated by ELK1 promotes cell proliferation in glioma via regulating miR-574-5p/RAB10. Biomed. Pharmacother. **122**(1), 109658 (2020)
5. Ridge, P.G., Karch, C.M., Hsu, S., et al.: Linkage, whole genome sequence, and biological data implicate variants in RAB10 in Alzheimer's disease resilience. Genome Med. **9**(1), 100 (2017)
6. Tavana, J.P., Rosene, M., Jensen, N.O., et al.: RAB10: an Alzheimer's disease resilience locus and potential drug target. Clin. Interv. Aging **14**(1), 73–79 (2019)
7. Good, R.G., Müller, M.P., Wu, Y.: Mechanisms of action of Rab proteins, key regulators of intracellular vesicular transport. Bio. Chem. **398**(5–6), 565–575 (2017)
8. Pylypenko, O., Hammich, H., Yu, I., et al.: Rab GTPases and their interacting protein partners: Structural insights into Rab functional diversity. Small GTPases **9**(1–2), 22–48 (2018)
9. Wang, J., Chou, K.: Insight into the molecular switch mechanism of human Rab5a from molecular dynamics simulations. Bio. Biophys. Res. Commun. **390**(3), 608–612 (2009)
10. Rai, A., Oprisko, A., Campos, G., et al.: bMERB domains are bivalent Rab8 family effectors evolved by gene duplication. eLife **5**(1), e186475 (2016)
11. Berman, H.M., Westbrook, J., Feng, et al.: The Protein Data Bank. Nucleic Acids Res. **28**(1), 235–242 (2000)
12. Sali, A., Blundell, T.L.: Comparative protein modelling by satisfaction of spatial restraints. J. Mol. Biol. **234**(1), 779–815 (1993)
13. O'Boyle, N.M., Banck, M., James, C.A., et al.: Open Babel: An open chemical toolbox. J. Cheminformatics **3**(1), 33 (2011)
14. Trott, M., Olson, A.J.: AutoDock Vina: Improving the speed and accuracy of docking with a new scoring function, efficient optimization, and multithreading. J. Comput. Chem. **31**(1), 455–461 (2010)
15. Schrödinger Release 2020–3.: Maestro. New York NY (2020)
16. Spoel, D.V.D., Lindahl, E., Hess, B., et al.: GROMACS: Fast, flexible, and free. J. Comput. Chem. **26**(1), 1701–1718 (2005)
17. Huang, J., MacKerell Jr., A.D.: CHARMM36 all-atom additive protein force field: Validation based on comparison to NMR data. J. Comput. Chem. **34**(1), 2135–2145 (2013)
18. Vanommeslaeghe, K., MacKerell Jr., A.D.: Automation of the CHARMM General Force Field (CGenFF) I: Bond Perception and Atom Typing. J. Chem. Inf. Model. **52**(12), 3144–3154 (2012)
19. Paul, M., Panda, M.K., Thatoi, E.H.: Developing Hispolon-based novel anticancer therapeutics against human (NF-κβ) using in silico approach of modelling, docking and protein dynamics. J. Biomol. Struct. Dyn. **37**(15), 3947–3967 (2019)

Importance of Meta-analysis in Studies Involving Plant Responses to Climate Change in Brazil

Janaina da Silva Fortirer[1], Adriana Grandis[1],
Camila de Toledo Castanho[2], and Marcos Silveira Buckeridge[1]([✉])

[1] Laboratory of Physiology and Ecology of Plants (Lafieco), Department of Botany, Biosciences Institute, University of São Paulo, São Paulo, Brazil
`msbuck@usp.br`
[2] Departamento de Ciências Ambientais,
Universidade Federal de São Paulo - UNIFESP, São Paulo, Brazil

Abstract. Meta-analysis synthesizes individual research results on the same subject and provides information that indicates bottlenecks in research. Due to the massive production of data, integrative analyzes are necessary, presenting more consistent views of biological phenomena. Broad themes such as plants' response to climate change have been the subject of meta-analyses since 1996, as there is a global concern about the effect of elevated CO_2 on plants and forests. We propose using meta-analysis to compile existing data, including studies related to the effects of high CO_2 on Brazilian biomes' vegetation. For that, we found 36 articles on the theme after a systematic review. Physiological parameters such as photosynthesis, leaf area, and non-structural carbohydrates are essential to understand the plant's responses to elevated CO_2 using meta-analysis. However, these parameters are not present in a considerable portion of the literature, decreasing the statistical power of meta-analytical strategies. The meta-analysis of plants' biological responses is usually performed with several species, although there are also studies with single species. The use of many species increases the variance of the effects, highlighting the need for multilevel modeling to consider the dependence among data on the same species. We discuss how to carry out studies considering the variables needed in future meta-analyses to contribute to better data integration relevant to national reports. In this way, we expect that meta-analytical strategies could be essential for national decision-making and complement global analyses such as those made by the IPCC.

Keywords: Elevated CO_2 · High temperature · Plant physiology · Tropical species

Supported by Capes. Fellowship: 88882.461730/2019-01.

J. C. Setubal and W. M. Silva (Eds.): BSB 2020, LNBI 12558, pp. 221–234, 2020.
https://doi.org/10.1007/978-3-030-65775-8_21

1   Introduction

The number of scientific publications is increasing exponentially, and there is a need for revisions that compile data produced to guide several study areas [34]. For many years these reviews were carried out by narrative reviews [12]. However, the narrative review is subjective and often not reproducible, as it can be skewed depending on the point of view and preferences of each author [36]. To answer this demand, the reviews have been applying the meta-analysis methodology that incorporates a previous systematic review to extract information, following the studies' inclusion criteria, where all the steps are documented [28]. Meta-analysis is a set of statistical methods that quantitatively compares the results of different studies that address a common issue [10, 22]. Meta-analysis is the grandmother of 'big data' and 'open science'. The implementation of meta-analytic techniques was the first effort to collect and synthesize pre-existing data to determine patterns, make predictions, and make evidence-based decisions[11]. Furthermore, qualitative data presentation that indicates gaps in current knowledge and new research needs to be performed [33]. The meta-analysis is the analysis of the analyzes [5]. After Glass [10] first used the term "meta-analysis" in 1976, the method has been widely applied and developed in the areas of medicine and sociology [41]. In the 1990 s, meta-analysis has been used in ecology and evolutionary biology [33] and is not yet widely known in the biological sciences [33].

Several studies show terrestrial ecosystems' responses to climate change, mainly due to increased atmospheric CO_2. The concentration of carbon dioxide (CO_2) in the atmosphere has increased from ~ 280 ppm (parts per million) to ~ 410 ppm from the industrial revolution to the present [19]. This increase in atmospheric CO_2 is due to fossil fuels, forest burning, and land-use changes [18, 19]. The projection for the 2100 s of the Intergovernmental Panel on Climate Change (IPCC) is that the CO_2 levels will increase to 1300 ppm [18]. However, CO_2 is not only one of the leading gases responsible for the greenhouse effect (GHG) [19, 38], but it is also an essential component for photosynthesis, leading to growth and higher productivity of the ecosystem [2, 23]. Plant's dry mass consists of 40% carbon fixed by photosynthesis [25]. The CO_2 concentration increases lead to a rise in temperature, and together they induce drastic changes in the terrestrial ecosystem, such as changing the pattern of rainfall in certain regions [23], or they can also cause tree mortality due to water losses or the forest burning [29]. Thus, most studies performed in this century try to understand how plants can respond to the increase in CO_2. Indeed, several individual publications have focused on the three variables that mostly affected the climate: CO_2, temperature, and water stress.

Meta-analyses on climate change have been primarily applied in studies with temperate climate species and have proven to be a valuable tool in this field [2, 6, 47]. Curtis et al. [6] used the meta-analysis to summarize more than 500 studies on the high CO_2 effects. They conclude that there is a 28% increase in tree biomass allocation. Wand et al. [47] showed that biomass increased by 33% and 44% when submitted to high CO_2 in plants holding C3 and C4 photosynthesis, respectively. In another meta-analysis covering 120 individual studies, the main effects on the plant physiology under high CO_2 in the Free Air Carbon Enrichment (FACE) system were described [2]. However, as mentioned above, these studies were performed on plants from the temperate climate, and it is necessary to include studies from tropical climate [2]. Tropical and subtropical forests contribute significantly to carbon assimilation and storage on the planet. However, until 2009, research on tropical species represented less than 10% of the studies [24]. Despite representing 13% of the Earth's surface, tropical forests resemble, in addition to their enormous genetic diversity, the largest carbon reservoir (52%) in the world (340 Gt/C biomass) [24]. The Brazilian territory sits on a larger tropical area in South America. Displaying six different biomes (Amazon, Cerrado, Atlantic Forest, Caatinga, Pampas, and Pantanal), the neotropical region is often considered a hotspot of biodiversity [1]. Due to the tropical species richness found in Brazil, studies involving the effects of an increase of CO_2 on plants' growth and development in these biomes have been performed [3,35], contributing to the knowledge of the tropical plant responses to climate change. However, there is still no meta-analysis study that synthesizes the total effect of high CO_2 on Brazil's tropical plants from these published data. To address this problem, this work had the following objectives: 1) to approach the importance of the meta-analysis methodology with a systematic review for use in studies on climate change; 2) Provide an overview of Brazilian research on the effect of high CO_2 on tropical plants, and 3) discuss future studies on which species and biomes should be prioritized and some parameters that could better explain the variation in the high CO_2 effects on plants. This way of compiling the data may, in the future, provide a scientific basis for the adoption of the best public policy and a reliable database about the response of Brazilian forests to climate change.

2   Methodology

2.1   How to Get the Data to Perform the Meta-analysis

The steps for performing meta-analysis follow the scientific research procedures: problem, question, hypothesis, method, data collection, and data analysis [27]. Data collection in the meta-analysis studies is carried out employing a systematic review, which selects primary studies and, in this case, experimental studies on plant responses to the increase in CO_2 [28]. A systematic review is necessary to identify and describe all the steps performed in selecting studies and the extraction of data so that the general result will be reproducible [28](Fig. 1A). In this work, we describe the history of meta-analysis using studies on the effect of elevated CO_2 on plants. These studies were obtained from the search in the Web of Science. The keywords used in the search were: "meta-analysis" OR "meta-analytic" AND "CO2" AND "plant," on August 18, 2020. To verify the primary studies on the effect of increasing CO_2 in plants in Brazil, a search was carried out in three databases (Web of Science, Scielo, and Brazilian Digital Library of Theses and Dissertations). The words used in this search were: "elevated CO2"

OR "increas* CO2" OR "CO2 enrichment" OR "rising CO2" OR "high* CO2". This search was conducted in March 2019. The criteria of inclusion of the works found in these searches were as follows: (I) studies conducted in Brazil with the effect of elevated CO_2 on plant physiological responses; (II) experiments with Free Air Carbon Enrichment (FACE) or Open-Top Chamber (OTC) systems; (III) studies that added: sample size, mean and standard deviation/error of the control and treatment group (Fig. 1A).

After completing all stages of the systematic review, the full text of Brazilian primary studies included in the meta-analysis was screened to extract the following data: species name, sample number, average and standard deviation/standard error for the control group (ambient CO_2), and treatment (elevated CO_2) for each variable (see Fig. 3), experimental system (FACE or OTC) and experiment time. The moderating variables contribute to part of the variance observed in a meta-analysis. The moderator analysis can be conducted to determine the heterogeneity sources and the extent to which it contributes to the observed variability in *effect sizes* among studies [36]. In this sense, the moderating variables verified were the life habit (tree or herbaceous) and the photosynthetic pathway (with C3 or C4 metabolism) to check whether these moderating variables influence the plants' magnitude responses to elevated CO_2.

2.2   How to Apply a Meta-analysis

The first stage to perform a meta-analysis is determining the type of *effect size* [27] (Fig. 1B). *Effect size* is the basic unit of meta-analysis. It makes possible the standardization of individual studies' results by providing an average estimate of the effect of elevated CO_2 compared to ambient CO_2 [12,14], and the *effect size's* in logarithmic scale is generally used [27]. The choice of the *effect size* used for the dataset extracted from the primary studies raises the necessity to check whether there is dependence on the data. Research in biology may include more than one analysis in the same experiment, and there are also different studies for the same species or studies that investigate several species. When these data types are included in the meta-analysis, the study requires a hierarchical model approach because it allows dependence on the data set [33]. After adjusting to the models, it is necessary to perform the *heterogeneity test* to determine the variation among studies, which is not attributable to the sample variance [33].

The *publication bias test* should be applied in the meta-analysis. The *publication bias* occurs when the published studies are not representative of the totality of the studies performed. For example, significant results that confirm the expectations of the research are more likely to be published than non-significant results [13]. The methods commonly used to assess publication bias are funnel charts [39] and the Egger test [9]. Finally, it is recommended to perform the sensitivity test to check the data's consistency and possible outliers. For the present study, we used "metafor" package for analysis [46] and ggplot2 package for the graphics [48], both in the R version 3.6.0 program [42].

3 Results and Discussion

3.1 History of Climate Change Conferences and the Impact on the Meta-analytical Production on the Effect of Elevated CO_2 on Plants

Figure 2 presents historical highlights of the evolution of global awareness about climate change (Fig. 2A). These historical steps are contrasted with the evolution of the scientific publication regarding the effects of elevated CO_2 and temperature on plants (Fig. 2B). These steps in climate change awareness history (Fig. 2A) highlight society, scientists, and the government's concern with climate change. These and many other conferences (Conference of the Parties - COPs) and the Intergovernmental Panel on Climate Change (IPCC) reports have been gradually changing political decision-making worldwide. Nevertheless, in order to design effective public policies to mitigate and adapt to the effects of climate change, there is still an urgent need to produce more scientific knowledge about the impact of climate change worldwide.

Due to concerns about air pollution and acid rain, in 1972, the first major United Nations Organization (ONU) conference on the environment was held in Stockholm [37] (Fig. 2A). After 18 years, in 1990, the IPCC published the first report on the increase in GHG concentrations on Earth's atmosphere [16]. After that, Brazil started to participate actively in these discussions on climate and environment, and in 1992 it held the United Nations Conference on Environment and Development, called Rio 92 [32,37]. In 1995, the 1st Conference of the Parties (COP-1) was held in Germany, and the IPCC Second Report [17] was also published.

COP-3 was held in 1997 in Kyoto - Japan. It generated targets for a 5% reduction in Greenhouse Gas emissions (GHG) [43]. GHG is one of the main factors that increase the effect of climate change [18] (Fig. 2A). Around this period, the first individual publications relating plants and climate change appeared in the literature, and consequently, the employment of meta-analyzes to assess the common effect among all experiments became possible (Fig. 2B). Between 1996 (the first meta-analysis on this theme) and 2006, four meta-analytical publications per year were published (Fig. 2B). By 2007 twice as many meta-analysis studies were produced on the effect of elevated CO_2 on plants. This increase of meta-analytical reviews is concomitant with the fourth IPCC report. This is consistent with the previously mentioned need for large-scale data analysis studies to guide global reports such as the IPCC ones.

The fourth IPCC report, released in 2007, brought a strong message confirming that the increase in CO_2 results from human activities. One of the report's key messages was that global warming could be irreversible *if drastic measures were not implemented*. Some ways of mitigating and adapting to climate scenarios were suggested [19]. That year the IPCC won the Nobel Peace Prize for its achievement.

The GHG emissions, mainly CO_2, and the increase in temperature were reported in previous reports. However, only in 2009 at COP-15 in Copenhagen,

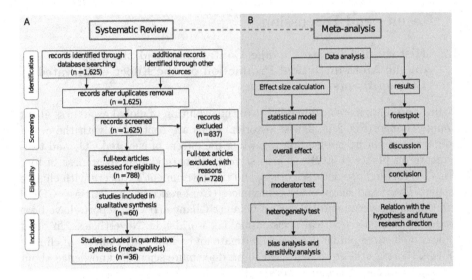

Fig. 1. (A) Flowchart showing the systematic review steps modified from Liberati et al. [28]. Numbers in parentheses represent the data obtained in the systematic review carried out in this work for publications on the CO_2 increases and its physiological responses in plants using individual Brazilian plant studies. After performing the systematic review and data compilation, (B) flowchart describing the steps to perform a meta-analysis. Based on Lei et al. [27].

it was proposed that there would be a 2 °C increase in global temperature [44]. This prediction was only possible because the individual studies were carried out very rapidly between 2007 and 2009, resulting in meta-analytical articles that compiled many studies and showed how the temperature could affect plants and forests' productivity (Fig. 2B).

Analyzing the meta-analytical articles (Fig. 2B), one year before and in the years following the IPCC reports (Fig. 2A), the number of reviews adopting this approach increased dramatically. One possible explanation is the adjustment of the timing of scientific literature publications on climate change and the publication of the IPCC reports. While an IPCC report is being prepared – what can take a few years – the scientific community is consulted and becomes aware of its ongoing work. As the IPCC sets limit dates for inclusion in the report's citations, many publications tend to be released before the IPCC deadline. As a result, many publications became available after the report's launch, increasing the amount of data available and turning new meta-analyses more viable. This relationship among chronology of events and world reports with scientific knowledge production shows the importance of both to the development of new technologies to adapt or mitigate the consequences of climate change. Due to the large number of scientific data generated, the projections have become more realistic and can be calculated using more complex mathematical models. In the fifth report from IPCC, it was estimated that from 2100, atmospheric CO_2

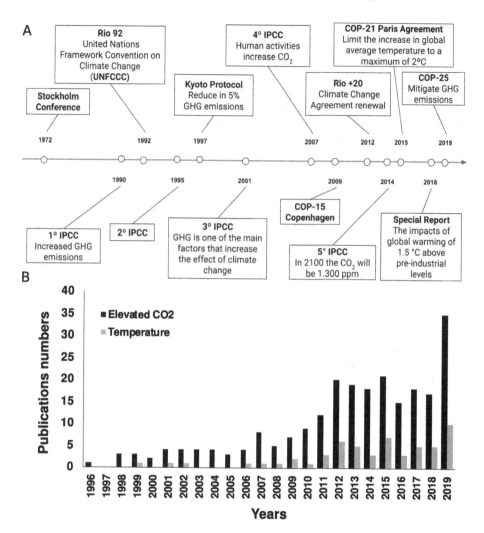

Fig. 2. (A) World history events ranging from meetings to IPCC reports that addressed global climate change. (B) Distribution over years of published manuscripts (from 1996 to 2019) of meta-analyses performed worldwide that address plant responses to high CO_2 (236 studies).

would reach an atmospheric concentration of about 1.300 ppm [18]. Moreover, the global temperature is likely to reach 1.5 °C around 2030 if there is no action to reduce GHG emissions [20,45].

From 1996 to 2019, we found 236 articles that performed meta-analyses related to high CO_2 effects on plants (Fig. 2B). The pioneering study that used the meta-analysis to assess elevated CO_2 responses on plants was conducted by Curtis [7] (Fig. 1B). In 2005, Ainsworth & Long [2] published a meta-analysis with 120 works showing physiological aspects and some species' productivity. This work was a milestone for studies researching climate change, with 3.157 citations (until October 20, 2020), demonstrating the power of quantitative synthesis of meta-analysis.

The meta-analytic studies addressed some ecological processes and relationships, including the effect of elevated CO_2 on photosynthesis and plant respiration, growth, competition, and interaction among plants, productivity, exchange and conductance of gases in the leaves, soil respiration, carbon, and nitrogen accumulation in the soil, and seed production [7,30,47]. This shows that the meta-analysis has been widely used in research containing experiments with elevated CO_2 in plants and comes with each new publication trying to couple the high CO_2 with other factors such as temperature and water conditions.

Alternatively, most meta-analyses, such as Ainsworth & Long [2], only include experiments performed with plants from temperate climate regions. This points to the need for studies involving tropical plant responses to climate change [2,21]. The importance of carrying out studies on species in a tropical climate is highlighted by the fact that from the 236 meta-analyses found in this work, only three studies on experiments with tropical species carried out in Brazil were included among the primary studies reviewed [4,31,40].

3.2 Integrative Analysis of Variables that Are Contemplated in Brazilian Studies Eligible for a Meta-analysis on Plant Responses to Climate Change

The systematic review carried out in the three databases returned 1.625 studies (Fig. 1A). After reading titles and abstracts, we excluded 837 studies because they were not experimental studies or were not conducted in Brazil. An evaluation of the texts was performed for the 788 eligible articles, and 728 articles were excluded for not meeting at least one of the following criteria: research conducted in Brazil on the elevated CO_2 effects on physiological responses in plants and experiments with FACE or OTC. In the next stage of the systematic review, 60 studies were included in the qualitative synthesis, but of these, only 36 displayed the required data for meta-analysis, which are sample size, mean, and standard deviation/error of the control and treatment groups (Fig. 1A). The small number of studies included in the meta-analysis demonstrated that the primary studies do not report all the information necessary to perform a meta-analysis. From the studies included in the quantitative synthesis, the information described in the methodology was extracted, and then the steps shown in Fig. 1B were followed.

Of the 36 studies performed in Brazil, 27 species were contemplated, 13 classified as trees, and 14 herbaceous species. Regarding the photosynthetic pathway, 20 species were classified with the C3 photosynthetic pathway, and seven species were from plants displaying the C4 photosynthetic pathway. Furthermore, there were relatively few studies with C4 plants compared to C3 (Fig. 3B). Despite some studies with maize [26], the elevation of CO_2 did not have any effect, confirming the hypothesis that species displaying C4 photosynthesis type do not respond or respond very weakly to CO_2 elevation. However, when sugarcane, also a tropical C4 grass, was subjected to elevated CO_2, an increase in photosynthesis and biomass was observed [8]. These results show that there seem to exist alternative physiological mechanisms by which C4 species could respond to elevated CO_2.

Species categorization, according to their habit, can provide information about the magnitude of the high CO_2 effects. In this sense, whether plants are trees or herbs and the photosynthesis metabolism (C3 or C4) seem to directly influence the size of the high CO_2 effect in the plants studied (Fig. 3B). However, when categorizing the different species, the studies number becomes smaller, as shown in Fig. 3, in which the herbs category C4 is left with only three observations (Fig. 3C), reducing the power of statistical analysis.

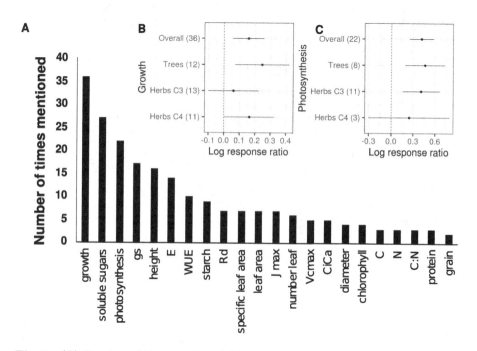

Fig. 3. (A) Number of times each variable was used within the 36 studies included in this meta-analysis. (B) Size of the high CO_2 effects on plant growth. (C) Size of the effect of high CO_2 on plant photosynthesis. Data represent means ±95% CI for each functional group analyzed. Values that do not overlap to zero have a significant difference ($\alpha < 0.05$). The numbers of observations are presented in parentheses.

Life's habit (trees or herbs) influences the magnitude of high CO_2 responses on plant biomass. From 1,625 studies included for the meta-analysis, 36 observations were obtained to estimate the effect of biomass (growth) (Figs. 1 and 3A). This higher growth variable number is due to the measurement facility during experiments since a single scale can be used to obtain the variable data. This representativeness of studies enables a more detailed analysis of the effect of high

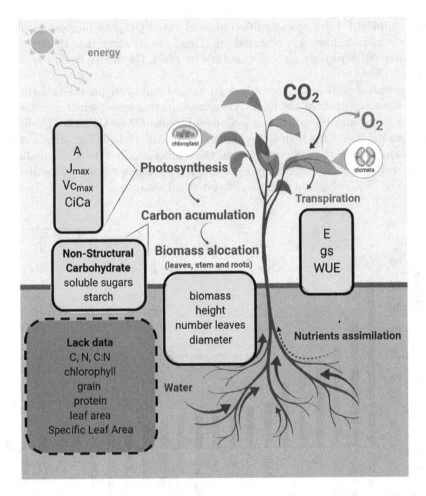

Fig. 4. Physiological variables more commonly measured in experiments with plants submitted to climate changes (CO_2, water stress, and temperature). The variables more frequently analyzed in Brazilian studies are in the yellow boxes (continuous line box), and variables in the red box (dashed line box) did not have enough studies (number of studies less than five) to perform a more consistent meta-analysis, needing to be measured in future studies. A = net CO_2 assimilation rate, J_{max} = maximum electron transport rate, Vc_{max} = maximum carboxylation of Rubisco, Ci:Ca = intercellular and environmental CO_2 rate, Rd = dark respiration, E = leaf transpiration, gs = stomatal conductance, WUE = water use efficiency, C:N = carbon and nitrogen ratio, C = carbon, and N = nitrogen. (Color figure online)

CO_2 on plants. However, if the photosynthesis variable, which represents how much the plant assimilates carbon, is observed, only 22 studies display this type of variable (Fig. 3A). A smaller number of observations reported photosynthesis measurements, which leads to a more complex level of plant responses. It is important to note that to measure photosynthesis, it is necessary to use expensive equipment to be operated by a highly qualified, well-trained professional. Altogether, these limitations may explain the lower number of publications reporting this kind of data in the literature.

A meta-analysis's reliability depends on having several individual studies and the relatively high number of variables within them. In the search for Brazilian studies, there was a diversity of variables reported. However, most variables were not common to all studies found (Fig. 3A). In this sense, it was necessary to standardize the number of variables that could be evaluated. The most studied variables in the studies were growth (plant biomass - 36), followed by soluble sugars (27), and A (photosynthetic assimilation - 22). The other variables were reported with a very low frequency in the selected studies (Fig. 3A). Few observations for each variable can influence the meta-analysis's statistical power since the higher the sample number, the greater the consistency of the result [15]. Our data show the diversity of variables that should be taken into account in the future for meta-analytical studies of plants in elevated CO_2 and particularly in studies of plant species in Brazilian biomes. Furthermore, standardization of the variables used in experiments with single species would allow future meta-analyzes to be more reliable, affording more consistent conclusions about how the effect of elevated CO_2 influences the plants' physiology.

This work suggests that some of the physiological parameters presented in Fig. 4, which contain the main variables that would bring more strength to Brazilian studies' future meta-analysis on plants' high CO_2 effects. The variables that are within the red (dashed line box) in Fig. 4 are those that have relatively little data. The variables in yellow boxes (continuous line box) are those that have a significant amount of data for a meta-analytical study. However, there are insufficient individual studies to perform a more detailed meta-analysis to verify moderators' influence on the magnitude of high CO_2 effects in tropical species.

Regarding the responses of tropical plant species to elevated CO_2, our systematic review, followed by classification criteria, generated 36 studies suitable for meta-analysis. Our results indicate significant differences in plant responses regarding habit (trees or herbs) and the photosynthesis type. However, as the biological implications were out of this chapter's scope, the more profound exploration of the meta-analysis performed with the Brazilian studies will be reported and discussed elsewhere.

4   Conclusion

In this study, we surveyed the history of global meta-analyses of the effects of CO_2 and temperature on plants. The purpose of the present work was not a complete biological analysis of the results of our meta-analysis but rather to bring a perspective on how important meta-analytical approaches are in the area of Climate Change. We also discussed the limitations inherent in the meta-analytical approach.

We found a close correlation (likely involving the release of the IPCC reports) between the rise in meta-analysis concerning the responses of plants to elevated CO_2 and the socio-political events that determined sustainable development goals on the planet during the last 50 years.

Given the importance of Brazilian biomes and agriculture for the planet and the importance of land use for the impacts of global climate change, we conclude that it is urgently needed that more studies would be performed with species from the neotropics. A meta-analysis will undoubtedly be crucial to understand such impacts and for decision making. However, most studies would have to bring a plethora of variables that can afford the use of this valuable statistical method.

References

1. Aguiar, S., Santos, I.D.S., Arêdes, N., Silva, S.: Redes-bioma: Informação e comunicação para ação sociopolítica em ecorregiões. Ambiente Soc. **19**(3), 233–252 (2016)
2. Ainsworth, E.A., Long, S.P.: What have we learned from 15 years of free-air co2 enrichment (face)? a meta-analytic review of the responses of photosynthesis, canopy properties and plant production to rising co2. New Phytol. **165**(2), 351–372 (2005). https://doi.org/10.1111/j.1469-8137.2004.01224.x
3. Arenque, B.C., Grandis, A., Pocius, O., de Souza, A.P., Buckeridge, M.S.: Responses of *Senna reticulata*, a legume tree from the amazonian floodplains, to elevated atmospheric CO_2 concentration and waterlogging. Trees **28**(4), 1021–1034 (2014). https://doi.org/10.1007/s00468-014-1015-0
4. Cleveland, C.C., et al.: Relationships among net primary productivity, nutrients and climate in tropical rain forest: a pan-tropical analysis. Ecol. Lett. **14**(9), 939–947 (2011). https://doi.org/10.1111/j.1461-0248.2011.01658.x
5. Cook, T.D.: The potential and limitations of secondary evaluations. In: Analysis and Responsibility, Educational Evaluation (1974)
6. Curtis, P.S., Wang, X.: A meta-analysis of elevated CO_2 effects on woody plant mass, form, and physiology. Oecologia **113**(3), 299–313 (1998). https://doi.org/10.1007/s004420050381
7. Curtis, P.: A meta-analysis of leaf gas exchange and nitrogen in trees grown under elevated carbon dioxide. Plant, Cell Environ. **19**(2), 127–137 (1996). https://doi.org/10.1111/j.1365-3040.1996.tb00234.x
8. De Souza, A.P., et al.: Elevated CO_2 increases photosynthesis, biomass and productivity, and modifies gene expression in sugarcane. Plant, Cell Environ. **31**(8), 1116–1127 (2008). https://doi.org/10.1111/j.1365-3040.2008.01822.x
9. Egger, M., Smith, G.D., Schneider, M., Minder, C.: Bias in meta-analysis detected by a simple, graphical test. Bmj **315**(7109), 629–634 (1997). https://doi.org/10.1136/bmj.315.7109.629
10. Glass, G.V.: Primary, secondary, and meta-analysis of research. Educ. Res. **5**(10), 3–8 (1976). https://doi.org/10.3102/0013189X005010003
11. Gurevitch, J., Koricheva, J., Nakagawa, S., Stewart, G.: Meta-analysis and the science of research synthesis. Nature **555**(7695), 175–182 (2018). https://doi.org/10.1038/nature25753
12. Harrison, F.: Getting started with meta-analysis. Methods Ecol. Evol. **2**(1), 1–10 (2011). https://doi.org/10.1111/j.2041-210X.2010.00056.x

13. Haworth, M., Hoshika, Y., Killi, D.: Has the impact of rising CO_2 on plants been exaggerated by meta-analysis of free air CO_2 enrichment studies? Front. Plant Sci. **7**, 1153 (2016). https://doi.org/10.3389/fpls.2016.01153

14. Hedges, L.V., Gurevitch, J., Curtis, P.S.: The meta-analysis of response ratios in experimental ecology. Ecology **80**(4), 1150–1156 (1999). https://doi.org/10.1890/0012-9658(1999)080[1150:TMAORR]2.0.CO;2

15. Hedges, L.V., Pigott, T.D.: The power of statistical tests for moderators in meta-analysis. Psychol. Methods **9**(4), 426 (2004). https://doi.org/10.1037/1082-989X.9.4.426

16. IPCC: Climate change: The ipcc 1990 and 1992 assessments (1990). https://www.ipcc.ch/report/climate-change-the-ipcc-1990-and-1992-assessments/. Accessed 25 Aug 2020

17. IPCC: Sar climate change 1995: Synthesis report (1995). https://www.ipcc.ch/site/assets/uploads/2018/05/2nd-assessment-en-1.pdf. Accessed 25 Aug 2020

18. IPCC: Tar climate change 2001: Synthesis report (2001). https://www.ipcc.ch/report/ar3/syr/. Accessed 25 Aug 2020

19. IPCC: Ar4 climate change 2007: Synthesis report (2007). https://www.ipcc.ch/report/ar4/syr/. Accessed 25 Aug 2020

20. IPCC: Global warming of 1.5 °c: Special report (2018). https://www.ipcc.ch/sr15/. Accessed 25 Aug 2020

21. Jones, A.G., Scullion, J., Ostle, N., Levy, P.E., Gwynn-Jones, D.: Completing the face of elevated CO_2 research. Environ. Int. **73**, 252–258 (2014). https://doi.org/10.1016/j.envint.2014.07.021

22. Koricheva, J., Gurevitch, J., Mengersen, K.: Handbook of meta-analysis in ecology and evolution. Princeton University Press, New Jersey (2013)

23. Körner, C.: Plant CO_2 responses: an issue of definition, time and resource supply. New Phytol. **172**(3), 393–411 (2006). https://doi.org/10.1111/j.1469-8137.2006.01886.x

24. Körner, C.: Responses of humid tropical trees to rising CO_2. Annu. Rev. Ecol. Evol. Syst. **40**, 61–79 (2009). https://doi.org/10.1146/annurev.ecolsys.110308.120217

25. Lambers, H., Chapin III, F.S., Pons, T.L.: Plant Physiological Ecology. Springer, New York (2008)

26. Leakey, A.D., et al.: Photosynthesis, productivity, and yield of maize are not affected by open-air elevation of CO_2 concentration in the absence of drought. Plant Physiol. **140**(2), 779–790 (2006). https://doi.org/10.1104/pp.105.073957

27. Lei, X., Peng, C., Tian, D., Sun, J.: Meta-analysis and its application in global change research. Chin. Sci. Bull. **52**(3), 289–302 (2007). https://doi.org/10.1007/s11434-007-0046-y

28. Liberati, A., et al.: The Prisma statement for reporting systematic reviews and meta-analyses of studies that evaluate health care interventions: explanation and elaboration. J. Clin. Epidemiol. **62**(10), e1–e34 (2009). https://doi.org/10.1016/j.jclinepi.2009.06.006

29. Lovejoy, T.E., Nobre, C.: Amazon tipping point: last chance for action (2019). https://doi.org/10.1126/sciadv.aba2949

30. Luo, Y., Hui, D., Zhang, D.: Elevated CO_2 stimulates net accumulations of carbon and nitrogen in land ecosystems: a meta-analysis. Ecology **87**(1), 53–63 (2006). https://doi.org/10.1890/04-1724

31. Moles, A.T., et al.: Which is a better predictor of plant traits: temperature or precipitation? J. Veg. Sci. **25**(5), 1167–1180 (2014). https://doi.org/10.1111/jvs.12190

32. MRE: Ministério das relações exteriores do brasil (2019). http://www.itamaraty.gov.br/pt-BR. Accessed 25 Aug 2020
33. Nakagawa, S., Noble, D.W., Senior, A.M., Lagisz, M.: Meta-evaluation of meta-analysis: ten appraisal questions for biologists. BMC biology **15**(1), 1–14 (2017). https://doi.org/10.1186/s12915-017-0357-7
34. Nakagawa, S., et al.: Research weaving: visualizing the future of research synthesis. Trends Ecol. Evol. **34**(3), 224–238 (2019). https://doi.org/10.1016/j.tree.2018.11.007
35. Palacios, C., Grandis, A., Carvalho, V., Salatino, A., Buckeridge, M.: Isolated and combined effects of elevated CO_2 and high temperature on the whole-plant biomass and the chemical composition of soybean seeds. Food Chem. **275**, 610–617 (2019). https://doi.org/10.1016/j.foodchem.2018.09.052
36. Quintana, D.S.: From pre-registration to publication: a non-technical primer for conducting a meta-analysis to synthesize correlational data. Front. Psychol. **6**, 1549 (2015). https://doi.org/10.3389/fpsyg.2015.01549
37. Ribeiro, W.C.: A ordem ambiental internacional. Editora Contexto (2001)
38. Ruddiman, W.F.: The anthropogenic greenhouse era began thousands of years ago. Climatic Change **61**(3), 261–293 (2003). https://doi.org/10.1023/B:CLIM.0000004577.17928.fa
39. Sterne, J.A., Egger, M.: Funnel plots for detecting bias in meta-analysis: guidelines on choice of axis. J. Clin. Epidemiol. **54**(10), 1046–1055 (2001). https://doi.org/10.1016/S0895-4356(01)00377-8
40. Stevens, N., Lehmann, C.E., Murphy, B.P., Durigan, G.: Savanna woody encroachment is widespread across three continents. Glob. Change Biol. **23**(1), 235–244 (2017). https://doi.org/10.1111/gcb.13409
41. Sutton, A.J., Higgins, J.P.: Recent developments in meta-analysis. Stat. Med. **27**(5), 625–650 (2008). https://doi.org/10.1002/sim.2934
42. Team, R.C., et al.: R: A language and environment for statistical computing (2013)
43. UNFCCC: Kyoto protocol to the united nations framework convention on climatechange (1997). http://unfccc.int/resource/docs/convkp/kpeng.pdf. Accessed 25 Aug 2020
44. UNFCCC: Statements on behalf of the group of g77 and china (2009). https://unfccc.int/resource/docs/2009/cop15/eng/11a01.pdf. Accessed 25 Aug 2020
45. UNFCCC: Adoption of the paris agreement. united nations framework convention on climate change (2015). http://unfccc.int/paris_agreement/items/9485.php. Accessed 25 Aug 2020
46. Viechtbauer, W.: Conducting meta-analyses in r with the metafor package. J. Stat. Softw. 36(3), 1–48 (2010). https://doi.org/10.18637/jss.v036.i03
47. Wand, S.J., Midgley, G.F., Jones, M.H., Curtis, P.S.: Responses of wild c4 and c3 grass (poaceae) species to elevated atmospheric CO_2 concentration: a meta-analytic test of current theories and perceptions. Glob. Change Biol. **5**(6), 723–741 (1999). https://doi.org/10.1046/j.1365-2486.1999.00265.x
48. Wickham, H.: Elegant Graphics for Data Analysis (ggplot2). Springer, New York (2009)

A Brief History of Bioinformatics Told by Data Visualization

Diego Mariano[1]([✉]) (ID), Mívian Ferreira[2] (ID), Bruno L. Sousa[2] (ID),
Lucianna H. Santos[1] (ID), and Raquel C. de Melo-Minardi[1] (ID)

[1] LBS - Laboratory of Bioinformatics and Systems, Department of Computer Science,
Universidade Federal de Minas Gerais - UFMG, Belo Horizonte, Brazil
diegomariano@ufmg.br
[2] LLP - Laboratory of Programming Languages, Department of Computer Science,
Universidade Federal de Minas Gerais - UFMG, Belo Horizonte, Brazil

Abstract. Bioinformatics is an interdisciplinary research field that aims to analyze biological data through computational approaches. In the last years, the evolution of technological resources has provided a tidal wave of biological data. Consequently, an unprecedented amount of studies using bioinformatics approaches
have been released, increasing peer-reviewed published papers. Here, we tell a
brief history of bioinformatics based on literature data analysis and visualization.
We collected abstracts and other metadata from papers published from 1998 to
2019 in four leading bioinformatics journals: (i) Oxford Bioinformatics; (ii) BMC
Bioinformatics; (iii) Briefings in Bioinformatics; and (iv) PLoS Computational
Biology. Our results show an increase in publication number and international
collaborations. We also observed an increase in publications by Chinese authors.
Latin America continues to have a low percentage of global scientific bioinformatics production. However, Brazil excels in this region, being responsible for almost
half of Latin America papers published. Our results also point out the recent trend
of using Python as the programming language for bioinformatics applications, followed by Perl, Java, and R. We hope these data visualizations can provide insights
to understand the recent changes and evolution in the bioinformatics field. The
developed interactive visualizations are available at http://bioinfo.dcc.ufmg.br/his
tory/.

Keywords: Bioinformatics · Computational biology · Data visualization

1 Introduction

Bioinformatics is an interdisciplinary research field whose principle is using models
and algorithms to analyze biological data and solve biologically related problems [1].
Bioinformatics' roots are in the early 1960s when computers, used for military purposes,
became available for universities and research institutes. At that time, researchers began
to use computers to try answering fundamental questions in life sciences [2].

Margaret Dayhoff was a pioneer in bioinformatics studies at that time. She proposed
the use of mathematical approaches for analyzing amino acid frequencies and mutation

© Springer Nature Switzerland AG 2020
J. C. Setubal and W. M. Silva (Eds.): BSB 2020, LNBI 12558, pp. 235–246, 2020.
https://doi.org/10.1007/978-3-030-65775-8_22

probabilities in biological sequences. Since the late 1950s, experimental approaches have allowed the sequencing of small protein structures, such as insulin [3]. This culminated in the creation of the first database of amino acid sequences and structures, the so-called "Atlas of Protein Sequence and Structure" [4]. Dayhoff and collaborators also proposed computational methods for sequence comparisons to detect homologous proteins using a substitution matrix called PAM (Percent Accepted Mutation), which contributed to the rising of the molecular evolution field [2].

Also, Needleman and Wunsch proposed a dynamic programming method for sequence alignment that is the base for the state-of-the-art methods used nowadays [5]. By the end of the 1960s, we observed the rising of structural bioinformatics, for example, with the construction of a three-dimensional model of a Cytochrome c protein [6]. At the beginning of the 1980s, the first methods for phylogenetic tree inference based on DNA sequences and the maximum likelihood were proposed [7]. Moreover, BLAST, a tool for local sequence alignments, was proposed in the 1990s mainly due to the increase of sequences availability [8].

However, three main milestones led to the modern bioinformatics field were: (i) the DNA sequencing methods, (ii) the genome projects, and (iii) the rise of supercomputers and the Internet [2]. DNA sequencing methods have existed since the 1970s, such as the Sanger chain-termination sequencing method [9]. However, these methods were slow and expensive. For instance, the human genome project (HGP) was an international research effort to map all of the genes and sequences of the human genome. This project started at the beginning of the 1990s but was only officially completed in 2003. The HGP gave us a genetic blueprint of the human being, but the sequencing costs were high.

The game changed when ingenious strategies were used in combination with computational approaches. Shotgun sequencing was introduced in the 1970s [10]. In 1995, a similar strategy was used to obtain a complete nucleotide sequence of the *Haemophilus influenzae* bacterium [11] and, in 2010, to get the complete genome of the fly *Drosophila melanogaster*, a eukaryote with a sequence length of ~ 120 Mb [12]. All these developments led to the emergence of Next-Generation Sequencing (NGS) platforms [13]. These technologies are characterized by high-throughput, with reduced processing time, and costs each time lower. This culminated in the diffusion of genome projects, even in small or middle research laboratories. Later, other high-throughput technologies appeared, such as microarray and RNA-Seq, used to analyze gene expression, and cryo-electron microscopy, which sped up the resolution of 3D macromolecular structures.

Thus, bioinformatics went from a tool for biological analysis to an interdisciplinary research field, mainly focused on the development of new models, algorithms, tools, and new types of analysis based on computational approaches to deal with biological data and get knowledge from them. In the last years, these evolutions of technological resources originated a tidal wave of data [14]. Consequently, an unprecedented amount of studies using bioinformatics approaches have been released, increasing peer-reviewed published papers in the field.

Here we tell the recent history of Bioinformatics based on metadata collected from scientific papers. We know that the history of bioinformatics has been told before in some publications [1, 15, 16]. However, we aimed to understand the recent state-of-the-art of the bioinformatics research, visualize changes motivated by technological evolution, and detect trends in published studies. We also performed an exploratory analysis of data collected from the bioinformatics literature and constructed visualizations to get insights into the bioinformatics field's brief history.

2 Materials and Methods

2.1 Data Collection

We empirically selected four journals for performing theses analyses: (i) Oxford Bioinformatics (ISSN 1367-4803); (ii) BMC Bioinformatics (ISSN 1471-2105); (iii) Briefings in Bioinformatics (ISSN 1467-5463); and (iv) PLoS Computational Biology (ISSN 1553-7358). We collected data using the PubMed API [17, 18] and in-house scripts coded using Python. Firstly, we collected a list of ID entries for all papers available at PubMed until October 7th, 2019, for the four journals. Then, we collected an XML file with PMID, date of publication, journal data, title, abstract, authors, and their affiliations for each entry. Citations were collected using Scopus Search API (https://dev.elsevier.com/).

2.2 Data Visualizations

Exploratory data analysis was performed using Python and R scripts. The web tool was developed using CodeIgniter Web framework (https://codeigniter.com), DataTables (https://datatables.net), jQuery (https://jquery.com), and other frameworks. Data were stored in a structured database using MySQL (https://www.mysql.com). Interactive visualizations were constructed using D3.js (https://d3js.org) and Flourish Data Visualization & Storytelling tool (https://flourish.studio). Static visualizations were constructed using RStudio (https://rstudio.com) and were refined using Adobe Photoshop CC 2019.

3 Results and Discussion

We collected 13,798 items from Oxford Bioinformatics, 1,418 items from Briefings in Bioinformatics, 9,265 items from BMC Bioinformatics, and 6,623 items from PLoS Computational Biology. In the publication timeline analysis for the last 22 years, we can observe the increase in the number of papers being published (Fig. 1A).

We can notice an expressive growth around 2007-2009 and the emergence of important open access journals in this period (BMC Bioinformatics and PLoS Computational Biology).

We also observed a substantial increase in international collaborations in publications (Fig. 1B), which agrees with the global trend, as corroborated by an analysis performed for other general science journals, such as Nature [19]. In 2019, the number of publications showing co-authors with different nationalities corresponded to approximately 25%. In contrast, the proportion of single authors' publications had a substantial drop.

It was also an evident rise in the numbers of papers published by Chinese authors and a reduction of the United Kingdom proportion in global production. Latin America has a low participation in global scientific production. However, Brazil leads the bioinformatics scientific production in the region, being responsible for almost half of Latin America publications (Fig. 1C-D).

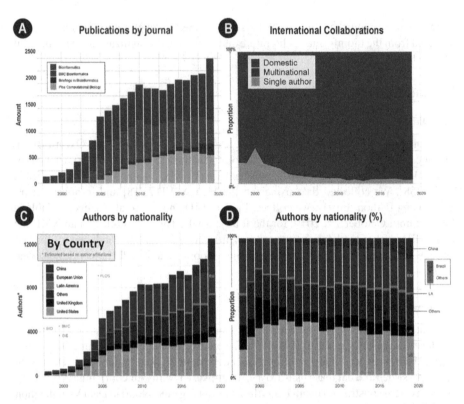

Fig. 1. (A) Publications by journal (1998–2019). (B) The proportion of international collaborations. "Domestic" corresponds to collaborations between researchers of the same institute or the
same country, "Multinational" from different countries, and "Single author" to individual papers.
(C) Nationality estimated based on author affiliation declarations (hence, this may not represent
the real nationality). (C) The amount and (D) the proportion. United Kingdom's (UK) data was
not included in the European Union (EU) group to ease the comprehension (even, before Brexit).

3.1 The History that the Keywords in Papers Tell Us

We analyzed the keywords reported in the papers to establish what are the major research
topics addressed in bioinformatics publications. We obtained 7,167 unique keywords
reported in the papers. From them, eleven were reported in the top five positions from
1998 to 2019 (Fig. 2).

Since 2011, only a restricted set of keywords has scored the top five list: humans,
computational biology, algorithms, models, and software. Interestingly, the keyword
"humans" that leads the top five groups since 2011, was less cited in the previous years,
except in 2000–2001 when it appeared in the fifth position. Even though most of the
bioinformatics research is more focused on developing new algorithms, models, and
software, a fact well illustrated by the other four keywords in the rank, it was expected
to find papers using keywords describing application areas for these approaches. Indeed,
we can observe that "humans" appear in the top five list in the years closer to the first

Keywords most used in Bioinformatics papers (1998-2019)

Plos Computational Biology | BMC Bioinformatics | Bioinformatics | Briefings in Bioinformatics

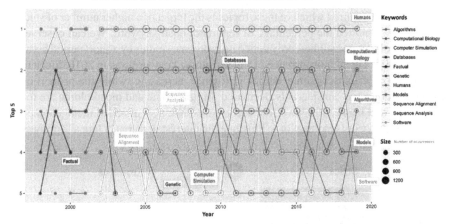

Fig. 2. Top five keywords listed in bioinformatics papers from 1998 to 2019.

announcements of the human genome draft (2001). We hypothesized that, in the following years, the bioinformatics research was characterized by considerable efforts to establish the architecture for data processing and storage, collection and organization of data, and development of desktop and web tools. Since many publications have accomplished these objectives, this provided scaffolds for more applicable studies focused on human data.

The top five ranking list only told us the main and general topics. To tell more about recent bioinformatics content history, it is necessary to analyze the importance of some keywords and correlate them with historical facts. To obtain these insights, we constructed word clouds with keywords with at least 50 occurrences by year (available on the website). Analyzing the word cloud, we can detect when some topics turned into trends or reduced, comparing the word size changes according to different periods. Although this analysis is quite limited to show specific changes in the keyword use (since it is hard to compare and find many words together through different word clouds), this could be used to identify targets for further analysis. Hence, based on this analysis, we raised some questions: what topic is the target for a higher number of studies: genome, transcriptome, or proteome? Are there keywords that are less used nowadays than in the past? What is related to the recent increasing in molecular dynamics publications? Why "artificial intelligence" was very reported in the 2000s, but fewer today? To try to answer these questions, we constructed several visualizations and discussed what they depict.

3.2 Omics

A Bioinformatics' fundamental consists of the use of computational approaches to analyze data related to the central dogma of molecular biology, *i.e.,* the process in which information flow from DNA to RNA to protein. Thus, genome, transcriptome, and proteome have been considered the main topics in this area, compounding the "Omics" study fields, such as genomics, transcriptomics, and proteomics (and new study areas such as metagenomics, metabolomics, and so on).

A priori, we inquired which of these topics is the target of a higher number of studies. In Fig. 3A, we plotted the percentage of reports of each keyword. We can observe the sovereignty of genome studies, only defied by proteome studies by a short period between 2007 and 2008. The first occurrence of transcriptome keyword was in 2010. Since then, the proportion of transcriptome studies is slightly increasing, while the proportion of proteome studies has considerably decreased. Metagenome and metabolome are recent study areas, and hence, still have few occurrences. We also included metagenome and metabolome keywords in this plot to illustrate the interest in recent "Omics" approaches.

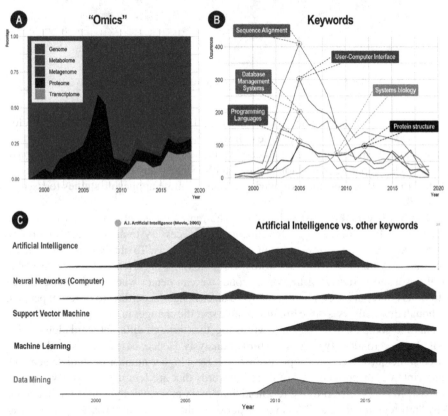

Fig. 3. Keywords analysis. (A) Percentage of citations of "Omics" keywords from 1998 to 2019. (B) Use of the keywords "Database Management Systems", "User-Computer Interface", "Programming Languages", and "Sequence Alignment", "Protein Structure", and "Systems Biology" from 1998 to 2019. (C) Use of "Artificial Intelligence" and other correlated keywords from 1998 to 2019.

3.3 Keywords in Disuse

We also observed a peculiar behavior in the usage of some keywords, such as: "Database Management Systems", "User-Computer Interface", "Programming Languages", and "Sequence Alignment" (Fig. 3B). These keywords correspond to essential topics in bioinformatics. So, why have they been less used nowadays?

Our hypothesis relies on the release of the HGP and the blast of the genome projects. At first, researchers did not know how to deal with high amounts of data. Besides, the scientific community interest was more focused on the development of novel models and algorithms to solve relevant problems than on data management. This naturally reflected in publications aiming to establish better ways to model, store, and access these data. Also, the high number of software executed by command lines created a gap of user-friendly tools that could be explored by publications. Nowadays, these necessities also persist, but because of the high number of studies, some topics present tools widely consolidated, such as "sequence alignment". Although this topic is essential, nowadays, the use of sequence alignment is considered trivial, with standard tools already established and few novel inventions.

We also observed a slight reduction in the use of "protein structure" and "systems biology" keywords. However, this could change in the next years due to the evolution in the Cryogenic electron microscopy methods used for determining protein structures and the increase of computational power that will allow simulations of larger systems.

Additionally, the interest in programming languages initially increased, impelled by projects that aimed to develop libraries for analyzing biological data, such as packages derived from the projects Bioconductor [20–24], and later Biopython [25, 26]. When a large number of packages were obtained, research topics began to focus, for example, on applications of use.

3.4 Artificial Intelligence and the Influence of the Pop Culture in Science

We analyzed the use of the keyword "Artificial Intelligence" and other related keywords. We observed that this keyword's use increased from 2001 until 2008, when reduced (Fig. 3C). Then, publications using "Artificial Intelligence" were made, but the authors started to use more specific descriptions, such as "neural networks", "support vector machine", "machine learning", and "data mining".

It is interesting to report that artificial intelligence is studied since the middle of the 20th century, as well as the other listed topics. Hence, what could explain the expressive increase in the use of this topic as a keyword and its subsequent reduction?

A peculiar explanation for this phenomenon is that in 2001 was released the famous movie "A.I. Artificial Intelligence" (directed by Steven Spielberg). The film's popularity may have influenced the keyword "artificial intelligence" in academic works. Later, this keyword's use decreased, and more specific descriptions of the A.I. techniques used were adopted. This suggests that pop culture could influence how studies are disseminated.

3.5 A Promising Future for Molecular Dynamics

Classical and quantum molecular dynamics are a set of techniques to simulate the dynamic behavior, movements, and interactions of molecules in a system across periods.

They have been used, for example, to perform computational predictions of cancer drug resistance [27], to understand allosteric immune escape pathways in the HIV-1 envelope glycoprotein [28], and to simulate the action of enzymes used in biofuel production [29, 30]. Although these methods are known to have high computational cost requirements, since 2011, the number of citations of "molecular dynamics simulation" has increased substantially (Fig. 4).

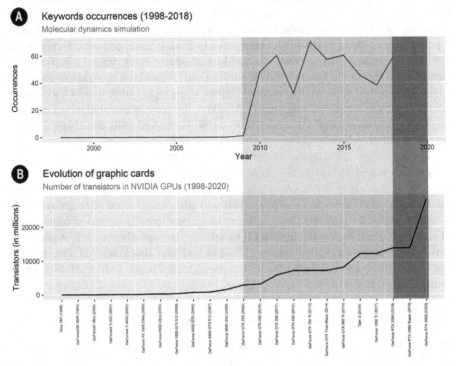

Fig. 4. (A) Use of "molecular dynamics" as a keyword (above) compared to (B) the number of transistors in NVIDIA GPUs (below). Also, it is important to highlight that molecular dynamics researchers prefer to publish their experiments in specialized journals, such as the Journal of Biomolecular Structure and Dynamics. Source: Adapted from https://vintage3d.org/dbn.php. Accessed on September 16th 2020.

We searched for connections in the number of reports with the number of transistors in NVIDIA graphic cards from 1998 until 2020. We observed that the number of citations of molecular dynamics simulation seems to correlate to the evolution of Graphic Processing Unities (GPUs). The evolution of GPUs is impelled by the game industry because of the necessity of more realistic games. However, researchers have adapted the GPU usage to process force fields calculus used in molecular dynamics simulations. Thus, molecular dynamics scientists have successfully used gamer graphic cards in research applications, which led to better results than those from CPU only supercomputers from a few years ago.

Based on this analysis, we foresee a promising future for molecular dynamics research making good use of GPUs evolution. Figure 4 shows that the new generation of graphic cards brought a huge increase in GPUs' number of transistors. The number of transistors in graphic cards can be related to their performance. The more quantity of transistors, the more the capacity of data processing. For instance, the graphic card GeForce256 SDR (released in 1999) presented 23,000,000 transistors. A new GeForce RTX 3090 (released in 2020) has 28,000,000,000 transistors ($\sim 1,200 \times$ more). This is possible due to the reduction in the size of the components. GeForce256 SDR used the chip NV10 with lithography of 220 nm. On the other hand, RTX 3090 uses the chip GA102 with lithography of only 8 nm. More powerful graphic cards will increase the number and the size of studies using these techniques.

3.6 Popular Programming Languages for Bioinformatics

Our results also point out a recent trend of using Python as the programming language for bioinformatics research and applications, followed by Perl (perhaps the first bioinformatic programming language), Java, and R. A recent large-scale analysis study of code from bioinformatics projects available at GitHub suggested that the languages Python, R, and Java are more succinct than lower-level languages [31]. This allows developers to construct advanced programs written in fewer code lines, which can contribute to the popularity of these languages for bioinformatics. Besides, initiatives for teaching Python to deal with biological data have contributed to expanding this language by bioinformaticians [32, 33].

3.7 Limitations of This Study

This study's major limitation is that it could include only a few of the interesting findings and correlations in the main text because of space restrictions. We expect to have solved this by developing an interactive web tool and turn it publicly available for users to explore data.

Furthermore, this research is restricted by the journals selected and the period used for data gathering. Frequent updates of data collected can be a challenge due to restrictions of the data sources APIs. Also, the keyword analysis topics were defined by an empirical analysis of the word clouds. In the future, we intend to improve these analyses using context-dependent string search engines, such as the natural language processing (NLP) API from Google Cloud.

In the popular programming language analysis, we found a lower number of results once we used text mining to detect occurrences in the abstract (many authors did not declare the programming language used). However, we expect to get a realistic sample of programming languages used. We would like to emphasize that it is not possible to define the "best programming language" for bioinformatics research applications. The choice of a programming language depends on the developer's familiarity with their syntax, the context of use, the requirements, and the algorithms' computational costs to be implemented. Besides, we consider the availability of packages and libraries for biological data analysis, and developers' community support is an essential item to be observed.

Last, we have tried to establish which authors contributed more to bioinformatics based on the number of publications and citations. However, this analysis was limited due to the challenges in disambiguating names. This could be solved by the adoption of unique codes for identifying scientists, such as ORCID. Unfortunately, these data were not available for most of the databases consulted. Even so, we constructed lists of authors that most collaborated in studies published in the main bioinformatics journals (stratified by publication number, citations, and countries) and made them available on the website. Also, PubMed API returned a limited number of citations in documents indexed there. We used the Scopus API to improve this analysis.

4   Conclusion

In this paper, we presented an overview of bioinformatics publications in the last 22 years based on four high-impact bioinformatics journals. We consider that this paper's main scientific contribution is to present the reports of the state-of-the-art of publications in bioinformatics (past and present), presenting our predictions based on what the data showed us. We show an increasingly collaborative world: data show an increase in the bioinformatics' scientific production with the increasing participation of several countries. However, there is still much to improve (see Latin America's low participation in global scientific production). The keyword analysis showed changes in the major topics addressed in bioinformatics papers. We also hypothesized a growth in molecular dynamic simulation publications due to the correlation between published works and recent evolutions in GPUs. The most cited programming languages analysis showed Python, Perl, Java, and R as the most popular programming languages for bioinformatics.

Additionally, we proposed a web tool for exploratory data analysis as supplementary material. Here, we present only an overview of what the data showed us, focusing on some details that caught our attention. Our readers can interact with the data in the web tool, obtain insights, and maybe reach conclusions that we may not even have imagined when writing this article. We also want to encourage (and provoke) our readers to think about bioinformatics' perspectives as a science (and their specific areas of activity). The scientist's role is to observe, question, and propose solutions that lead to society's improvement. For scientists to fulfill their roles, they should be able to adapt to changes and know the history and discuss the future is a fundamental step to this. We hope that these data visualizations can provide insights and raise more thought-provoking discussions about bioinformatics evolution, trends, and perspectives. We provided the data sets and other interactive visualizations at http://bioinfo.dcc.ufmg.br/history/.

Acknowledgments. The authors thank the funding agencies: CAPES, FAPEMIG, and CNPq. This study was financed in part by the Coordenação de Aperfeiçoamento de Pessoal de Nível Superior - Brasil (CAPES) - Finance Code 001. Project grant number 51/2013 - 23038.004007/2014-82.

References

1. Akalin, P.K.: Introduction to bioinformatics. Mol. Nutr. Food Res. **50**, 610–619 (2006)
2. Hagen, J.B.: The origins of bioinformatics. Nat. Rev. Genet. **1**, 231–236 (2000)
3. Moore, S., Spackman, D.H., Stein, W.H.: Automatic recording apparatus for use in the chromatography of amino acids, pp. 1107–1115 (1958)
4. Dayhoff, M.O.: Atlas of protein sequence and structure. National Biomedical Research Foundation (1972)
5. Needleman, S.B., Wunsch, C.D.: A general method applicable to the search for similarities in the amino acid sequence of two proteins. J. Molecular Biol. **48**, 443–453 (1970)
6. Levinthal, C.: Molecular model-building by computer. Sci. Am. **214**, 42–52 (1966)
7. Felsenstein, J.: Evolutionary trees from DNA sequences: a maximum likelihood approach. J. Mol. Evol. **17**, 368–376 (1981)
8. Stephen, F., Altschu, P., Warren, G., Webb, M., Eugene, W., Myers, David, J.L.: Basic Local Alignment Search Tool (1990)
9. Sanger, F., Nicklen, S., Coulson, A.R.: DNA sequencing with chain-terminating inhibitors. Proc. National Acad. Sci. **74**, 5463–5467 (1977)
10. Staden, R.: A strategy of DNA sequencing employing computer programs. Nucleic Acids Res. **6**, 2601–2610 (1979)
11. Fleischmann, R.D., Adams, M.D., White, O., Clayton, R.A., Kirkness, E.F., Kerlavage, A.R., et al.: Whole-genome random sequencing and assembly of Haemophilus influenzae Rd. Science **269**, 496–512 (1995)
12. Adams, M.D., Celniker, S.E., Holt, R.A., Evans, C.A., Gocayne, J.D., Amanatides, P.G., et al.: The genome sequence of drosophila melanogaster. Science **287**, 2185–2195 (2000)
13. Mariano, D.C.B., Pereira, F.L., Aguiar, E.L., Oliveira, L.C., Benevides, L., Guimarães, L.C., et al.: SIMBA: a web tool for managing bacterial genome assembly generated by Ion PGM sequencing technology. BMC Bioinform. **17**(Suppl 18), 456 (2016)
14. It's sink or swim as a tidal wave of data approaches. Nature, **399**, 517 (1999)
15. Gauthier, J., Vincent, A.T., Charette, S.J., Derome, N.: A brief history of bioinformatics. Brief. Bioinform. **20**, 1981–1996 (2019)
16. Hogeweg, P.: The Roots of Bioinformatics in Theoretical Biology. PLoS Comput. Biol. **7**, e1002021 (2011)
17. Canese, K., Weis, S.: PubMed: The Bibliographic Database. National Center for Biotechnology Information (US) (2013). https://www.ncbi.nlm.nih.gov/books/NBK153385/. Accessed 14 Sep 2020
18. NCBI Resource Coordinators: Database resources of the national center for biotechnology information. Nucleic Acids Res. **46**, D8–D13 (2018)
19. Monastersky, R., Noorden, R.V.: 150 years of nature: a data graphic charts our evolution. Nature **575**, 22–23 (2019)
20. Carey, V.J., Gentry, J., Whalen, E., Gentleman, R.: Network structures and algorithms in Bioconductor. Bioinformatics **21**, 135–136 (2005)
21. Chen, H., Lau, M.C., Wong, M.T., Newell, E.W., Poidinger, M., Chen, J.: Cytofkit: a bioconductor package for an integrated mass cytometry data analysis pipeline. PLoS Comput. Biol. **12**, e1005112 (2016)
22. Fournier, F., Joly Beauparlant, C., Paradis, R., Droit, A.: rTANDEM, an R/Bioconductor package for MS/MS protein identification. Bioinformatics **30**, 2233–2234 (2014)
23. Gådin, J.R., van't Hooft, F.M., Eriksson, P., Folkersen, L.: AllelicImbalance: an R/ bioconductor package for detecting, managing, and visualizing allele expression imbalance data from RNA sequencing. BMC Bioinform. **16**, 194 (2015)

24. Gentleman, R.C., Carey, V.J., Bates, D.M., Bolstad, B., Dettling, M., Dudoit, S., et al.: Bioconductor: open software development for computational biology and bioinformatics. Genome Biol. **5**, R80 (2004)

25. Talevich, E., Invergo, B.M., Cock, P.J., Chapman, B.A.: Bio.Phylo: a unified toolkit for processing, analyzing and visualizing phylogenetic trees in Biopython. BMC Bioinform. **13**, 209 (2012)

26. Cock, P.J.A., Antao, T., Chang, J.T., Chapman, B.A., Cox, C.J., Dalke, A., et al.: Biopython: freely available Python tools for computational molecular biology and bioinformatics. Bioinformatics **25**, 1422–1423 (2009)

27. Sun, X., Hu, B.: Mathematical modeling and computational prediction of cancer drug resistance. Brief. Bioinform. **19**, 1382–1399 (2018)

28. Sethi, A., Tian, J., Derdeyn, C.A., Korber, B., Gnanakaran, S.: A mechanistic understanding of allosteric immune escape pathways in the HIV-1 envelope glycoprotein. PLoS Comput. Biol. **9**, e1003046 (2013)

29. Costa, L.S.C., Mariano, D.C.B., Rocha, R.E.O., Kraml, J., da Silveira, C.H., Liedl, K.R., et al.: Molecular dynamics gives new insights into the glucose tolerance and inhibition mechanisms on β-glucosidases. Molecules **24**, 3215 (2019)

30. Lima, L.H.F., de Fernandez-Quintéro, M., Rocha, R.E.O., Mariano, D.C.B., Melo-Minardi, R.C., de Liedl, K.R.: Conformational flexibility correlates with glucose tolerance for point mutations in β-glucosidases – a computational study. J. Biomolecular Structure Dyn. 1–20 (2020)

31. Russell, P.H., Johnson, R.L., Ananthan, S., Harnke, B., Carlson, N.E.: A large-scale analysis of bioinformatics code on GitHub. PLoS ONE **13**, e0205898 (2018)

32. Ekmekci, B., McAnany, C.E., Mura, C.: An Introduction to Programming for Bioscientists: A Python-Based Primer. PLoS Comput. Biol. **12**, e1004867 (2016)

33. Mariano, D., Martins, P., Helene Santos, L., de Melo- Minardi, R.C.: Introducing Programming Skills for Life Science Students. Biochemistry and Molecular Biology Education (2019). https://doi.org/10.1002/bmb.21230

Computational Simulations for Cyclizations Catalyzed by Plant Monoterpene Synthases

Waldeyr Mendes Cordeiro da Silva[1,2,3]([envelope]) [ID], Daniela P. de Andrade[1],
Jakob L. Andersen[4][ID], Maria Emília M. T. Walter[2][ID], Marcelo Brigido[2][ID],
Peter F. Stadler[3,4,5,7,8][ID], and Christoph Flamm[6][ID]

[1] NEPBio - Federal Institute of Goiás, Formosa, Brazil
waldeyr.mendes@ifg.edu.br
[2] University of Brasília, Brasília, DF, Brazil
[3] Department of Computer Science, Bioinformatics Group, Interdisciplinary Center
for Bioinformatics, University of Leipzig, Leipzig, Germany
[4] Department of Mathematics and Computer Science, University of Southern
Denmark, Odense, Denmark
[5] German Centre for Integrative Biodiversity Research (iDiv) Halle-Jena-Leipzig,
Leipzig, Germany
[6] Institute for Theoretical Chemistry, University of Vienna, Vienna, Austria
[7] Max Planck Institute for Mathematics in the Sciences, Leipzig, Germany
[8] Santa Fe Institute, Santa Fe, USA

Abstract. Metabolic pathways collectively define the biochemical repertory on an organism exposing the steps of its production. *In silico* metabolic pathways have been reconstructed using a wide range of computational methods. The reconstructed metabolic pathways can vary in some aspects, among which, in the context of this work, it is relevant to remark the data structure and granularity of biochemical details. Inferring chemical reactions using graph grammar rules is a method that exposes the initial, intermediates, and final products by modeling pathways over graphs and hypergraphs. Plant monoterpenes are volatile compounds with applications in industry, biotechnology, and medicine. They also play a vital ecological role. The last enzymatic reaction in the plant monoterpenes production chain is plentiful of possibilities due to the promiscuous nature of the terpene synthases (TPS), a case in which the application of inferring chemical reactions using graph grammar rules is suitable. In this work, we designed graph grammar rules that express cyclization reactions catalyzed by plant monoterpene synthases. As a result, it was generated a reachable chemical space of potential plant monoterpene blend, which can be computationally exploitable. In addition, these graph grammar rules were added to the 2Path, and a graphical interface was provided to aid the simulation code outlining.

Keywords: Plant · Monoterpene · Biosynthesis · Computational · Simulation

© Springer Nature Switzerland AG 2020
J. C. Setubal and W. M. Silva (Eds.): BSB 2020, LNBI 12558, pp. 247–258, 2020.
https://doi.org/10.1007/978-3-030-65775-8_23

1   Introduction

Terpenes are the oldest and various plant natural products as essential oils, and some of them, as volatile compounds, play a crucial role in response to herbivores, interaction with other plants, and attracting pollinators [33,36]. The benefits of terpenes properties for humans spread through their use as flavors in agricultural and industrial products, or as fragrances in foods and cosmetics, plus pharmaceuticals and biofuels [2,39].

The wide range of terpenes and its "chemodiversity" is expected as a characteristic of life, taking into account the considerable biodiversity of plants and their interactions with other organisms [16,36]. This chemodiversity is related to the chemical mechanisms catalyzed by Terpene synthases/cyclases (TPS) influencing the terpenes' variety. This variety may be related to their biological function by adjusting the mixture and amount of terpenes to the specificity of the target, both in communication relations and concerning protection against numerous predators, parasites and competitors [16,32,36].

Terpenes are named according to the number of C_5 isoprenoid units incorporated into their carbon skeletons as mono- (C_{10}), sesqui- (C_{15}), di- (C_{20}), sester- (C_{25}), tri- (C_{30}) and sesquarterpenes (C_{35}) [37]. The isoprenoid units can be Isopenthenyl Diphosphate (IPP) or its allylic isomer Dimethylallyl Diphosphate $(DMADP)$ that are condensed by prenyltransferases to produce larger prenyl diphosphates, such as the monoterpene precursor Geranyl Diphosphate (GDP), which is the minimum-length cyclization substrate terpene biosynthesis [10]. Commonly, after the GPP diphosphate loss and a $C_1 - C_6$ bond formation, the monoterpene formation proceeds through the α-terpinyl cation (Fig. 1) by means of a cascade of reactions that include $C - C$ bonds, Wagner-Meerwein rearrangements, allyl- and methyl-shifts caused by conformational changes of intermediate cations, and carbocation capture by water and hydride [33].

There are two classes of TPSs: Class I and Class II, defined by catalytically essential amino acid motifs [7,27]. TPSs I convert linear, all-trans, isoprenoids, geranyl (C10)-, farnesyl (C15)-, or geranylgeranyl (C20)-diphosphate into numerous varieties of monoterpenes, sesquiterpenes, and diterpenes. The TPSs I bind their substrate by coordinating a trinuclear divalent metal ion catalytic site (generally a Mg^{2+}), consisting of a central cavity formed by mostly antiparallel α-helices. This catalytic site has an aspartate-rich $DDxxD/E$ motif, and often another NSE/DTE motif in the C-terminal portion [24]. TPSs Class II act by triggering GGPP protonation, which results in successive carbocations and cyclizations to form, for example, copalyl-diphosphate (CPP) [27]. In the Class II TPSs, the $DxDD$ motif (distinct from the TPS I $DDxxD/E$ motif) catalyzes the reaction, also using a Mg^{2+} cofactor to assist substrate binding and positioning [15]. The terpenes diversity also can be influenced by nucleotide changes in the alleles of TPS genes [26]. In plants, the production of terpenes can be compartmentalized (Fig. 2), and monoterpenes, for example, can be produced in specialized structures: plastids [31].

Enzyme function prediction is particularly challenging when dealing with TPS because of their capability to produce numerous carbon skeletons by

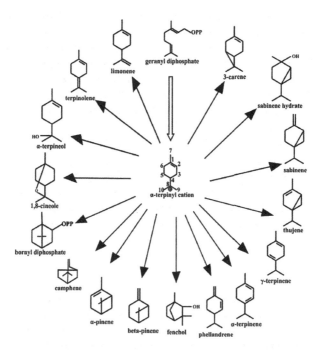

Fig. 1. Examples in the blend of monoterpenes produced by plants through Geranyl Diphosphate (GPP) cyclizations.

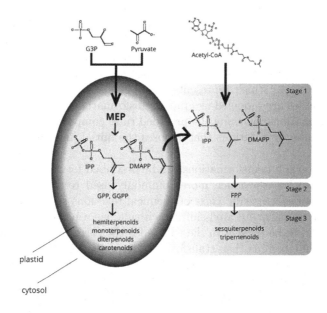

Fig. 2. Plant compartmentalization of terpenes biosynthesis.

catalyzing complex carbocation rearrangements. Thus, beyond the primary sequence homology, metadata, such as environmental or experimental conditions and cheminformatics approaches could help annotate the TPS accurately. Considering that the TPS annotation may be enhanced by exploring the feasible terpene blend produced by cyclization of their precursors, this work presents a computational approach to generate it by simulating carbocations of GPP catalyzed by *in vitro* or *in vivo* patterns of plant monoterpene synthases that have already been described in the literature.

The work is organized as follows. Section 2 introduces the method, explaining how the chemical network is produced, its data structure, and traversing. Section 3 discusses the results and compares this approach with others, highlighting the contributions of our results for molecular biology and cheminformatics. Finally, Sect. 4 presents the conclusion and an outline of the next steps to consolidate the approach as a tool for the scientific community.

2   Method

Degenhardt *et al.* [11] described reaction mechanisms for plant mono- and sesquiterpene synthases identifying several monoterpenes as products of these chemical rearrangements. Based on Degenhardt *et al.* [11] review and other sources of plant enzymatic GPP cyclizations [6,13,17,18,36], we extended the approach presented by Silva *et al.* [33] for plant sesquiterpenes biosynthesis by including monoterpenes.

The method formally models the chemical reactions on a mechanistic level, building pathways assembled as a chemical network. The chemical network is abstracted as a directed multi-hypergraph, where the vertices correspond to molecules and hyperedges to reactions. Each vertex represents a molecule, which is abstracted by an undirected graph, where atoms are vertices, and the bonds are edges. In other words, each vertex of the chemical network is composed of an undirected graph, which represents a molecule, and the chemical reactions on these molecules are modeled as graph transformations. The accumulation of graph transformations following the provided rules compose a network obtained in a given number of iterations that exposes the initial, intermediate, and final compounds.

Rule-based graph transformations can be described formally by graph grammars that generalize the much more commonly used term-rewriting systems. Each rule describes a specific *class* of chemical reactions such as a ring closure from C_1 to C_6 atom, or an allyl-shift. Each rule has a pattern L of atoms and bonds that need to be present in the educts for the corresponding reaction. The matched part of educts is then transformed as specified by the rule.

We used the *double pushout* (DPO) formalism [29] for graph rewriting because it is particularly suitable to model chemistry: it ensures reversibility of transformation and supports well-defined atom maps [3]. Here, each rule has the form $p = (L \xleftarrow{l} K \xrightarrow{r} R)$ where L is the left graph, R is the right graph, and K is a context graph. Graph morphisms l and r describe the embedding of the context into the L and R by connecting these graphs by $l\colon K \to L$ and $r\colon K \to R$. If a rule p is applied to a graph G, it is mandatory that L "matches" a part of G. The existence of another graph morphism ($m\colon L \to G$) captures this, and together with the rule p and the matching morphism m, define the transformation of the substrate G to the product H, written as $G \stackrel{p,m}{\Longrightarrow} H$. Figure 3 shows an example of a rule for quenching by water, where the molecules (H_2O and α-terpynol), the rule, and the matching morphism are exposed according to the DPO graph transformation.

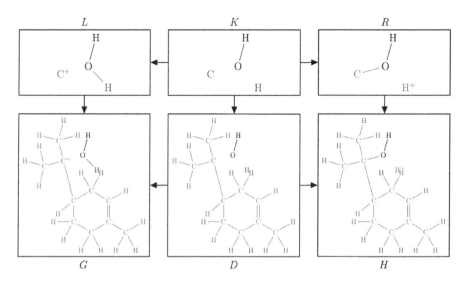

Fig. 3. Example of rule for quenching by water, and its application to the molecules H_2O and α-terpynol.

MedØlDatschgerl [4], or MØD for short, is a software package for chemically inspired graph transformation that can target a set of initial molecules and generate the reaction networks automatically, applying rule-based graph transformations. Some of the rules from Silva *et al.* [33] were revised to compose a more comprehensive new set of rules able to generating mono- and sesquiterpenes. A total of 17 graph transformation rules were manually designed in GML format [21] to represent the literature-based chemical mechanisms that lead to the compounds shown in the Fig. 1.

These rules transform the initial set of molecules (GPP and H_2O), matching them in a first iteration, and generating a new set of chemically feasible molecules. Then, in a new iteration, the set of rules is again applied to the new set of compounds, generating a third set of compounds. The same process can be repeated for an arbitrary number of iterations. It is possible to customize both the set of rules and the number of iterations for each simulation. We have built a Web interface to facilitate this task, which allows us to generate a simulation source code file according to the chosen parameters. A graphical explanation of the proposed method is shown in Fig. 4 and the graphical representaion of the rules is shown in Fig. 5.

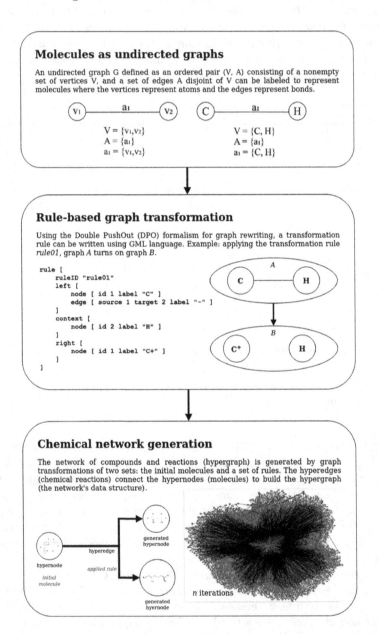

Fig. 4. Method summary with examples of abstractions/codes, and their applications for the chemical network generation based on graph transformations.

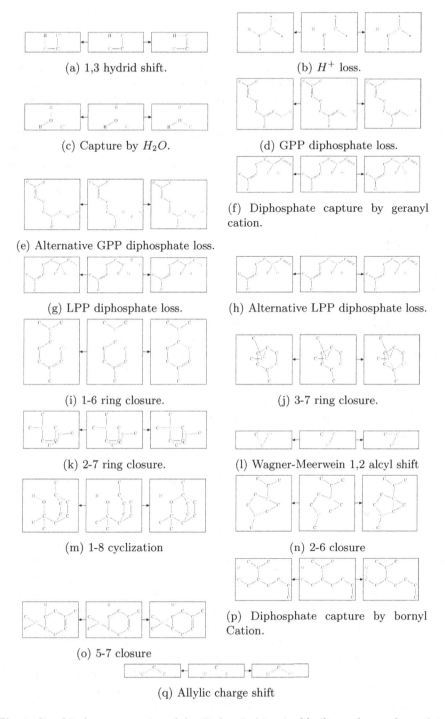

(a) 1,3 hydrid shift.

(b) H^+ loss.

(c) Capture by H_2O.

(d) GPP diphosphate loss.

(e) Alternative GPP diphosphate loss.

(f) Diphosphate capture by geranyl cation.

(g) LPP diphosphate loss.

(h) Alternative LPP diphosphate loss.

(i) 1-6 ring closure.

(j) 3-7 ring closure.

(k) 2-7 ring closure.

(l) Wagner-Meerwein 1,2 alcyl shift

(m) 1-8 cyclization

(n) 2-6 closure

(p) Diphosphate capture by bornyl Cation.

(o) 5-7 closure

(q) Allylic charge shift

Fig. 5. Graphical representation of the 17 chemical inspired built graph transformation rules for the GPP cyclizations.

3   Results and Discussion

This work provides material and method for computer simulations capable of reproducing the blend of monoterpenes produced by plants using chemical mechanisms reported in the literature. Based on the presented seventeen graph grammar rules, which embedded the enzymatic cyclizations of Geranyl Diphosphate (GPP) described in [6,11,13,17,18,36], the simulations were able to generate a massive universe of monoterpenes as final products, including those shown in Fig. 1. Along with them, the simulations also produced all the cyclization steps, explicitly exposing the final and intermediates compounds, forming an extensive chemical network of potential monoterpene products. The entire chemical network, a hypergraph, can be computationally accessed, processed for further analysis, and conveniently exported in a PDF report.

Combining the graph transformation rules and the number of iterations allows exploring a vast blend of feasible monoterpenes, exposing all the enzymatic mechanisms of cyclization of its precursor, the GPP. Notably, in this chemical space, there are three groups of monoterpenes: i) those that are found in nature and are known and deposited in scientific databases such as CHEBI [19], PubChem [25], and KEGG [23]; ii) those that are found in nature but remain unknown; iii) those that are physically feasible but probably will not be found in nature, maybe due to the high energy cost of their production. Scientific advances in technologies such as mass spectrometry combined with gas chromatography have allowed allocating more and more monoterpenes in the first group, the known ones. For all these groups, the results presented here allow us to clarify the chemical mechanisms of their biosynthesis.

Along with this, the analysis of omic data has increasingly provided evidence based on the primary sequence of amino acid residues for functional protein annotation. Like other enzymes, the TPS sequence of amino acid residues influences its three-dimensional structure [28], and mutations in these amino acids can affect their structure, and hence their function, causing changes in the efficiency, specificity, or concentration of their products [8,10,40]. Also, different blends of terpenes are produced under different scenarios atop their primary amino acid sequence. [5]. These scenarios can be, for example, the availability of GPP, the subcellular compartmentalization, the pH, developmental stage, and organ, abiotic and biotic stress, and fitness costs [1,2,5,20,38]. The dependency on conditions beyond sequence homology makes it challenging to annotate TPS accurately. Thus, the results presented here could compose a computational annotation system that considers the following elements: the biological sequence of the TPS, a blend of feasible terpenes produced by it, and a weighted subset of this blend produced by a particular TPS under a particular scenario.

Although simulations can be customized, source code writing may be nontrivial for non-computational researchers. Looking to find a way to make the simulation environment more friendly, we created a Web interface that relieves this condition, making customization of the simulation more intuitive and less laborious as some drag and drop and clicks are enough. Such an interface (Fig. 6, combined with a Docker Image with all the environment ready, is available in the GitHub[1].

Fig. 6. On-line interface to build the simulation code.

3.1 Comparison with Related Works

There are some important computational metabolic pathway prediction initiatives that consider intermediate transformations in chemical reactions such as RetroRules [14], Biotransformer [12], Isegawa *et al.* [22], and Tian *et al.* [37]. Although providing comparable results, these approaches differ by using distinct methods and data structures to represent the molecules and their transformations. RetroRules [14] and Biotransformer [12] uses chemical reaction descriptions encoded by SMARTS [35] and SMIRKS [34]. Isegawa *et al.* [22] used AFIR [30] method to computationally predict pathways for terpene formation. Tian *et al.* [37] presented a computational approach to generate all possible carbocations of monoterpene synthases defining and organizing the product chemical space.

This work has similar objectives and results to RetroRules [14], Biotransformer [12], and Tian *et al.* [37], but quite a different methodology, presenting itself as a versatile alternative. Chow *et al.* [9] used Tian *et al.* [37] approach to characterize a sesquiterpene synthase from *Streptomyces clavuligerus*, which corroborates the idea that the chemical networks can help the functional assignment of TPS.

4   Conclusion and Future Work

This work presents a way to generate and explore the monoterpenes biosynthesis and their diversity. It is done through a computationally tractable data structure

---

[1] https://github.com/waldeyr/2PathTerpenes.

representing a chemical network that exposes the inputs, intermediates, and final compounds at an atom-bond level. The results presented here, allied to biological scenarios and the primary sequences of TPS, can enhance their functional annotation. Moreover, plant metabolic engineering could take advantage of this approach to help designing genes that produce monoterpenes' desired assortment for ecological and industrial purposes. These monoterpene results are added to the previous results of sesquiterpenes [33] to compose a terpene biosynthesis simulation system called 2Path. Also, it brings a Web interface that makes writing the simulations more intuitive and less laborious as some drag and drop, and clicks are enough. All the work is available at https://github.com/waldeyr/2PathTerpenes.

Acknowledgments. This research was funded in part by CAPES through a sandwich scholarship to W.M.C.d.S. It was additionally supported in part by the Independent Research Fund Denmark, Natural Sciences, grant DFF-7014-00041. M.E.M.T.W. and M.B. has been continuously supported by productivity fellowship from CNPq.

References

1. Abbas, F., Ke, Y., Yu, R., Yue, Y., Amanullah, S., Jahangir, M.M., Fan, Y.: Volatile terpenoids: multiple functions, biosynthesis, modulation and manipulation by genetic engineering. Planta **246**(5), 803–816 (2017). https://doi.org/10.1007/s00425-017-2749-x

2. Aharoni, A., Jongsma, M.A., Bouwmeester, H.J.: Volatile science? metabolic engineering of terpenoids in plants. Trends Plant Sci. **10**(12), 594–602 (2005)

3. Andersen, J.L., Flamm, C., Merkle, D., Stadler, P.F.: Inferring chemical reaction patterns using graph grammar rule composition. J. Syst. Chem. **4**, 4 (2013)

4. Andersen, J.L., Flamm, C., Merkle, D., Stadler, P.F.: A Software Package for Chemically Inspired Graph Transformation. In: Echahed, R., Minas, M. (eds.) ICGT 2016. LNCS, vol. 9761, pp. 73–88. Springer, Cham (2016). https://doi.org/10.1007/978-3-319-40530-8_5

5. Block, A.K., Vaughan, M.M., Schmelz, E.A., Christensen, S.A.: Biosynthesis and function of terpenoid defense compounds in maize (zea mays). Planta **249**(1), 21–30 (2019)

6. Bohlmann, J., Steele, C.L., Croteau, R.: Monoterpene synthases from grand fir (Abies grandis): cDNA isolation, characterization, and functional expression of myrcene synthase, (-)-(4S)- limonene synthase, and (-)-(1S,5S)-pinene synthase. J. Biol. Chemistry **272**(35), 21784–21792 (1997)

7. Chen, F., Tholl, D., Bohlmann, J., Pichersky, E.: The family of terpene synthases in plants: a mid-size family of genes for specialized metabolism that is highly diversified throughout the kingdom. Plant J. **66**(1), 212–229 (2011)

8. Chen, H., Li, G., Köllner, T.G., Jia, Q., Gershenzon, J., Chen, F.: Positive darwinian selection is a driving force for the diversification of terpenoid biosynthesis in the genus oryza. BMC Plant Biol. **14**(1), 1–12 (2014)

9. Chow, J.Y., et al.: Computational-guided discovery and characterization of a sesquiterpene synthase from Streptomyces clavuligerus. Proceedings of the National Academy of Sciences of the United States of America **112**(18), 5661–6 (2015)

10. Christianson, D.W.: Structural and chemical biology of terpenoid cyclases. Chemical Rev. **117**(17), 11570–11648 (2017)
11. Degenhardt, J., Köllner, T.G., Gershenzon, J.: Monoterpene and sesquiterpene synthases and the origin of terpene skeletal diversity in plants. Phytochemistry **70**(15–16), 1621–1637 (2009)
12. Djoumbou-Feunang, Y., Fiamoncini, J., Gil-de-la-Fuente, A., Greiner, R., Manach, C., Wishart, D.S.: BioTransformer: a comprehensive computational tool for small molecule metabolism prediction and metabolite identification. Journal of Cheminformatics **11**(1), 1–25 (2019). https://doi.org/10.1186/s13321-018-0324-5
13. Dong, L., Jongedijk, E., Bouwmeester, H., Van Der Krol, A.: Monoterpene biosynthesis potential of plant subcellular compartments. New Phytologist **209**(2), 679–690 (2016)
14. Duigou, T., du Lac, M., Carbonell, P., Faulon, J.L.: Retrorules: a database of reaction rules for engineering biology. Nucleic Acids Res. **47**(D1), D1229–D1235 (2018)
15. Gao, Y., Honzatko, R.B., Peters, R.J.: Terpenoid synthase structures: a so far incomplete view of complex catalysis. Nat. Prod. Rep. **29**(10), 1153 (2012)
16. Gershenzon, J., Dudareva, N.: The function of terpene natural products in the natural world. Nat. Chem. Biol. **3**(7), 408–414 (2007)
17. Godard, K.A., White, R., Bohlmann, J.: Monoterpene-induced molecular responses in Arabidopsis thaliana. Phytochemistry **69**(9), 1838–1849 (2008)
18. Gutensohn, M., et al.: Cytosolic monoterpene biosynthesis is supported by plastid-generated geranyl diphosphate substrate in transgenic tomato fruits. Plant J. **75**(3), 351–363 (2013)
19. Hastings, J., et al.: Chebi in 2016: Improved services and an expanding collection of metabolites. Nucleic Acids Res. **44**(D1), D1214–D1219 (2016)
20. Heinig, U., Gutensohn, M., Dudareva, N., Aharoni, A.: The challenges of cellular compartmentalization in plant metabolic engineering. Current Opinion Biotechnol. **24**(2), 239–246 (2013)
21. Himsolt, M.: Gml: a portable graph file format. Html page under http://www.fmi.uni-passau.de/graphlet/gml/gml-tr.html, Universität Passau (1997)
22. Isegawa, M., Maeda, S., Tantillo, D.J., Morokuma, K.: Predicting pathways for terpene formation from first principles-routes to known and new sesquiterpenes. Chem. Sci. **5**(4), 1555–1560 (2014)
23. Kanehisa, M., et al.: The kegg database. In: Novartis Foundation Symposium, pp. 91–100. Wiley Online Library (2002)
24. Kempinski, C., Jiang, Z., Bell, S., Chappell, J.: Metabolic engineering of higher plants and algae for isoprenoid production. In: Schrader, J., Bohlmann, J. (eds.) Biotechnology of Isoprenoids. ABE, vol. 148, pp. 161–199. Springer, Cham (2015). https://doi.org/10.1007/10_2014_290
25. Kim, S., et al.: Pubchem 2019 update: improved access to chemical data. Nucleic Acids Res. **47**(D1), D1102–D1109 (2019)
26. Köllner Schnee, C., Gershenzon, J., Degenhardt, J.: The Variability of Sesquiterpenes Emitted from Two Zea mays\nCultivars Is Controlled by Allelic Variation of Two Terpene\nSynthase Genes Encoding Stereoselective Multiple\nProduct Enzymes. The Plant Cell **16**(May), 1115–1131 (2004)
27. Liu, W., et al.: Structure, function and inhibition of ent-kaurene synthase from bradyrhizobium japonicum. Sci. Rep. **4**, 612 (2014)
28. Liu, W., et al.: Structure, function and inhibition of ent-kaurene synthase from bradyrhizobium japonicum. Sci. Rep. **4**, 6214 (2014)

29. Löwe, M.: Algebraic approach to single-pushout graph transformation. Theor. Comput. Sci. **109**(1), 181–224 (1993)

30. Maeda, S., et al.: Artificial Force Induced Reaction (AFIR) method for exploring quantum chemical potential energy surfaces. Chemical Record **16**(5), 2232–2248 (2016)

31. Nagegowda, D.A.: Plant volatile terpenoid metabolism: biosynthetic genes, transcriptional regulation and subcellular compartmentation. FEBS Letters **584**(14), 2965–2973 (2010)

32. O'maille, P.E., et al.: Quantitative exploration of the catalytic landscape separating divergent plant sesquiterpene synthases. Nat. Chemical Biol. 4(10), 617–623 (2008)

33. da Silva, W.M.C., et al.: Exploring plant sesquiterpene diversity by generating chemical networks. Processes **7**(4), 240 (2019)

34. Systems, D.C.I.: A reaction transform language. http://daylight.com/dayhtml/doc/theory/theory.smirks.html. Accessed 30 Jan 2019

35. Systems, D.C.I.: SMARTS - a language for describing molecular patterns. http://www.daylight.com/dayhtml/doc/theory/theory.smarts.html (2008). Accessed 30 Jan 2019

36. Talapatra, S.K., Talapatra, B.: Biosynthesis of terpenoids: the oldest natural products. Chemistry of Plant Natural Products, pp. 317–344. Springer, Heidelberg (2015). https://doi.org/10.1007/978-3-642-45410-3_5

37. Tian, B., Poulter, C.D., Jacobson, M.P.: Defining the product chemical space of monoterpenoid synthases. PLoS Comput. Biol. **12**(8), 1–13 (2016)

38. Vattekkatte, A., Garms, S., Brandt, W., Boland, W.: Enhanced structural diversity in terpenoid biosynthesis: enzymes, substrates and cofactors. Organic Biomolecular Chemistry **16**(3), 348–362 (2018)

39. Zhang, Y., Nielsen, J., Liu, Z.: Engineering yeast metabolism for production of terpenoids for use as perfume ingredients, pharmaceuticals and biofuels. FEMS Yeast Res. 17(8), fox080 (2017)

40. Zhuang, X., et al.: Dynamic evolution of herbivore-induced sesquiterpene biosynthesis in sorghum and related grass crops. The Plant J. **69**(1), 70–80 (2012)

Oncogenic Signaling Pathways
in Mucopolysaccharidoses

Gerda Cristal Villalba Silva[1,2,3] (iD), Luis Dias Ferreira Soares[3,5] (iD),
and Ursula Matte[1,2,3,4(✉)] (iD)

[1] Postgraduate Program in Genetics and Molecular Biology, UFRGS,
Porto Alegre 91501970, Brazil
umatte@hcpa.edu.br
[2] Gene Therapy Center, HCPA, Porto Alegre 90035903, Brazil
[3] Bioinformatics Core, HCPA, Porto Alegre 90035903, Brazil
[4] Department of Genetics, UFRGS, Porto Alegre 91501970, Brazil
[5] Graduation Program on Biotechnology/Bioinformatics, UFRGS,
Porto Alegre 91501-970, Brazil

Abstract. Cancer cells depend on several signaling pathways and organelles, such as the lysosomes. Defects in the activity of lysosomal hydrolases involved in glycosaminoglycan degradation lead to a group of lysosomal storage diseases called Mucopolysaccharidoses (MPS). In MPS, secondary cell disturbance affects pathways common to cancer. This work aims to identify oncogenic pathways related to cancer in the different MPS datasets available in public databases and compare the ontologies across the different types of MPS. For this, we used 12 expression datasets of 6 types of MPS. Statistical analysis was based on hypergeometric distribution followed by FDR correction. We found several enriched pathways across the 12 MPS studies, among being 57.65% were KEGG pathways, 32.5% of GO Biological Process, 2.5% GO Celular Component, and 7.35% GO Molecular Function. Hippo signaling pathway and MAPK signaling pathway appear in all datasets. Proteoglycans in cancer, Rap1 signaling pathway, and Cytokine-mediated signaling pathway appears in 11 of 12 datasets. The lysosome participates in several biological processes, like autophagy, cell adhesion and migration, and antigen presentation. These processes also may affect in several types of cancer and Lysosomal Storage Diseases. Studying the tumor ontology signature in lysosomal disorders may help understand lysosomal storage diseases and cancer's underlying mechanisms. This may help amplify therapeutic approaches for both types of diseases.

Keywords: Cancer pathways · Gene ontology · Lysosomal storage diseases

1 Introduction

Several metabolic pathways are deranged in cancer cells. The proliferation ability of tumors depends on a cascade of signaling pathways in several cancer cells' organelles, such as the lysosomes [1]. Lysosomes are cellular compartments responsible, among

© Springer Nature Switzerland AG 2020
J. C. Setubal and W. M. Silva (Eds.): BSB 2020, LNBI 12558, pp. 259–264, 2020.
https://doi.org/10.1007/978-3-030-65775-8_24

other functions, for the degradation of macromolecules through acid hydrolases contained within them. Defects in these enzymes culminate in the lysosomal accumulation of intermediate metabolites or macromolecules, known as lysosomal storage diseases [2]. Lysosomal Storage Diseases (LSD) are a group of more than 50 rare metabolic diseases, among which we can highlight the Mucopolysaccharidoses (MPS). In MPS, secondary cell disturbance affects pathways common to cancer.

This work aims to identify oncogenic pathways related to cancer in the different datasets of MPS available in public databases and to compare the ontologies across the different types of MPS.

2 Methods

Gene expression analysis considered 12 datasets available at GEO (https://www.ncbi. nlm.nih.gov/geo), from six different MPS types. For RNA-seq data, we used edgeR, and for microarray data, we used R packages according to the experiment's platform. Furthermore, the data present in this work are available in the MPSBase (https://www.ufrgs. br/mpsbase/). Statistical analysis was based on hypergeometric distribution followed by FDR correction. We perform the enrichment analysis in Cytoscape, with Bingo and ClueGo plugins. We search the child terms with QuickGo. We selected 12 datasets, being 2 of MPS I; 1 from MPS II; 1 from MPS IIIA; 3 MPS IIIB; 1 MPS VI; and 4 from MPS VII. These datasets comprise an RNA-seq data of Illumina HiSeq 2500 platform of human iPSC-derived Neuronal Stem Cell (MPS I, GSE111906); Agilent-021193 Canine (V2) microarray of Ascending Aorta, Descending Aorta and Carotid Aorta (MPS I, GSE78889); AB SOLiD 3 Plus System (Mus musculus) of Brain samples (MPS II, GSE95224); Agilent-028005 SurePrint G3 Mouse GE 8×60K Microarray of Brain and Blood samples (MPS IIIA, GSE97759); Agilent-012694 Whole Mouse Genome G4122A of Lateral entorhinal cortex and Medial entorhinal cortex (MPS IIIB, GSE15758); Affymetrix Human Exon 1.0 ST Array of iPSC-derived Neuronal Stem Cell (MPS IIIB, GSE23075); Affymetrix Human Exon 1.0 ST Array of HeLa depleting NAGLU (MPS IIIB, GSE32154); Affymetrix Mouse Gene 1.0 ST Array of ARSB null mouse hepatic cells (MPS VI, GSE77689); Illumina Mouse-8 Expression BeadChip of Descending aorta (MPS VII, GSE30657); Affymetrix Mouse Genome 430A 2.0 Array of six brain regions (MPS VII, GSE34071); Affymetrix Mouse Exon 1.0 ST Array of iPS embryo-derived ES cells with controls derived from B6 Blu ES cells and Mouse embryonic fibroblast (MPS VII, GSE36017); and Affymetrix Mouse Genome 430A 2.0 Array of hippocampus (MPS VII, GSE76283).

3 Results

We found 680 oncogenic enriched ontologies across the 12 MPS studies, among being 392 were KEGG pathways (57.65%), 221 GO Biological Process (32.5%), 17 GO Cellular Component (2.5%), and 50 of GO Molecular Function (7.35%). Hippo signaling pathway and MAPK signaling pathway appears in all datasets. Proteoglycans in cancer, Rap1 signaling pathway, and Cytokine-mediated signaling pathway appears in 11 of 12 datasets (see Fig. 1).

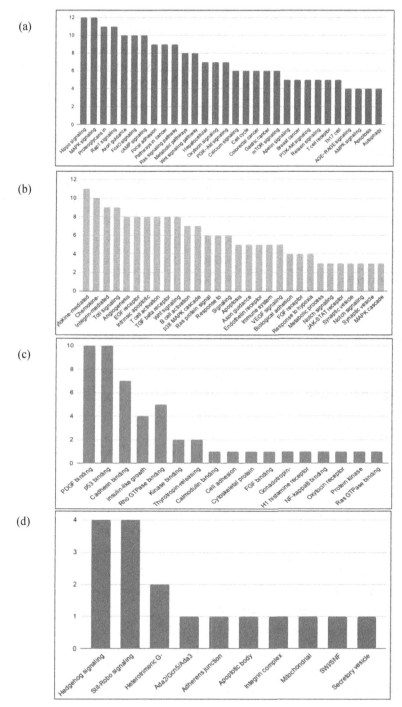

Fig. 1. Top Gene ontology of oncogenic terms of MPS. (a) KEGG pathways; (b) GO Biological Process; (c) Molecular Function; (d) Cellular Component.

Considering only the MPS types, the ontologies Axon guidance, Focal adhesion, Hippo signaling pathway, MAPK signaling pathway, Metabolic pathways, Pathways in cancer, PI3K-Akt signaling pathway, Proteoglycans in cancer, Rap1 signaling pathway, and Ras signaling pathway are present in all the MPS types found in the GEO. The following Table 1 gives a summary of the most frequent oncogenic ontologies according to the MPS type.

Table 1. Prevalent oncogenic enriched pathways of datasets analyzed. In bold, the ontology appears in all MPS types.

Term	Ontology	MPS I	MPS II	MPS IIIA	MPS IIIB	MPS VI	MPS VII
Apelin signaling pathway	KEGG	X	X	X			X
Apoptosis	KEGG	X		X	X	X	X
Autophagy	KEGG	X		X	X		X
Axon guidance	KEGG	X	X	X	X	X	X
Calcium signaling pathway	KEGG	X	X		X	X	X
cAMP signaling pathway	KEGG		X	X	X	X	X
Focal adhesion	KEGG	X	X	X	X	X	X
FoxO signaling pathway	KEGG	X	X	X	X		X
Hepatocellular carcinoma	KEGG	X	X	X			X
Hippo signaling pathway	KEGG	X	X	X	X	X	X
MAPK signaling pathway	KEGG	X	X	X	X	X	X
Metabolic pathways	KEGG	X	X	X	X	X	X
mTOR signaling pathway	KEGG	X		X	X	X	X
Oxytocin signaling pathway	KEGG	X	X	X	X	X	
Pathways in cancer	KEGG	X	X	X	X	X	X
PI3K-Akt signaling pathway	KEGG	X	X	X	X	X	X
Proteoglycans in cancer	KEGG	X	X	X	X	X	X
Rap1 signaling pathway	KEGG	X	X	X	X	X	X
Ras signaling pathway	KEGG	X	X	X	X	X	X
Wnt signaling pathway	KEGG	X	X	X	X	X	X
Chemokine-mediated signaling pathway	GO_BP	X		X	X	X	X
Cytokine-mediated signaling pathway	GO_BP	X		X	X	X	X
EGFR signaling pathway	GO_BP	X	X	X	X		X
Slit-Robo signaling complex	GO_CC			X	X	X	X
p53 binding	GO_MF			X	X	X	X

The dataset with the most enriched pathways is GSE32154 (MPS IIIB, *Homo sapiens*) with 90 ontologies (see Fig. 2). The dataset with the most enriched KEGG terms is GSE30657 (MPS VII, *Mus musculus*) with 60 KEGG terms. The GSE32154 (MPS IIIB, *Homo sapiens*) have the most GO Biological Process enriched terms, with 45 terms. For GO Cellular Component, the dataset with more enriched terms in this category is GSE32154 (MPS IIIB, *Homo sapiens*) with 5 terms. Lastly, in the GO Molecular Function, the GSE32154 is the dataset with more enriched terms (10 terms found).

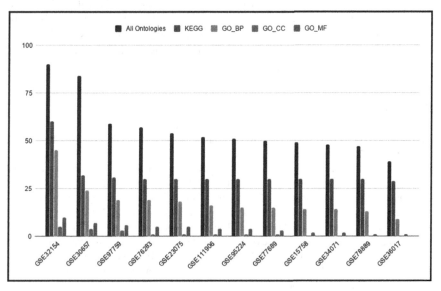

Fig. 2. Number of enriched cancer ontologies across the MPS datasets.

4   Discussion

Oncogenic activation can lead to the destabilization of lysosomal membranes and an increase of lysosomal hydrolases into the cytosol, where they can contribute to the demise of the cancer cell [3].

Axon guidance and Wnt signaling pathway are related to the neurological impairment found in several MPS types [4]. Alterations in autophagy are frequently found in MPS patients with neurodegenerative symptoms [4]. In cancer, autophagy are related to cancer initiation, proliferation, and survival [5].

The ontologies Cell cycle, Hippo, Notch, PI-3-Kinase/Akt, RAS, TGFβ signaling, P53 and β-catenin/WNT signaling pathway are considered canonical oncogenic pathways. Unfortunately, 89% of tumors found in the TCGA consortium had at least one driver alteration in these pathways, and 57% percent of the tumors had at least one alteration potentially targetable by currently available treatments [6]. We hypothesize that these signaling pathways are altered because glycosaminoglycans play an essential role in the composition of the extracellular matrix [7], helping to regulate processes such as metabolic signaling, apoptosis, cell migration, adhesion, and antigen presentation, in both cancer and MPS.

5 Concluding Remarks

The available public data is essential for amplified the multi-omic knowledge of complex and rare diseases. Bioinformatic approaches, such as gene enrichment analysis, may help us understand the complexity of processes deranged in several diseases. Studying the tumor ontology signature in lysosomal disorders may help understand lysosomal storage diseases and cancer's underlying mechanisms. This may help amplify therapeutic approaches for both types of diseases.

References

1. Cairns, R.A., Harris, I.S., Mak, T.W.: Regulation of cancer cell metabolism. Nat. Rev. Cancer **11**(2), 85–95 (2011). https://doi.org/10.1038/nrc2981
2. Matte, U., Pasqualim, G.: Lysosome: the story beyond the storage. J. Inborn Errors Metab. Screen. **4**, e160044 (2016). https://doi.org/10.1177/2326409816679431
3. Kallunki, T., Olsen, O.D., Jäättelä, M.: Cancer-associated lysosomal changes: friends or foes? Oncogene **32**(16), 1995–2004 (2012). https://doi.org/10.1038/onc.2012.292
4. Fiorenza, M.T., Moro, E., Erickson, R.P.: The pathogenesis of lysosomal storage disorders: beyond the engorgement of lysosomes to abnormal development and neuroinflammation. Hum. Mol. Genet. **27**(R2), R119–R129 (2018). https://doi.org/10.1093/hmg/ddy155
5. Martinez-Carreres, L., Nasrallah, A., Fajas, L.: Cancer: linking powerhouses to suicidal bags. Front. Oncol. **7**, 204 (2017). https://doi.org/10.3389/fonc.2017.00204
6. Sanchez-Vega, F., et al.: Oncogenic signaling pathways in the cancer genome atlas. Cell **173**(2), 321–337 (2018). https://doi.org/10.1016/j.cell.2018.03.035
7. Davidson, S.M., Vander Heiden, M.G.: Critical functions of the lysosome in cancer biology. Ann. Rev. Pharmacol. Toxicol. **57**(1), 481–507 (2017). https://doi.org/10.1146/annurev-pharmtox-010715-103101

Natural Products as Potential Inhibitors for SARS-CoV-2 Papain-Like Protease: An in Silico Study

Jesus Alvarado-Huayhuaz[1] , Fabian Jimenez[2] , Gerson Cordova-Serrano[3] ,
Ihosvany Camps[4] , and Wilmar Puma-Zamora[1,5(✉)]

[1] Laboratorio de Investigación en Biopolímeros y Metalofármacos, Universidad Nacional de Ingeniería, Lima, Peru
jalvaradoh@uni.pe
[2] Department of Technological Operations, Alluriam HealthCare Inc., Orlando, FL, USA
[3] Escuela de Farmacia y Bioquímica, Universidad María Auxiliadora, Lima, Peru
[4] Laboratorio de Modelagem Computacional, Universidade Federal de Alfenas, Minas Gerais, Brazil
[5] Escuela de Farmacia y Bioquímica, Universidad Nacional de San Antonio Abad del Cusco, Cusco, Peru
070994@unsaac.edu.pe

Abstract. In December 2019 in Wuhan, China, the first case of a patient infected with a new coronavirus was reported, later called SARS-CoV-2 by the World Health Organization, and so far, there is no approved drug or vaccine against this new virus. Considering this, previous studies identified several essential SARS-CoV-2 viral proteins, among these to the papain-like protease (PL^{pro}), an enzyme with an important role in viral spread and evading the host immune response, making it an attractive drug target. For this reason, using a library of 213,038 structures of natural products and virtual screening, we identified 10 molecules with high affinity for residues of the SARS-CoV-2 PL^{pro} allosteric site, which could be tested in vitro against the virus or as lead-compounds to develop inhibitors that are more effective.

Keywords: SARS-CoV-2 · Natural products · Virtual screening

1 Introduction

SARS-CoV-2 is an RNA virus whose genome has 10 open reading frames that encode various structural and non-structural proteins [23]. Among these proteins, one of the most studied together with the main protease and the protein spike, is the papain-like protease (PL^{pro}) which is necessary to facilitate the spread of the virus and help to evade the host immune response [7]. One of the most outstanding aspects of SARS-CoV-2 PL^{pro} is its role as an ubiquitin-specific protease, especially the ubiquitin-like protein ISG15, which is a known

© Springer Nature Switzerland AG 2020
J. C. Setubal and W. M. Silva (Eds.): BSB 2020, LNBI 12558, pp. 265–270, 2020.
https://doi.org/10.1007/978-3-030-65775-8_25

regulatory agent of human immune response pathways. In this way, it is proposed that blocking SARS-CoV-2 PL^{pro} could stop the viral infection and improve the immune response [18].

Since to date there is no officially an effective vaccine or antiviral against this coronavirus [2], we highlight the importance of rapid search techniques like virtual screening (VS) and molecular docking to identify possible inhibitors of this enzyme, using the evaluation of molecular interactions in its active and allosteric sites. In this work, we focus on natural products (NPs) because in these organic molecules we can find a large surface area and multiple functional groups capable of interacting by hydrogen bonds or van der Waals forces.

2   Methods

We carried out two stages. First, a VS with AutoDock Vina (referred to as Vina from here on) [19] and 213,038 chemical structures of NPs collected in the freely available Universal Natural Product Database - In Silico MS/MS Database (UNPD-ISDB) [1].

To ensure the reliability of the VS results, the second stage consisted of molecular docking divided into three sub-stages with Glide [5] from the suite Schrödinger. The best hits of the VS were docked with standard precision (Glide-SP), extra precision (Glide-XP), and then with the Induced Fit Docking (IFD) protocol, that allows flexibility to the residues from the pocket. Finally, the binding free energy (ΔG_b) of the best 10 complexes from IFD protocol was calculated using the physics-based MM-GBSA (Molecular Mechanics-Generalized Born Surface Area) method with the Prime [10] module of the suite Schrödinger.

Preparation of the Receptor. From Protein Data Bank we retrieve the structure of the SARS-CoV-2 PL^{pro} with PDB ID 6W9C (resolution 2.7 Å). For VS, all water molecules and ions were removed and polar hydrogens were added to the receptor using AutoDock Tools v1.5.6 [14]. For molecular docking, the protease was prepared by removing unnecessary water molecules and ions with the Protein Preparation Wizard [13] module and considering the protonation states of the residues at pH 7.4 with Epik [17].

Ligand Preparation. The structures of NPs in InChI strings were obtained from the UNPD-ISDB. 3D coordinates were assigned to them and then they were geometrically minimized with 20 thousand steps using the Open Babel software (v3.0.0) [16]. For VS and molecular docking, we add appropriate hydrogens for pH 7.4.

Validation of Docking Protocol. For the validation we used the crystallized structure of SARS-CoV PL^{pro} (PDB ID 3E9S) to check the suitability and accuracy of Vina and Glide as appropriate docking tools. The protein and its native ligand were prepared under the same conditions detailed above. The box size and the coordinates of the centroid of the binding site were calculated in the same way as for VS and molecular docking stages.

Virtual Screening. Ligands and protein in PDBQT format were used as input files. The centroid of the binding site (x = −50.1, y = 15.1 and z = 34.4) and the optimal box size (40 × 40 × 40 Å), which cover the active and allosteric sites, were obtained with the POCASA server [25] and the eBoxSize script [4] respectively. Vina was used with an exhaustiveness level of 24 in the Quinde 1 supercomputer.

Molecular Docking and MM-GBSA. The best 1000 ligands were selected from VS according to their binding affinity values. These were docked with Glide-SP in the SARS-CoV-2 PLpro active and allosteric sites. From the results with Glide-SP, the best 100 hits were selected and docked with Glide-XP. From the last step with Glide-XP, the best 10 hits were selected and again they were docked with the IFD protocol, allowing flexibility to the residues close to the active and allosteric sites. For these steps, the OPLS_2005 force field was used, and the same conditions of binding site coordinates and box size from the VS were used. Finally, the ΔG_b of the best ligand-receptor complexes from IFD was estimated, according to Eq. 1, with the Prime module using an implicit solvation model VSGB and the OPLS_2005 force field. In Eq. 1, $G_{complex}$, $G_{protein}$, and G_{ligand} are the free energies of the complex, protein, and ligand, respectively.

$$\Delta G_b = G_{complex} - (G_{protein} + G_{ligand}) \tag{1}$$

3   Results and Discussion

Validation of Docking Protocol. Using the Open Babel obrms tool, the RMSD was calculated between the native co-crystallized ligand (TTT; 5-amino-2-methyl-N-[(1R)-1-naphthalen-1-ylethyl] benzamide) of the SARS-CoV PL^{pro} and its best re-docked conformations obtained with Vina (0.47 Å), Glide-SP (0.57 Å) and Glide-XP (0.81 Å). Since RMSD values are less than 2.0 Å with respect to the native ligand, it is considered that the protocols used in the present study are reliable [21].

Virtual Screening. According to Vina, 88,848 NPs obtained a binding affinity between −6 and −7 kcal/mol, which represents 41.7% approximately of all ligands (Fig. 1). The 1000 NPs selected for the following molecular docking tests had a binding affinity between −8 to −13 kcal/mol.

Molecular Docking and MM-GBSA. We identify 10 molecules with the best ΔG_b values. Hydrogen bonds (H-bonds) interactions were found with many residues from the subsites S3 and S4 (underlined in Table 1). These molecules

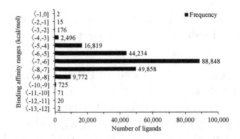

Fig. 1. Frequency of ligands according to binding affinity, calculated by Vina.

could block the enzyme activity through non-covalent interactions (between 6 to 9 H-bonds and lipophilic interactions) with many residues of the SARS-CoV-2 PL^{pro} allosteric site, inducing the closure of the loop [3]. These hits are relatively large glycosides and tannins with previously reported biological properties.

Table 1. ΔG_b values, interacting residues, classification and biological source of the 10 molecules with the best affinity for SARS-CoV-2 PL^{pro}

ISDB ID	ΔG_b (kcal/mol)	H-bonds interactions	Type of compounds	Biological source
208044	−131.251	K157, E161, D164, R166, E167, Y264, Y268, Q269, T301	Steroidal glycoside	M. pubescens [22]
139907	−129.517	C155, N156, D164, R166, E167, Y273, D302	Steroidal glycoside	O. japonicus [12]
113050	−127.701	C155, K157, E161, Y171, P248, G266, Q269	Steroidal glycoside	A. filicinus [26]
066371	−121.304	E161, R166, Y171, E203, P248, Y264, Y273, T301	Hydrolyzable tannin	Q. phillyraeoides [15]
141146	−118.245	K157, L162, Y171, P248, G266, N267, Y268, Q269	Terpene glycoside	P. quinquefolium [20]
058326	−105.371	C155, N156, E167, Y171, P248, Y264, G266, Y273, T301	Steroidal glycoside	A. asphodeloides [11]
150945	−86.977	E161, L162, D164, E167, Y264, Y273, T301	Steroidal glycoside	A. amurensis [9]
035549	−75.780	K157, E161, L162, D164, E167 G266, N267, Q269	Steroidal glycoside	D. inarticulata [24]
171225	−73.354	E161, L162, E167, S245, Y273, D302. *Pi-cation: R166	Hydrolyzable tannin	L. speciosa [6]
122929	−56.530	E161, L162, D164, R166, Y273, T301	Hydrolyzable tannin	J. nigra [8]

4   Conclusion

In this study, we identified 10 molecules that can interact favorably with residues of the allosteric site of PL^{pro}, being considered stable complexes according to their ΔG_b values. This search was done using a library of 213,038 structures of NPs collected in the UNPD-ISDB, and techniques like virtual screening and molecular docking. These molecules are mainly glycosides and tannins, and their molecular structures are relatively large, presenting between 6 to 9 H-bonds and lipophilic interactions with residues of SARS-CoV-2 PL^{pro} allosteric site. However, still are needed prediction of pharmacokinetic properties and molecular dynamics simulations to evaluate the possibility that these molecules can be used for in vitro PL^{pro} inhibition assays, or lead optimization processes.

Acknowledgments. The team gives special thanks to Siembra Empresa Pública (Imbabura, Ecuador) for their support with Quinde 1 supercomputer. Also thanks to Dr. Meng Guo and Dr. Li Wang (National Supercomputing Center in Jinan, China), for their support with the implementation of Vina and helpful discussions, and to Dr. Ana Valderrama (Universidad Nacional de Ingeniería, Perú) for her personal financial contribution to the project. Finally, remembering that #SinCienciaNoHayFuturo.

References

1. Allard, P.M., et al.: Integration of molecular networking and In-Silico MS/MS fragmentation for natural products dereplication. Anal. Chem. **88**, 3317–3323 (2016)
2. Arnold, C.: Race for a vaccine. New Sci. **245**, 44–47 (2020)
3. Estrada, E.: COVID-19 and SARS-CoV-2. Modeling the present, looking at the future. Phys. Rep. **869**, 1–51 (2020)
4. Feinstein, W.P., Brylinski, M.: Calculating an optimal box size for ligand docking and virtual screening against experimental and predicted binding pockets. J. Cheminform. **7**(1), 1–10 (2015). https://doi.org/10.1186/s13321-015-0067-5
5. Friesner, R.A., et al.: Extra precision glide: docking and scoring incorporating a model of hydrophobic enclosure for protein-ligand complexes. J. Med. Chem. **49**, 6177–6196 (2006)
6. Guo, S., et al.: The anti-diabetic effect of eight Lagerstroemia speciosa leaf extracts based on the contents of ellagitannins and ellagic acid derivatives. Food Funct. **11**, 1560–1571 (2020)
7. Harcourt, B.H., et al.: Identification of severe acute respiratory syndrome coronavirus replicase products and characterization of papain-like protease activity. J. Virol. **78**, 13600–13612 (2004)
8. Ho, K.V., et al.: Identifying antibacterial compounds in black walnuts (Juglans nigra) using a metabolomics approach. Metabolites **8**, 58 (2018)
9. Ishii, T., Okino, T., Mino, Y., Tamiya, H., Matsuda, F.: Plant-growth regulators from common starfish (Asterias amurensis Lütken) waste. Plant Growth Regul. **52**, 131–139 (2007)
10. Jacobson, M.P., Friesner, R.A., Xiang, Z., Honig, B.: On the role of the crystal environment in determining protein side-chain conformations. J. Mol. Biol. **320**, 597–608 (2002)
11. Ji, D., Huang, Z.y., Fei, C.h., Xue, W.w., Lu, T.l.: Comprehensive profiling and characterization of chemical constituents of rhizome of Anemarrhena asphodeloides Bge. J. Chromatogr. B **1060**, 355–366 (2017)
12. Kang, Z.Y., Zhang, M.J., Wang, J.X., Liu, J.X., Duan, C.L., Yu, D.Q.: Two new furostanol saponins from the fibrous root of Ophiopogon japonicus. J. Asian Nat. Prod. Res. **15**, 1230–1236 (2013)
13. Madhavi Sastry, G., Adzhigirey, M., Day, T., Annabhimoju, R., Sherman, W.: Protein and ligand preparation: parameters, protocols, and influence on virtual screening enrichments. J. Comput. Aided. Mol. Des. **27**, 221–234 (2013)

14. Morris, G.M., et al.: AutoDock4 and AutoDockTools4: automated docking with selective receptor flexibility. J. Comput. Chem. **30**, 2785–2791 (2009)

15. Nonaka, G.i., Nakayama, S., Ishioka, I.N.: Tannins and related compounds. LXXXIII. Isolation And Structures of Hydrolyzable Tannins, Phillyraeoidins A-E From Quercus Phillyraeoides. Chem. Pharm. Bull. **37**, 2030–2036 (1989)

16. O'Boyle, N.M., Banck, M., James, C.A., Morley, C., Vandermeersch, T., Hutchison, G.R.: Open babel: an open chemical toolbox. J. Cheminform. **3**, 33 (2011)

17. Shelley, J.C., Cholleti, A., Frye, L.L., Greenwood, J.R., Timlin, M.R., Uchimaya, M.: Epik: a software program for pK a prediction and protonation state generation for drug-like molecules. J. Comput. Aided. Mol. Des. **21**, 681–691 (2007)

18. Shin, D., et al.: Papain-like protease regulates SARS-CoV-2 viral spread and innate immunity. Nature (2020)

19. Trott, O., Olson, A.J.: AutoDock Vina: Improving the speed and accuracy of docking with a new scoring function, efficient optimization, and multithreading. J. Comput. Chem. **31**, 455–461 (2009)

20. Tung, N.H., Shoyama, Y.: Eastern blotting analysis and isolation of two new dammarane-type saponins from american ginseng. Chem. Pharm. Bull. **60**, 1329–1333 (2012)

21. Wang, R., Lu, Y., Wang, S.: Comparative evaluation of 11 scoring functions for molecular docking. J. Med. Chem. **46**, 2287–2303 (2003)

22. Zhao, W., Xu, R., Qin, G., Vaisar, T., Lee, M.S.: Saponins from mussaenda pubescens. Phytochemistry **42**, 1131–1134 (1996)

23. Wu, F., et al.: A new coronavirus associated with human respiratory disease in China. Nature **579**, 265–269 (2020)

24. Yang, W.L., Tian, J., Peng, S.L., Guan, J.F., Ding, L.S.: Chemical constituents of Diuranthera inarticulata. Yao Xue Xue Bao **36**, 590–4 (2001)

25. Yu, J., Zhou, Y., Tanaka, I., Yao, M.: Roll: a new algorithm for the detection of protein pockets and cavities with a rolling probe sphere. Bioinformatics **26**, 46–52 (2009)

26. Zhou, L., Cheng, Z., Chen, D.: Simultaneous determination of six steroidal saponins and one ecdysone in Asparagus filicinus using high performance liquid chromatography coupled with evaporative light scattering detection. Acta Pharm. Sin. B **2**, 267–273 (2012)

Author Index

Printed in the United States
By Bookmasters